SIGNALS AND IMAGES

A C.I.P. Catalogue record for this book is available from the Library of Congress

ISBN 0-7923-4466-9

Published by Kluwer Academic Publishers,
P.O. Box 17, 3300 AA Dordrecht, The Netherlands.

Kluwer Academic Publishers incorporates
the publishing programmes of
D. Reidel, Martinus Nijhoff, Dr W. Junk and MTP Press.

Sold and distributed in the U.S.A. and Canada
by Kluwer Academic Publishers,
101 Philip Drive, Norwell, MA 02061, U.S.A.

In all other countries, sold and distributed
by Kluwer Academic Publishers Group,
P.O. Box 322, 3300 AH Dordrecht, The Netherlands.

Printed on acid-free paper

Printed in the Netherlands

Signals and Images

Selected Papers from the 7th and 8th GIRI Meeting, held in Montpellier, France, November 20–21, 1993, and Jerusalem, Israel, December 10–11, 1994

Edited by

Madeleine Bastide

Faculty of Pharmacy,
University of Montpellier I,
Montpellier, France

Groupe International de Recherches sur l'Infinitésimal

GIRI

International Research group on Very Low Dose and High Dilution Effects

GIRI

KLUWER ACADEMIC PUBLISHERS
DORDRECHT / BOSTON / LONDON

Scientific Committees

GIRI Meeting, Montpellier, France, 20-21 November 1993

A.J. Gabius
Professor of Pharmaceutical Chemistry, University of Marburg, Germany

M. Manfait
Professor of Physics, University of Reims, France

J. Michel
Professor of Medicine, Hebrew University and Hadassah, Jerusalem, Israel

GIRI Meeting, Jerusalem, Israel, 10-12 December 1994

Z. Bentwitch
Professor of Immunology, Kaplan Hospital, Rehovot, Israel

P.Y. Turpin
Professor of Physics, Curie Institute and Paris VI University, Paris, France

R. van Wijk
Molecular Biology, Utrecht University, The Netherlands

Organizing Committee

M. Bastide
Professor of Immunology , University of Montpellier, France

J. Cambar
Professor of Cellular Biology, University of Bordeaux, France

R. Halm
Doctor, Pharmacist, Nice, France

M. Manfait
Professor of Physics, University of Reims, France

M. Oberbaum
MD, Kaplan Hospital, Rehovot, Israel

H. Wagner
Professor of Pharmacology, University of Munich, Germany

CONTENTS

PREFACE

The GIRI is designated by the letters of its french name "Groupe International de Recherches sur l'Infinitésimal" which is actually translated as the "International Research Group on Very Low Dose and High Dilution Effects". This group was created in 1987 by 15 researchers in response to the need for a platform where scientists working with very dilute solutions including homeopathic ones could communicate and compare their results. The aim of the GIRI was to bring together pharmacologists, biologists, physicists, chemists and physicians in order to exchange experiences and develop joint research projects. The distinctive feature of the research activities of the group was the study of very low or ultra low dose impulses or very high dilutions similar to those which are used in homeopathic therapy and considered as devoid of any molecule.

The group is also interested in the possible medicinal and therapeutic relevance of these doses or dilutions as well as the mechanism of action of the very diluted solutions of active principles on biological systems.

These last ten years were very fruitful thanks to the possibility of organizing an annual meeting. The first meetings were very unproductive because of the small number of experiments. However each year, the number of research teams increased, different models were proposed, serious discussions took place and the proceedings of the 3rd and 4th meetings was edited in 1990 (*Signals and Images: New Pharmacological Approaches and their concepts, M.Bastide ed., Ateliers Alpha Bleue Publisher, Paris*). Only five full papers were published corresponding to the questions raised by this kind of research. The 5th and 6th meetings were not published.

We are now able to propose a book which includes 23 full papers from more than fifty contributors corresponding to the two last GIRI meetings: the 7th (held in Montpellier, France, on 20-21 November 1993) and the 8th (held in Jerusalem, Israel, on 10-11 December 1994). The GIRI gets bigger every year and now includes more than 100 members from 22 different countries. This evolution stimulates more and more researches on this subject often considered as scientific nonsense or a scientific error. All these papers have been reviewed according to the standards of scientific publications and many of them are now published in regular scientific journals. They illustrate perfectly the evolution of the ideas and the new experimental and theoretical approaches of this uncommon research. It becomes obvious that different hypothesis can enlighten the interpretations of these different papers. Part of them can be interpretated according to the classical way of thinking; but the true interpretation of the similia law of homeopathic medicine is quite different from the mechanistic approach.

To help the reader, the papers will be organized into four chapters; each chapter will be introduced by a short analysis of the papers included. The *"Introduction"* debates the question of scientific evolution and revolution in the context of modern science. The 1st chapter *"Hormesis"* gathers together all the papers related to this concept: these models are often evoked to demonstrate or to explain the similia law although hormesis is always based on a relationship of identity. The 2nd chapter *"In Vitro and in Vivo Experimental Models"* includes many experiments which demonstrate low dose or high dilution activity and is introduced by a summary of classical receptology. The 3rd chapter *"Therapeutics and Provings "* discusses the question of analysis of the symptoms in a systemic way of thinking and also includes pathogenetic studies (provings) as well as homeopathic therapeutic studies in humans or in animals. The last chapter presents an *"Epistemologic Approach"* which become necessary in order to enlarge the possibility of interpretation of the law of similarity and the high dilution effectiveness considering that these dilutions are above the Avogadro number.

This new field of research is very exciting and introduces new scientific concepts supported by experimental results. Above all, we observe that this nascent science is totally concerned by "living" organisms and as such, it becomes necessary to define what we design as "information" brought by non-molecular high dilutions.

This book presents the brain-storming work of this research group and is one of the starting points of a scientific evolution: " *The proliferation of concurrent variants of the conventional paradigm, the fact of being willing to try anything, the expression of marked discontent, recourse to philosophy and discussion on the theoretical foundations, all signs are many symptoms of a passage from normal research to extraordinary research"*. T.S.Kuhn, 1962, *The Structure of Scientific Revolutions*, The University of Chicago Press, Chicago.

Madeleine BASTIDE

LIST OF CONTRIBUTORS

ANDRIOLI G. Institute of Chimica Microscopia Clinica, University of Verona, Laboratorio Centrale, Ospedale Policlinico, 37134 Verona, Italy.

BAROUILLET M.P. Renal Physiology and Physiopathology Research Group, Faculty of Pharmacy 146 av Leo Saignat, 33076 Bordeaux ,France.

BASTIDE M. Laboratory of Immunology and Parasitology, Faculty of Pharmacy, Montpellier I University, 15 av Charles Flahault, 34060 Montpellier cedex 1, France.

BAYLE J.D. Laboratory of General Physiology, Faculty of Sciences, Montpellier II University, Pl. E. Bataillon, 34090 Montpellier cedex 5,France.

BELLAVITE P. Institute of Chimica Microscopia Clinica, University of Verona, Laboratorio Centrale, Ospedale Policlinico, 37134 Verona, Italy.

BENTWICH Z. R. Ben Ari Institute of Clinical Immunology, Kaplan Hospital, Rehovot, Israel.

BIASI D. Institute of Patologia Medica, University of Verona, Laboratorio Centrale, Ospedale Policlinico, 37134 Verona, Italy.

BLOSTIN R. Dr Veterinary, 42 rue F. Poulenc 77430 Champagne sur Seine, France.

BONNE C. Laboratory of Physiology, Faculty of Pharmacy, 15 Avenue C.Flahault, 34060 Montpellier Cedex 01, France.

BONNE C. Laboratory of Physiology, Faculty of Pharmacy, Montpellier I University, 15 av Charles Flahault, 34060 Montpellier Cedex 1, France.

BOUDARD F. Laboratory of Immunology and Parasitology, Faculty of Pharmacy, Montpellier I University, 15 av Charles Flahault, 34060 Montpellier cedex 1, France.

BRANDS M. MD, Homeopaths without Borders Netherlands, Laurier Gracht 47 B, 1016 RG Amsterdam, The Netherlands.

CAMBAR J. Renal Physiology and Physiopathology Research Group, Faculty of Pharmacy 146 av Leo Saignat, 33076 Bordeaux ,France.

CARRÉE H. MD - Lecturer at the Faculty of Medecine of Marseilles, Résidence Ste Victoire, Av. de Craponne, 13100 Aix-en-Provence, France.

CHAGNAUD J.L. Laboratoire PIOM, ENSCPB, BP 108, Avenue Pey Berland , 33402 Talence cedex, France.

CHIRUMBOLO S. Institute of Chimica Microscopia Clinica, University of Verona, Laboratorio Centrale, Ospedale Policlinico, 37134 Verona, Italy.

CITRO M. Research Institute Alberto Sorti (IDRAS), Via Passalacqua 19, 10122 Turin, Italy.

CRISTEA A. Laboratory of Pharmacology, Faculty of Pharmacy, Str Av. Sanatescu 43, 71324 Bucharest, Romania.

DARIE V. Laboratory of Pharmacology, Faculty of Pharmacy, Str Av. Sanatescu 43, 71324 Bucharest, Romania.

DELBANCUT A. Renal Physiology and Physiopathology Research Group, Faculty of Pharmacy 146 av Leo Saignat, 33076 Bordeaux ,France.

DEMANGEAT J.L. Nuclear Medicine Department, General Hospital, Haguenau, France.

DORFMANN P. Laboratoires Dolisos, 71 rue Beaubourg, 75003 Paris, France.

ENDLER C. Ludwig Boltzman Institut für Homöopathie, Durerg.4, 8010 Graz, Austria.

ERNST-HIEBER E. University of Freiburg, Department of Psychology, 79085 Freiburg, Germany.

FAIDERBE S. Laboratoire PIOM, ENSCPB, BP 108, Avenue Pey Berland , 33402 Talence cedex, France.

GEFFARD M. Laboratoire PIOM, ENSCPB, BP 108, Avenue Pey Berland , 33402 Talence cedex, France.

GRANDMONTAGNE Y. LTP-Dolisos - BP 20 - 51051 Reims Cedex France.

GRIES P.B. Nuclear Medicine Department, General Hospital, Haguenau, France.

GUELATTI M. Laboratory of General Physiology, Faculty of Sciences, Montpellier II University, Pl. E. Bataillon, 34090 Montpellier cedex 5, France.

HARPAZ N. R. Ben Ari Institute of Clinical Immunology, Kaplan Hospital, Rehovot, Israel.

HIEBER S. Department of Psychology, University of Freiburg, 79085 Freiburg Germany.

KALINKOVICH A. R. Ben Ari Institute of Clinical Immunology, Kaplan Hospital, Rehovot; The Hebrew University Hadassah Medical School, Jerusalem, Israel.

LAGACHE A. PhD, 5 Rue Ste Anastase, 75003 Paris.

LE MOIGNE J.L. G.R.A.S.C.E. URA CNRS 935, Université d'Aix-Marseille, Centre Forbin 15-19, Allée Claude Forbin, 13627 Aix-en-Provence Cedex 1, France.

LUCKEY T.D. 1009 Sitka Ct, Loveland, CO 80538, USA.

LUSSIGNOLI S. Institute of Chimica Microscopia Clinica, University of Verona, Laboratorio Centrale, Ospedale Policlinico, 37134 Verona, Italy.

MEKAOUCHE M. Laboratory of General Physiology, Faculty of Sciences, Montpellier II University, Pl. E. Bataillon, 34090 Montpellier cedex 5 ,France.

NICULA S. Laboratory of Pharmacology, Faculty of Pharmacy, Str Av. Sanatescu 43, 71324 Bucharest, Romania.

OBERBAUM M. R. Ben Ari Institute of Clinical Immunology, Kaplan Hospital, Rehovot, Israel.

OVELGÖNNE J.H. Department of Molecular Cellular Biology, Utrecht University, PO Box 80.056, 3508 TB Utrecht, The Netherlands.

POITEVIN B. French Association for Homeopathy Research, Les Segauds, 03110 Vendat, France.

PONGRATZ W. Ludwig Boltzman Institut für Homöopathie, Durerg.4, 8010 Graz, Austria.

RIAUCOURT A. Filiaporc-Dynal, 69, rue H. Le Vezouet BP 471 - 22 Loudéac Cedex France.

RUBIK B. Institute for Frontier Sciences, 7711-D Mc Callum St, Philadelphia, PA 19118, USA.

SANTONASTASO C. Institute of Chimica Microscopia Clinica, University of Verona, Laboratorio Centrale, Ospedale Policlinico, 37134 Verona, Italy.

SCHULTE J. National Superconducting Cyclotron Laboratory, Michigan State University, East Lansing, MI 48824-1321, USA.

SENEKOWITSCH F. Institute of Bioinformatics, Graz, Austria.

SMITH C.W. Department of Electric and Electronic Engineering, University of Salford, 827221 Salford, UK.

SOUREN J.E.M. Department of Molecular Cellular Biology, Utrecht University, PO Box 80.056, 3508 TB Utrecht, The Netherlands.

TEODORESCU-NEGRES S. Laboratory of Pharmacology, Faculty of Pharmacy, Str Av. Sanatescu 43, 71324 Bucharest, Romania.

TOPPER R. R Ben Ari Institute of Clinical Immunology, Kaplan Hospital, Rehovot, Israel.

VAN ERP V.M.A. Sc, Vrije Universiteit, Laurier Gracht 47 B, 1016 RG Amsterdam, The Netherlands.

VAN WIJK R. Department of Molecular Cellular Biology, Utrecht University, PO Box 80.056, 3508 TB Utrecht, The Netherlands.

WALACH H. Department of Psychology, University of Freiburg, 79085 Freiburg Germany.

WEISMAN Z. R. Ben Ari Institute of Clinical Immunology, Kaplan Hospital, Rehovot; The Hebrew University Hadassah Medical School, Jerusalem, Israel.

WIEGANT F.A.C. Department of Molecular Cellular Biology, Utrecht University, PO Box 80.056, 3508 TB Utrecht, The Netherlands.

YOUBICIER-SIMO B.J. Laboratory of Immunology and Parasitology, Faculty of Pharmacy, Montpellier I University, 15 av Charles Flahault, 34060 Montpellier Cedex 1, France.

INTRODUCTION

B. RUBIK.
The Perennial Challenge of Anomalies at the Frontiers of Science.

In the context of the History of Science, the scientists who challenge the dominant paradigm are ignored or attacked by the scientific mainstream. Modern science is totally devoted to the mechanistic paradigm and biology is totally linked to a physico-chemical reductionism: in this context, scientists are very resistant to new discoveries which like homeopathy or high dilution effects chip away the molecular dogma. But "considering that major changes in Science have never been brought about by isolated findings, but by collective evidence", it is fitting that such a text should introduce this book of "anomalies" at the frontier of science.

M.B.

THE PERENNIAL CHALLENGE OF ANOMALIES AT THE FRONTIERS OF SCIENCE *

Anomalies at the Frontiers of Science

B. RUBIK
Institute Frontier Science, 7711-D Mc Callum St, Philadelphia, PA 19118, USA

1. Introduction

Scholars have documented the resistance to novel scientific discovery by various groups, such as economic and religious groups. However, there has been less attention given to the resistance of the scientific community itself to challenging scientific discoveries (Barber, 1961). Nonetheless, we find it in the history, philosophy, and sociology of science and especially in the writings of those scientists who have personally suffered obstacles due to this resistance. Whereas the scientific community believes that it deals with novel controversial discoveries in a rational manner, this is rarely the case.

The history of science, medicine, and technology is full of rejections of novel discoveries that seemed anomalous in their time. Contemporary scientists laughed when Benjamin Franklin proposed that lightening was a form of electricity. Semmelweiss, a Viennese physician who documented that washing one's hands before obstetrical assistance would prevent childbed fever, was scorned and rejected by his contemporaries. William Crookes, the noted British scientist and member of the Royal Society who discovered the element thallium, was bitterly attacked by his scientific colleagues for his research in parapsychology. Lord Kelvin said that X-rays were a hoax. Helmholtz, who was not a physicist, but a medical doctor who formulated the theory of energy conservation and who was opposed by the physicists of his time noted how the "greatest benefactors of mankind usually do not obtain a full reward during their lifetime" (Murray, 1825). Lister warned medical students against blindness to new ideas in science, such as he had encountered against his own theory of antisepsis. Long after their time, many of these scientists whose ideas were rejected were regarded as formative thinkers who made significant contributions or even launched new scientific paradigms.

2. The Scientific Paradigm

In 1962 Thomas Kuhn published a seminal work, "The Structure of Scientific Revolutions", which addresses the manner in which science advances. Kuhn's main

* Plenary lecture presented at the 7th GIRI Meeting, November 1993, Montpellier, France.

M. Bastide (ed.), Signals and Images, 15–28.

thesis is that science is not a slowly growing body of knowledge approaching a true description of the world. Instead, science is characterized by periods of quiet research activity leading to a crisis, which may last for years to decades. During this transition period, scientific problems appear that cannot be resolved within the given paradigm. Scientific anomalies, experimental results that cannot be reconciled with current theory, may occur. Such anomalies are critical to progress in science. In fact, each new major advance in science starts with an anomaly that is unacceptable at first*. Therefore, anomalies are valuable because they inspire new ways of thinking. Conventional scientists attempt to explain the anomalies within the framework of the dominant paradigm, while a smaller, usually younger group of scientists develop an alternative paradigm. The crisis is resolved by a dramatic change of perspective, a paradigm shift. A struggle typically ensues that may result in the overthrow of the old paradigm. After the triumph of the new paradigm, the old paradigm eventually disappears in a time frame necessary to provide stability and confidence in the new paradigm. What was an anomaly earlier now becomes the expected result. Textbooks are rewritten such that they even disguise the very existence of the revolution that generated them. Eventually, new research uncovers problems with the new paradigm. Then the process repeats itself.

Kuhn notes how unconsciously ingrained the dominant paradigm is. He wrote, "Scientists often work from textbook models acquired through education and through subsequent exposure to the literature without knowing or needing to know they are accepting a community paradigm" (Kuhn, 1970). They work to fit their data into the ruling paradigm. The usual peer review process in science provides an adequate forum for evaluating new ideas and discoveries, but this is only true if those ideas and discoveries do not challenge the paradigm. As was mentioned previously, those considered incomprehensible or too challenging to current scientific understanding are typically rejected. Michael Polanyi, in defending this conservative nature of science, wrote, "There must be at all times a predominantly accepted scientific view of the nature of things, in the light of which, research is jointly conducted by members of the scientific community". Any evidence which contradicts this view has to be disregarded, even if it cannot be accounted for, in the hope that it will eventually turn out to be false and irrelevant" (Truzzi, 1990). Although the neglect of other possible conceptual categories is not malicious in intent, it can become malicious in effect because the dominant paradigm discourages and poorly tolerates competitors. That is, scientists prefer their work to appear as an integral, growing body of knowledge under the auspices of a single paradigm. Perhaps this is because scientists are encouraged to demonstrate what they know rather than to raise truly novel questions that challenge what they think they know.

Kuhn recognized an "essential tension" within science because it must preserve its accumulated knowledge by acting cautiously and conservatively, and on the other hand, remain an open system ready to take in novel, potentially revolutionary data and concepts (Kuhn, 1977). This balance is maintained in a number of ways. First, science places the burden of proof on those who claim to discover scientific anomalies or otherwise make revolutionary scientific claims. Second, the proof must be commensurate with the claim; that is, extraordinary claims require stronger than usual proof. (This relates to the principle of parsimony in science in which the simplest adequate theory is the most acceptable.)

---

* In this regard, it is interesting to note that in Chinese, the character for "crisis" also means opportunity

It is interesting to note that Kuhn (1970) believes that science generally progresses in a positive direction, but that some paradigm shifts have reversed concepts such that aspects of an even older paradigm may return in the form of new input reshaping old models. It is a common conviction that the world is progressing in one direction scientifically and socially, but as Kuhn points out, very often the clock is turned back with new scientific developments. For example, relativity and quantum theory, two of the most significant scientific paradigm shifts in the twentieth century, both turned back the clock in certain ways. The gravitational aspects of Einstein's general relativity reflect back to Newton's predecessors, and quantum mechanics has reversed some of the methodological prohibitions that had occurred in the earlier chemical revolution. Needless to say, the reshaping of older views into a new paradigm would have significance for homeopathy and low dose bioeffects. Many scientists today have the attitude that these phenomena from an era predating modern molecular biology have been overthrown, or that at best they represent a placebo effect. These scientists are victims of historicism who refuse to accept anything from an earlier time as bearing any modicum of truth.

3. Scientific Anomalies

According to science sociologist Marcello Truzzi, an anomaly is something that: (1) actually occurs (that is, something both perceived and validated); (2) is not explained by some accepted scientific theory; (3) is perceived to be something which is in need of explanation; (4) contradicts what we might expect from applying our accepted scientific models. I would suggest that the anomaly's lack of fit with accepted theory is the necessary element common to any real anomaly. It is a fact in search of an explanation (Truzzi, 1987).

In the field of anomalistic observations, or anomalistics (Wescott, 1980), that is, inquiry into anomalies and their role in science, there are different types of scientific anomalies, at least in retrospect. There are those that are recognized in their time by the scientific mainstream that become the subject of legitimate research activity, and those that go ignored by them because they are apparently too threatening. Many of the latter come from the "frontier sciences," that is, whole areas of scientific inquiry that have not yet been incorporated into conventional science. These areas are ignored or even considered irrelevant by the mainstream, in some cases, because they are often residues of older systems of knowledge that have been denounced as pseudo-science, as, for example, parapsychology and astrology.

The history of science shows that the most challenging anomalies, those that seriously challenge the dominant paradigm, are ignored by the scientific mainstream until they are explained, and only then are they recognized in retrospect. The term retrorecognition has been given to this type of recognition only after there is a compelling explanation for the anomaly (Lightman and Gingerich, 1991). Such anomalies make the scientific community uncomfortable, as it likes to think of science as an integral body of knowledge that is nearly complete. These unexplained facts are either ignored, reduced in importance, or merely accepted as "givens". Several factors are behind this attitude, such as the sheer intellectual difficulty in recognizing anomalies, the tendency to ignore a problem that cannot be easily solved, and the conservatism of science. But there is something more. The recognition of what were once anomalies under an older paradigm only after they are reconciled with a new paradigm clearly shows that the scientific community is unable to live with ambiguity and cognitive dissonance (psychological inconsistency). However, frontier scientists whose work challenges the paradigm appear to be of a different psychological makeup, with a higher tolerance for ambiguity and cognitive dissonance. It is interesting to note that such tolerance correlates highly with creativity scores in psychological testing (Barron, 1963). Furthermore, frontier scientists may be working from dimensions other than rationality and logic, for Kuhn has written, "The man who embraces a (new) paradigm at an early stage must often do it in defiance of the evidence. A decision of that kind can only be made on faith" (Kuhn, 1970).

4. The Role of Skepticism

Indeed, it is rare to find those scientists who are true skeptics, that is, without prejudice, open, and tolerant of uncertainty. It is unfortunate that the term "skeptic" is being used by many who are disbelievers or debunkers whose aim is to remove the anomaly, rather than true nonbelievers (Truzzi, 1987). This appears to be particularly the case for organized so-called skeptics groups such as the Committee for the Scientific Investigation of Claims of the Paranormal (CSICOP), which sponsors unusual critiques and other activities to discredit anomalous scientific claims, undermining the usual processes of replication attempts and peer review. In some cases this has involved members outside of the scientific community such as professional magicians in a process analogous to inquisitors for a dogmatic church (Maddox *et al.*, 1988). Unfortunately, this has the effect of creating fear among those who would have an interest in trying to replicate the anomaly, thereby blocking real scientific inquiry.

Where there are anomalies and frontier areas of science that seriously challenge the paradigm, the scientific community is often polarized into two categories: believers and disbelievers. Although the scientific community may consist largely of disbelievers, sometimes the frontier scientists or proponents of an anomaly act as "true believers". In some cases there are societies of "true believers" centered around maverick scientific claims that do not welcome open dialogue. In my opinion, they are no better than some of the mainstream scientists they criticize. Sometimes the discoverer of a challenging fact overstates his claims, jumping to conclusions about the importance of his discovery without adequate data. On the other hand, the "essential tension" of the scientific process renders it very difficult to find the right balance in reporting anomalous claims. If the discoverer understates his claim, it may go ignored; if he stresses its revolutionary character, it may gather more attention and resources for further study. From my own work aiming to facilitate new research and greater open-mindedness in frontier areas of science, I find that it is a difficult position to stand firm on the fine line that separates the believers from the disbelievers. In my opinion this is the best viewpoint to encourage an attitude of nonbelief that stimulates new questions and further experimentation. Apparently this viewpoint is not well understood or liked by most, as I am often accused of being "the enemy" of one group or the other. However, openness and a healthy level of skepticism are crucial in order to avoid pathological science.

5. The Power of New Questions and Approaches in Science

Scientists must approach nature by asking questions of her, and it is impossible to pose a question without some expectation or anticipation. Clearly, from the analysis of

Kuhn and numerous other scientific historians and sociologists, science is not context-independent. Scientific objectivity does not reside in theory-free perception. It lies in the flexibility to reject a cherished theory when an anticipated observation cannot be confirmed, and a contrary event or fact is perceived instead. Scientists may say that they see the data with their own eyes, but in fact, they see it through their brains. They cannot bypass this central focus and filter full of biases, products of both evolution and society. It is very difficult to "see" scientifically beyond the context of theory or expectations.

As an example, consider the following. Before Darwinism, the paradigm that preceded evolutionary theory was natural theology, in which each creature was considered to be perfectly adapted to its environment and designed for full functionality. While natural theology dominated, no one noticed that some organisms were less well adapted to their environment. Natural theology would not permit such questions. Ducks were webbed feet that could not swim, birds with wings that could not fly, and bats with eyes that could not see, went unnoticed. Darwin asked new questions and noticed that some animals were less well-adapted for their environment. He explained these anomalies on the basis of natural selection, an ongoing evolutionary process. The point here is to show the power of asking new questions that take us outside of the present scientific theory or paradigm. These offer the possibility of a breakthrough to a new way of seeing nature. As physicist Werner Heisenberg noted, "What we observe is not nature itself, but nature exposed to our method of questioning."

Another historical example of this goes back to microscopy of the 17th and 18th centuries. The great microscopist Van Leeuwenhoek and his contemporaries claimed they saw minute forms of complete babies inside sperm under the microscope. Their observations were shaped by the 2000-year-old idea that women contributed nothing to conception but the womb as an incubator. In this case, too, preconceived ideas determined what was scientifically observed.

In another historical example involving microscopy, different methodological approaches of observation based on different philosophies led to a scientific debate. In the 1940's the bacteriologist Adrianus Pijper maintained that bacterial flagella are not true motor organs, but are essentially insignificant, being merely cell wall byproducts of bacterial motility (Strick, 1994). From his observations of live bacteria under the dark-field microscope, he claimed that he saw small changes in the forms of the bodies of the bacteria, a slight undulating motion, which he proposed as a theory of bacterial motility. As it turned out, his view was unpopular because he was far outnumbered by those who fixed and stained dead bacteria for light microscopy or electron microscopy, which was newly introduced at that time. The majority of scientists then claimed that flagella were indeed the organelles of motility and showed evidence via microphotography of sites of flagellar attachment to the cell body. Pijper rejected these physical approaches, emphasizing that studies on the living state itself were critical to understanding cellular motility, and that the approaches using dead cells might yield artifacts. This lead to an ongoing debate, as both schools refused to "see" any evidence beyond their own viewpoints. In the end, Pijper lost the debate. His refusal to acknowledge the "superiority" of the electron microscope was held against him by the scientific majority.

Beyond the specifics of this historical debate, the latter case is important for us to consider because it reveals the perennial struggle between the naturalist and the mechanist in biology. It shows how naturalists' observations of living systems were replaced by a modern biology tightly linked to physico-chemical reductionism as new powerful, expensive, prestigious, technological tools came into being. These new physical methods require an often insensitive manipulation of organisms that distorts or even kills them in order to study them. The naturalists' approach came to be regarded as old-fashioned and even reminiscent of vitalism to the new biologists, who were led by several physicists-turned-biologists in the 1940's and 1950's. These were the people who ushered in a new scientific era, the revolution that became the dominant paradigm of molecular biology and biotechnology in recent decades.

6. The Resistance of Scientists to New Discoveries

Studies on the psychology of science suggest that scientists have a resistance to acknowledging data that contradicts their own hypothesis (Truzzi, 1990). In one study on falsifiability, a simple experiment was set up to compare the performance of a group of scientists and a group of clergymen. A false hypothesis was given to all of the participants. The means was provided for them to test the hypothesis, which they did not know was false. The results showed that most of the scientists refused to declare the hypothesis false, clinging to it longer despite the lack of evidence. The clergymen, however, more frequently recognized that the hypothesis was false. This and other studies show that scientists are at least as dogmatic, authoritarian, and irrational as nonscientists in resisting unexpected findings.

The historical examples cited earlier illustrate only a few reasons why resistance to novel discoveries in the scientific community occurs. Analysis of many other examples shows numerous ways in which scientists resist discoveries that are old paradigm breaking and new paradigm making. One of these mentioned earlier is the loathing of ambiguity. Most scientists prefer to elaborate what they think they know than rather focus on what they do not know; perhaps this is simply human nature. Along with that is fear of novelty. New discoveries require restructuring older ideas and ways of doing science. Change, whether it is personal, social, or intellectual is difficult, and may even cause a lifetime of work to become unimportant and obsolete. Related to this is the fact that older scientists have a tendency to resist the novel work of the younger. Innovative "outsiders" may also be rejected by the "insiders," especially if the new discovery comes from outside the field, as in the case of cold fusion**. There is also a faithfulness to old models, reflecting a belief in scientific concepts or simply conservatism. When Thomas Young proposed a wave theory of light, the scientific community remained faithful to the older corpuscular theory for some time. This tendency sometimes reveals a dogmatism or scientism. Paul Feyerabend accuses contemporary science of being a "church" in which scientists play a role that is in many respects similar to the role bishops and cardinals played not too long ago (Feyerabend, 1980). Another mode of resistance, also illustrated by the example cited earlier of Van Leeuwenhoek and his colleagues, is blindness due to preconceptions. It is extraordinarily difficult to "see" what may lie beyond one's paradigm, which delimits all questions posed of nature and ways of perceiving her. Anomalies without "causes" or an adequate explanatory model are rejected because they do not fit neatly into the body of science. If an anomalous claim pertains to an area reminiscent of mysticism, religion, older

** The Princeton Plasma Fusion physicists said of cold fusion, when it was first announced, "What would you do if you were working to develop a propeller airplane that did not yet fly and somebody else from outside the field suddenly invented a rocket ship?"(Mallowe, 1993).

paradigms that have been overthrown, or pseudo-science, this may be grounds for rejection by those who feel threatened by these associations. Along with that, occasionally conflicting personal religious ideas may be a reason for rejection. That was the case for both Galileo and Copernicus, and it also appears to be a factor in the debate between Creationists and Evolutionists. Scientists evaluating an anomalous finding sometimes take into account the relative professional standing of the discoverer as well as the number of prestigious followers of the new claim, and these are primarily political concerns. Concerned about their reputation, scientists are reluctant to take the lead in helping to advance a new claim. In relation to that, publications about the new scientific claim in other than the most prestigious peer-reviewed journals are taken less seriously and may be grounds for rejection or simply neglect. Finally, and perhaps most important to contemporary science is that where substantial funding is involved, patronage to those ideas endorsed and funded to the exclusion of others is overwhelming.

Today, because of large economic interests in science, biomedicine, and technology, and the increasing overlap between academia and industry, the resistance to new discoveries or ideas that challenge the dominant paradigm goes well beyond ideological concerns. Challenging ideas can be seen as threatening to big business interests, including the interests of those industries waging war against cancer or AIDS. Anyone who is a proponent of ideas that threaten large-scale economic interests can expect even harsher backlash from the scientific community, which in mainstream biology and medicine, is now closely linked to pharmaceutical and biotechnology firms. Surely that is one of the most significant reasons for rejection of novelty in biology and medicine today. Moreover, the many different fields of biology with their varied orientations to life that existed before big business science are presently extinct, at least in the U.S.

It is simply taboo to challenge seriously the dominant paradigm, and those who propose such maverick ideas or findings suffer extraordinary obstacles. Similar to the acceptance of novel discoveries, the obstacles are especially severe for those whose work threatens big economic interests that are now coupled to mainstream science.

7. Obstacles Faced by Scientists who Challenge the Paradigm

There are a number of serious, even extraordinary obstacles that scientists presently face as proponents of paradigm-challenging discoveries or where their reputation becomes associated with research on unconventional topics. These obstacles are not characteristic of a particular culture; they appear worldwide. (1) There is difficulty in obtaining funding, as there are simply no usual sources. (2) There is difficulty in publishing, and there is no real peer review. (3) There is loss of camaraderie. Colleagues fear a loss of reputation by association with a scientist who is deemed an outcast. (4) There is loss of reputation in the scientific community regardless of one's stature. (5) There are obstacles to promotion, retention, and tenure. (6) There is possible critical backlash from the scientific community. (7) There is a possible loss of employment and future employment opportunities.

The pursuit of research in frontier science areas such as homeopathy and extremely high dilution bioeffects, novel medical therapies or diagnostics, new energy technologies, and consciousness studies--research in any area that challenges the dominant paradigm--poses extraordinary hardships for scientists. Merely expressing an interest in these can affect one's reputation as a serious member of the scientific community. Whether one is a postdoctoral researcher, a junior professor, a member of a distinguished national academy of science, or a Nobel laureate, essentially the same obstacles remain. For those who have seemingly overcome these hurdles, publications of challenging scientific results may bring about unforeseen backlash in the form of discrediting the discoverer or the claim without really disproving it, prohibiting it from being tested by others. Moreover, this may prevent consideration of similar challenging claims in the scientific literature, textbooks, and education. The proponent of the anomalous claim is thus isolated from further debate and interaction with rest of the scientific community.

Many people associate such repressiveness with earlier times, but there are living examples today. One illustrious example--a case where big economic interests in biotechnology and medical testing are threatened--is that of Peter Duesberg, professor of molecular biology at the University of California at Berkeley. His work identifying the first oncogene to cause cancer and also decoding the first retrovirus genes earned him an outstanding international reputation as a molecular biologist and virologist. However, because of his recent criticism of the oncogene theory of cancer and especially his criticism of HIV as the cause of AIDS, he has essentially been silenced by the scientific community. No one will debate his arguments either in writing or in person. Duesberg is unable to publish in prestigious peer-reviewed journals, not even the Proceedings of the U.S. National Academy of Science, despite his stature as a member of the National Academy, because they rewrote the rules especially to prevent him from publishing. He lost his annual $300,000 Special Investigator Grant from the U.S. National Institutes of Health, which was expressly for the purpose of asking novel questions, and as a result, his students and technicians have had to leave. Duesberg has been excommunicated from the scientific community. Needless to say, the review panel who refused to renew his grant included scientists who earn their living from the theories that Duesberg is undermining, and many others in the mainstream also earn their living from these theories.

8. Strategies Toward Progress in the Frontier Sciences

With all of the obstacles and resistances, how can we help facilitate rational, objective criticism and fair peer review of anomalous claims? What strategies can we implement to bring progress to a frontier science area such as homeopathy and low dose bioeffects?

(1) We must recognize that there is no single critical experiment that can prove an anomaly. This is ridiculous from the scientific viewpoint, as the history and philosophy of science has shown that there is no such thing as a critical experiment.

(2) More empirical studies need to be undertaken by more researchers, and we need to work together at least to provide peer review of each others' work, if not outright collaboration. All too often, the work of pioneering frontier scientists represents isolated, individual efforts. By contrast, most quality science involves collaborative efforts. It is important to build on one another's work. Just as cooperative or collective phenomena in nature have unusual stability, there is also a strength in collective scientific efforts that is harder to dismiss.

(3) An interdisciplinary approach to anomalies is absolutely necessary, because we do not know ultimately where an anomaly will fit. In the case of homeopathy or high dilution bioeffects, interdisciplinary group collaboration with experiments performed in tandem on the same high dilution would be worthwhile, because for the first time it would reveal physical, chemical, and biological information about a single preparation.

This could develop into an international task force, a global cooperation, to address the problem.

(4) We must produce well-designed experiments that are well-communicated in the scientific literature, which will presumably continue to demonstrate the effect in a wide variety of biological systems.

(5) We must show replication of phenomena, especially by skeptics.

(6) We must also discover and document where no such anomalous effects are observed, so that the boundary conditions of the effect are clear.

(7) Conceptual work toward achieving a theoretical explanation for the effect is crucial for its recognition.

(8) We must keep the communication flowing between those working in the field who don't agree on the details. A diversity of opinions is extremely important because it drives the formation of new questions. Good science requires good and effective criticism. Furthermore, failures in communication from splinter groups in frontier areas of science only weaken the case, as their presence makes a statement to the scientific community that there is weakness or irrational behavior associated with the anomaly.

(9) One of our best strategies would be serve as mentors and inspire younger scientists to conduct research in novel areas of science. For one, it is most likely that presently established scientists will have to retire before a paradigm shift is completed, and most of them will not change their viewpoint. As physicist Max Planck sadly noted," A new scientific theory does not triumph by convincing its opponents and making them see the light, but rather because its opponents eventually die, and a new generation grows up that is familiar with it " (Kuhn, 1970). Niels Bohr put it somewhat differently: "Science advances--funeral by funeral."

(10) Retired scientists, who have less to lose in terms of their reputation or funding, are occasionally more open to new ideas or discoveries. Moreover, they may still wield political power in the scientific community. Therefore, communications with or other involvement of retired colleagues may be a viable strategy.

(11) Another strategy that may be used to advance scientific recognition of a challenging anomaly is to identify and align with whatever social, political, or economic interests that would very much like this particular piece of scientific unorthodoxy to be true, or at least to be highly interested in resolving the issue. When Robert O. Becker, medical researcher in bioelectromagnetics had the unorthodox idea in the 1970's that electromagnetic fields from power lines might be a health risk, he found no sympathetic ears in the scientific community or the electric power industry. However, he communicated the issue clearly in his popular writings and launched a public campaign in which the people demanded unbiased research to test his ideas. Within less than two decades, substantial U.S. government funds became available for this purpose.

(12) Another approach related to this strategy is to develop a successful application of the anomaly that will bypass the scientific community altogether. Once the application is adopted, scientists will be naturally drawn to the fundamental discovery underlying it.

(13) Finally, we should attempt to foster true skepticism--neither denial nor disbelief, but a balanced state of openness. The best way to do this is by personal example, by maintaining a level of healthy skepticism ourselves, with an emphasis on further questions. This is crucial to keeping science an open system of inquiry.

9. The Role of Homeopathy and Low Dose Bioeffects in the Future of Science

The observations of low dose biological effects challenge the dominant paradigm of mechanical reductionism, of viewing life as a collection of biomolecules responding to molecular stimuli. The enhanced potency of very low doses as in homeopathy appears to challenge molecular theory, one of the pillars of modern chemistry. On the other hand, it may demonstrate that something else is occurring at these very low doses that does not involve molecules.

Biological effects of low doses have been demonstrated in a growing number of studies worldwide, and we are now in the midst of a paradigm struggle. As Kuhn predicts, an intellectual and emotional battle is occurring: there have been nasty editorials, tenure battles, debates and arguments, splinter groups, the rejection of papers, frequent denial on the part of the scientific community, and many questions that have been raised for further research. From an historical perspective, the accretion of anomalies or numbers of anomalous observations in themselves are not enough to product a paradigm shift. Further effort is required. Conceptual work towards new theories and a paradigm that would reconcile them is critical to their recognition by the scientific community. No one other than the proponents of the anomalies will accomplish this. It remains for us, the frontier scientists, to design the theories, elaborate the new paradigm, and show how they explain our anomalies.

One of the best examples of a conceptual revolution is found in a 19th century science fiction classic: E.A. Abbott's Flatland. The inhabitants of Flatland live on a two-dimensional surface and have no concept of our third dimension. When a sphere visits Flatland, he is perceived as an anomaly: a circle that first grows bigger and then smaller. The sphere then lifts the leader of Flatland into the third dimension where he can see his whole world. This novel perspective not only clears up the anomaly, but offers a new perspective for everything. We need a similar major conceptual breakthrough for homeopathy and low dose bioeffects. When it occurs, it may reframe our ideas of matter, energy, life, and information in a radically new perspective.

Presently the greatest challenge to those working on homeopathy or low dose bioeffects is to develop a proper theoretical context for their observations. We need a theory of very high dilutions in the context of the organism. This would enable us to form testable questions that move the research from an accumulation of anomalous observations to a sequence of facts that fit together like pieces of a puzzle. It is becoming more apparent that molecular theory offers nothing but conceptual limitations for this field of inquiry, and that an alternative that goes beyond it must be sought. Moreover, I anticipate that a breakthrough toward a radically new view of chemistry is in the making, and it is long overdue. Quantum chemist, H. Primas, wrote (Primas, 1982), "The richness of chemical phenomena renders it impossible to discuss them exhaustively from a single point of view. The molecular view is just one of these views and has no privileged status.... While the molecular theory fell on fertile ground, the further development of a theory of chemical substances was deprived of intellectual incentive. Even today, chemical thermodynamics and chemical kinetics are still in a rudimentary state of development achieved at the turn of the century....The molecular idea flourished and degenerated into a dogma, requiring unqualified faith".

He also wrote, "Our vision of the world will be severely limited if we restrict ourselves to the molecular view. Molecular theories describe some aspects of matter, but it is not wise to think that they give us a description of reality 'as it is.' If questions of a different kind can be asked, nature will then respond in a new language."

ANOMALIES AT THE FRONTIER OF SCIENCE

As to the future of science, research on homeopathy and other low dose bioeffects offers the gift of new questions to the greater scientific community--not only for homeopathy and solution chemistry, but for the entire theory of condensed matter with ramifications for biology, chemistry, and physics. Chipping away at the molecular dogma and raising uncertainty about what scientists thought was bedrock truth should be seen as healthy for science. As physicist Louis de Broglie warned us, "The advances of science have always been frustrated by the tyrannical influences of certain preconceived notions that were turned into unassailable dogmas, and for that reason scientists must periodically reexamine their basic principles." Research on homeopathy and low dose bioeffects may lead to a revision or a refinement of molecular theory, or it may show that something other than molecular theory is involved at these low doses.

There is theoretical work in physics towards a new theory of matter that may hold promise for application to homeopathy and low dose bioeffects. Del Giudice (1991) and Preparata (1992) propose a novel theory of condensed matter based on quantum electrodynamics in which collective or cooperative phenomena are critical to its structure and properties. They show that conventional molecular theory works well for gases, but falls short in explaining the phenomena of liquids and solids. A system of molecules kept together by purely static forces becomes dynamically unstable beyond a certain density threshold. Therefore the system enters a lower energy configuration where molecules oscillate in tune with a self-produced coherent electromagnetic field. The energy gain is proportional to the particle density, and then matter is forced to condense. The theory predicts the appearance of coherence domains in solids and liquids such as water. Because the living cell and its structural subcomponents have dimensions of the same order of size of the calculated coherence domains in liquid water, it is expected that electrodynamical coherence may be relevant to the living state, in terms of enhanced stability and novel energy and information transactions. Such novel energy and information transactions, if they exist, may be relevant for homeopathy.

The results of many low dose experiments suggest new features of matter such as information that may be conveyed by more subtle properties of matter than molecules. It comes as no surprise that living systems, which are well known to involve many levels of order and different types of informational exchange, appear to be sensitive to what may be "informational" properties of very high dilutions of bioactive substances. Experiments from another frontier area of biology suggest that there may be subtle non-chemical bio-informational transfer in cellular systems (Kaznacheev, 1976; Kirkin, 1981). Still other experiments suggest that the zero point energy of the quantum vacuum may be involved in subtle informational transfer in biology (Reid, 1989). Perhaps an appropriate explanation for low dose bioeffects awaits us in a biophysics that is yet to be invented.

Whereas conventional science maintains that biological information is stored and transferred via biomolecular structures such as DNA, there is some indication that more subtle informational signals may elicit biological effects. In bioelectromagnetics there are many observations that extremely low-level nonionizing electromagnetic fields whose energy content is below the physical thermal noise limit can produce biological effects, sometimes robust. There is no agreed upon molecular mechanism for these effects. It has been postulated by some that these may act on the organism in such a way that they affect the organisms's endogenous electromagnetic field, which may be bio-regulatory. That is, they act at the level of the whole organism to provide bio-information or disrupt it rather than at the level of energy or power intensity directed to molecular receptors. Furthermore, it is possible that several other phenomena that elicit biological effects such as very high dilutions, homeopathy, healer treatments, acupuncture, and other types of "energy medicine" may mediate their effects by means of coherent excitations, forms of electromagnetic bio-information that might interact primarily with the organism's endogenous fields. Endogenous electromagnetic fields, which are properties of the entire organism rather than specific biomolecules, may be involved in self-regulation of the whole organism, and sensitive to a variety of subtle informational signals from the environment. These speculations not only challenge the concept of molecular mechanisms, but also the dogma that mechanical reductionism is the fundamental principle underlying the living state. However, much work needs to be done to develop these speculations into testable hypotheses and theories.

There are a number of other attacks on the mechanistic view of life that those working on homeopathy or low dose bioeffects should be aware of. Richard Strohman, a leading molecular biologist and Professor Emeritus at the University of California has recently presented some serious challenges to the genetic paradigm. He argues that the information for cellular activity is not in the individual genes, but is holistically located (Strohman, 1993). In his view, biological research is presently missing this integral program. The creativity of the organism, which is perhaps life's most salient feature, involves the interplay of the integral design and function of the organism with its environment. He raises the argument for an epigenetic rather than a genetic view of life, whereby environmental interactions produce hereditable changes. This means that a nonlinear interaction between the organism and its environment takes place, where the temporal sequence of events determines the complexity that unfolds even in the simplest organism. Of course, it is much easier to ask questions within the mechanistic reductionist framework by studying the fragments of a dead organism. It is much more difficult to study the interaction of genetic and environmental factors in a living organism and develop a science of life at this level. However, most biologists fail to see the limitations of their paradigm and the importance of aiming for this larger context.

There is a popular anecdote based on a Sufi story of a drunk who lost his keys somewhere on a dark street and is groping for them only under the street lamp. When asked where he lost them, he replies that he doesn't know, but he is looking there because the light is good. Similarly, the dominant paradigm of mechanical reductionism has prevailed because the biology community has asked only those questions where the "light is good," and the results are clear cut and reproducible.

Biologists explore, for the most part, those dynamical possibilities for life only where organisms "obey" the paradigm. They have missed the enormous creative potential of life in its subtle interactions and interrelationships. Furthermore, the genetic approach has not permitted "other" questions to be addressed, which, in fact, challenge the conventional approach and the dominant paradigm. Moreover, there is a terrible confusion in contemporary biology between the ontology of life, its epistemology, and the methodology. That is, the methodology used (mechanical reductionism) has frequently been equated with life itself or the model of how it functions. This is particularly true in the U.S. where higher education in science does not typically include coursework in the history or philosophy of science.

The whole organism may be a biological fundamental that cannot be reduced to its parts; the whole may be self-governing by virtue of its long-range electromagnetic fields that are the summation of many electrically charged component species and their interactions. This is reminiscent of the words of Claude Bernard, "The vital force directs phenomena that it does not produce; the physical agents produce phenomena that they do not direct." In 1839, when Bernard wrote this statement, the "vital force" was taken to mean a metaphysical concept beyond the scope of science. However, the "vital force" may indeed be a property of the whole organism, a time-varying electromagnetic field summation of all the electrically charged molecular events occurring within it. Subtle biological effects may be mediated through this subtle informational network at the level of the whole.

10. Conclusions

The dominant paradigm of mechanical reductionism that shaped science for the past few centuries, but was overthrown by developments in modern physics earlier this century, still governs modern biology and medicine. Mechanical reductionism, which was developed for the inanimate physical world, determines the scope of questions that can be posed for living organisms, and conventional biology is the collection of theory and results based on those questions. However, frontier scientists are exploring other features of life by asking new questions that go beyond the dominant paradigm. Their questions come from various frontier areas of science and medicine such as epigenetic heredity, bioelectromagnetics, homeopathy, and low dose bioeffects. The results of their investigations, which may be regarded as individual anomalies by the mainstream, may be taken together as evidence for the need of a bigger paradigm to accommodate them. Biology, it appears, may be entering a crisis.

Not only do these "anomalies" challenge our present view of life, but collectively they point to the necessity for a holistic view of life to complement the reductionistic view. Whereas conventional science maintains that biological information is stored and transferred via biomolecular structures such as DNA, the anomalies show that other informational signals not stored in chemical structures may elicit biological effects by possibly altering the subtle informational signals involved in biological regulation of the whole organism.

Major changes in science have never been brought about by isolated experimental findings, but by collective evidence. Thus, it is crucial for scientists who dare to venture into tributaries of the mainstream or uncharted terrain to come together to dialogue and share their data, to find that what may seem as isolated anomalies fit together to form the rudiments of an emerging paradigm. It is important to look at the problems of our science and the gaps in our knowledge. We must continually ask new questions, to never be satisfied with the old ones or the answers that have come to pass. Scientists must continually be motivated by the "mother" of all questions: what facets of nature remain undiscovered because what we consider to be theoretical certainties prevent the posing of new challenging questions?

11. References

Abbott, E.A. (1963) *Flatland*, Harper and Row Publishers, New York.

Barber, B. (1961) Resistance by scientists to scientific discovery, *Science* **134**, 596-602.

Barron, F. (1963) The disposition toward originality, in C.W. Taylor and F. Barron, (eds) *Scientific Creativity: Its Recognition and Development*, Wiley Pubmisher, New York, pp. 139-152.

Bernard, C. (1859) *Des Liquides de l'Organisme*, Tome III, Bailliere Publisher, Paris.

Del Giudice, E. and Preparata, G. (1991) Superradiance: Towards an understanding of the ground states of QED in condensed matter, in T.D. Clark *et al.* (eds)*Macroscopic Quantum Phenomena*, World Scientific Publisher, Singapore, pp. 167.

Feyerabend, P. (1980) Comments by Paul Feyerabend, *Zetetic Scholar* **6**, 52-54. I

Kaznacheev, V.P., Shurin, S.P., *et al.* (1976) Distant intercellular interactions in a system of two tissue cultures, *Psychoenergetic Syst.* **1**, 141-142.

Kirkin, A.F. (1981) Non-chemical distant interactions between cells in culture, *Biofizika* **26**, 839-843.

Kuhn, T.S. (1970) *The Structure of Scientific Revolutions*, Second Edition, University of Chicago Press Publisher, Chicago.
Kuhn, T.S. (1977) *The Essential Tension.*, University of Chicago Press Publisher, Chicago.
Lightman, A. and Gingerich, O. (1991) When do anomalies begin?, *Science* **255**, 690-695.
Maddox, J., Randi, J., and Stewart, W.W. (1988) "High dilution" experiments a delusion, *Nature* **334**, 287-290.
Mallove, E. (1993) Personal communication.
Murray, R.H. (1825) *Science and Scientists in the Nineteenth Century*, Sheldon Press, London.
Preparata, G. (1992) Coherence in QCD and QED, in T. Bressani *et al.* (eds) *Problems and Ideas of Modern Physics*, World Scientific Publisher, Singapore, pp. 3.
Primas, H. (1982) Chemistry and complementarity, *Chimia* **36**, 293-300.
Reid, B.L. (1989) On the nature of growth and new growth based on experiments designed to reveal a structure and function for laboratory space, *Medical Hypothesis* **29**, 105-127.
Strick, J. (unpublished paper) Adrianus Pijper and the debate over bacterial flagella: morphology and electron microscopy in bacteriology, 1946-1956. Presented April 1, 1994 at Joint Atlantic Seminar for the History of Biology at MIT.
Strohman, R. (1993) Ancient genomes, wise bodies, unhealthy people: limits of a genetic paradigm in biology and medicine, *Perspectives in Biology and Medicine* **37**, 112-145.
Truzzi, M. (1987) Zetetic ruminations on skepticism and anomalies in science, *Zetetic Scholar* **12**, 7-20.
Truzzi, M. (1990). Reflections on the reception of unconventional claims in science, *Frontier Perspectives* **1**, Fall/Winter, pp. 13-25.
Wescott, R.W. (1980). Introducing anomalistics: a new field of interdisciplinary study, *Kronos* **5**, 36-50.

CHAPTER I

HORMESIS

T.D.LUCKEY
Radiation Hormesis.
F.A.C.WIEGANT *et al.*
Stimulation of Self-recovery by Low Doses of Arsenite-intoxicated Cells.
R.VAN WIJK *et al.*
Stimulation of Cellular Self-recovery by Application of the Similia-principle.
S.FAIDERBE *et al.*
Preventive Effect of a Monoclonal Anti-idiotypic Antibody Compared to a Hormetic Model on Rat Sarcoma.
A.DELBANCUT *et al.*
Evidence and Mechanistic Approach of the Protective Effects of Heavy Metal High Dilutions in Rodent and in Renal Cell Cultures.

The modern notion of hormesis originated with the observations of Southam and Erlich (*Phytopathology*, **33**, pp 515-524, 1948) and was developpped by Stebbing (*Aquatic Tox.*, **1**, pp 227-238, 1981). Hormesis is a very well known phenomenon which demonstrates the reverse toxic action of a substance which becomes a stimulating agent at a lower concentration. This was demonstrated with ponderal doses, the reversion of the effect being obtained only by a small decrease of the concentration (10-100 times lower). Models have been created with various poisons and on all possible organisms whatever their level in evolution, from procaryotic cells to plants, from eucaryotic cells to mammals This reverse effect is studied here by Luckey who observed that cancer mortality rates were lowered by exposure to low doses of ionizing radiations.

Conversely, the administration of low doses of a poison to an organism increases the resistance towards that poison. The famous king Mithridate, using small daily doses of various poisons, could not be killed by poison and eventually died by a sword-thrust.

This model is presented here by Faiderbe *et al.* in a tumoral induction model provoked in rats by a chemical agent by comparison with the internal image of this toxic agent built by the immune system (anti-idiotypic antibody). A molecular hormetic protective effect was observed although the protection was

M. Bastide (ed.), Signals and Images, 29–30.

much more effective when the animals were pretreated by the internal image of the chemical agent.

A "post intoxication" treatment was also tested by Wiegant *et al.* and van Wijk by administration of ponderal lower concentrations of the poisons. They could observe a protective mechanistic effect by increasing the heat shock protein production. This effect was not strictly specific to the poison and cross-reactions were observed. It resembles an amplification of the defense process thus demonstrating a "mechanistic" similia-law which is rather different from the similia-law of homeopathy based on the observation of the "symptoms" (see the next chapter). On the contrary, Delbancut *et al.* protected rats or cultured cells by *successed* (or *potentized*) high dilutions 10^{-30} or 10^{-40}M of mercury chloride (non-molecular successed solutions) against the same chemical (mercury chloride). In this case, the protection seems to be strictly specific to the same toxic (no cross reaction) and the protective metallothionein is not released directly by the treatment but only at the intoxication time.

It seems that pretreatment or post-treatment by molecular doses only amplifies the "protective process" whatever the inducing agent (cross-reaction) while pretreatment by successed high dilutions (non molecular) induces a learning process in the living organism aimed to establish a protection against the "waited danger". Specific appropriate protective tools are therefore conceived and performed by the living organism only at the poisoning time by the "same poison" (law of identity). We can propose an informative function of the successed high dilution differs from the molecular function.

M.B.

RADIATION HORMESIS *

Biopositive Effect of Radiations

T.D. LUCKEY
1009 Sitka Ct, Loveland, CO 80538, USA

1. Introduction

My purpose is to promote harmony with nature and to improve our quality of life with the knowledge that cancer mortality rates decrease following exposure to low dose irradiation. Hormesis (Greek HORMO = I excite) is the stimulation of any system by low doses of any agent. Hormology is the study of excitation. Low doses of many agents evoke a biopositive effect ; large doses produce a bionegative effect. The message is simple : *small and large doses induce opposite physiologic results.*

Radiation hormesis implies stimulation by ionizing radiation. Cancer induction is the most feared action of large doses of ionizing radiation. Therefore, cancer mortality rates will be used to illustrate radiation hormesis in humans. Large doses of ionizing radiation increase cancer mortality rates ; this is considered to be harmful. Since small doses decrease cancer mortality rates, low dose irradiation is beneficial. Although small doses of radiation can stimulate cell an cancer growth, the stimulation of different components of our complex immune system more than compensates for simple cellular effects. The net effect is a decreased cancer mortality.

This report will a) review the development of the concept of hormesis, b) present an overview of radiation hormesis in humans, c) compare and contrast hormesis and homeopathy, and d) suggest a surprising mechanism. Constraints of time and space dictate that each of these will be very brief.

2. Development of the Hormesis Concept

Antibiotics became available to non-medical investigators toward the end of World War II. It was just 50 years ago that I received some streptomycin from the Eli Lilly Company. Our working hypothesis was : feeding antibiotics should prevent intestinal microbes from supplying unknown vitamins to the host ; the animal should then exhibit a new vitamin deficiency. The expected did not happen ; chicks fed streptomycin grew better than controls (Moore *et al*, 1946). The mechanism of action of dietary antibiotics was complex (Luckey, 1959). The use of germfree animals showed that part

* Plenary lecture presented at the 7th GIRI Meeting, November 1993, Montpellier, France.

M. Bastide (ed.), Signals and Images, 31–39.

of the action was directly on the host ; part was on the intestinal microbes (Luckey *et al*, 1956). This began the use of dietary antibiotics for animals all over the world.

A literature search revealed much information suggesting that large and small doses evoke opposite effects : Hippocrates, *Similia similibus curentur* or "likes are cured by likes" ; Paracelsus, the father of infinitesimal doses, "The dose makes the poison" ; Hahnemann, "Drugs have a dynamic effect when used in small doses" ; Cannon, "Adaptation to perturbations is the basis for homeostasis" ; Selye, "The General Adaptation Syndrome" ; and Arndt-Schultz, "Poisons are stimulants in small doses". These were supported by the remarkable findings of Richet (1906). He understood the oligodynamic action of metals as proposed by Nageli (1893) : very dilute solutions of metallic ions are toxic. Richet quantified the toxic effects by using serial dilutions of several metallic ions. He found that each metal ion exhibited a threshold and was stimulatory at subharmful doses. This linked out antibiotic response to classic science.

Hormesis was the word suggested by Southam and Erlich (1943) who determined the effective concentration of phenolic compounds in wood which protects trees from fungi decay. They diluted their extracts to determine the minimum amount needed for fungal retardation. The unexpected increase in growth rate with dilute solutions was called *hormesis*. They agreed to a broad definition of the word.

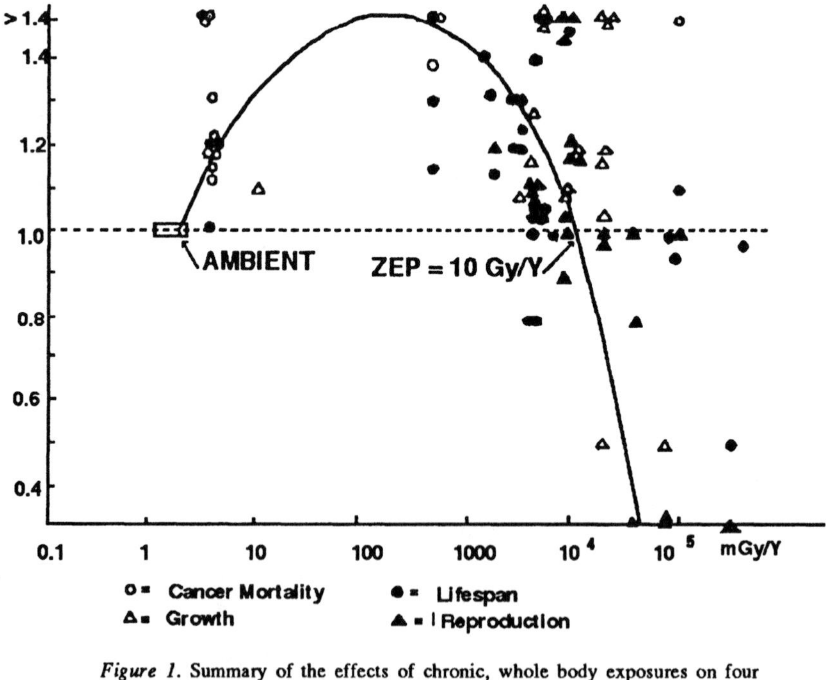

Figure 1. Summary of the effects of chronic, whole body exposures on four physiologic functions. Radiation hormesis is represented by the defined area above the horizontal line. When compared with the controls, represented by the horizontal line, large dose rates exert a negative effect.

Many physiologic functions show radiation hormesis : growth, neuromuscular development, hearing and visual acuity, learning and memory, fecundity, immune competence, cancer mortality and average lifespan (Luckey, 1990; 1993). The effects of chronic, whole body exposure to low doses of ionizing radiation upon four physiologic functions shows radiation hormesis (Figure 1). The controls, at ambient levels of radiation, are set at one, the horizontal line. Hormesis is displayed by values >1. Large doses decrease these functions. There is a definite threshold, the zero equivalent point (ZEP), at the transition between biopositive to bionegative effects. Note that the ZEP is more than 1,000 times ambient radiation levels. Note also that cancer mortality is displayed as an inverse function : the less cancer mortality, the higher the relative value.

The results from human experiences are grouped near ambient levels of radiation (Figure 1). Results from animal experiments comprise the large group of results on the right side. Since these were more than 100 times ambient levels of radiation, the two components leave a gap in our knowledge about low dose irradiation, an area for future research. When both chronic and acute exposures were considered, over 100 experiments showed statistically valid, p<0.01, radiation hormesis. There is no comparable set of results which show harm from low dose irradiation (Luckey, 1990; 1993). The results displayed in figure 1 provide background for this critical review of recent literature on cancer mortality in humans exposed to whole body irradiation.

3. Acute Exposure to Ionizing Radiation

Acute exposure to ionizing radiation is exemplified by the 86,520 Japanese survivors of atomic explosions ; 41,372 were in the exposed cohort (Shimizu *et al.*, 1992).
The total cancer mortality and leukemia mortality rates of lightly exposed survivors were consistently less than that of the control population (Figure 2).

The chisquare statistic indicates that these data are not significant because the sample size is too small. If the sample size were to be increased by a factor of ten, the data would show statistically significant decreased mortality rates for both total cancer and leukemia in the lightly exposed population, p<0.01.

Radiation hormesis in cancer mortality was found in 32,000 United States and 22,000 British military observers of atmospheric nuclear explosions (Robinette *et al.*, 1985; Darby *et al.*, 1988). The cancer mortality rate of Canadian military observers was only 88% of carefully selected military controls (Raman *et al.*, 1987). The leukemia mortality rate of the Canadian observers was only 40% of that of the unexposed controls. In each study the cancer motality rate of exposed personnel was lower that that of the general population.

The cumulative data represents about 100,000 acutely exposed persons in four countries. This is convincing evidence that whole body exposure to low doses of ionizing radiation do not cause increased mortality. The supporting animal data showed that both acute and chronic exposure to low dose irradiation decreased cancer mortality (Luckey 1990, 1993). The combined animal and human results provide impressive evidence that cancer mortality is decreased by acute exposure to low doses of ionizing radiation.

4. Chronic Exposure to Ionizing Radiation

Chronic whole body exposure to ionizing radiation is pertinent to the welfare of all societies. Epidemiologic studies of chronic exposure to ionizing radiation consistently showed a negative correlation between ambient (background) radiation and leukemia, cancer and total mortality rates (Luckey 1990, 1993). These are exemplified by a study in which lung cancer mortality rates were compared with radon in homes (Figure 3) (Cohen, 1993). The dashed line represents a popular fallacy. Unfortunately, many agencies use this erroneous concept for official recommendations and regulations. The solid line represents the mean values, with 5% errors bars, from 1700 counties of the United States in which 90% of the population reside. The negative slope of the solid line shows that the optimun concentration must be greater than the 4 pCi l^{-1} which some agencies consider to be excessive. Cohen's data confirm the results of Haynes from England and Wales (Haynes, 1988). Large radon clinics in Russia are used to treat a variety of minor disease problems (Bogoljubov, 1988).

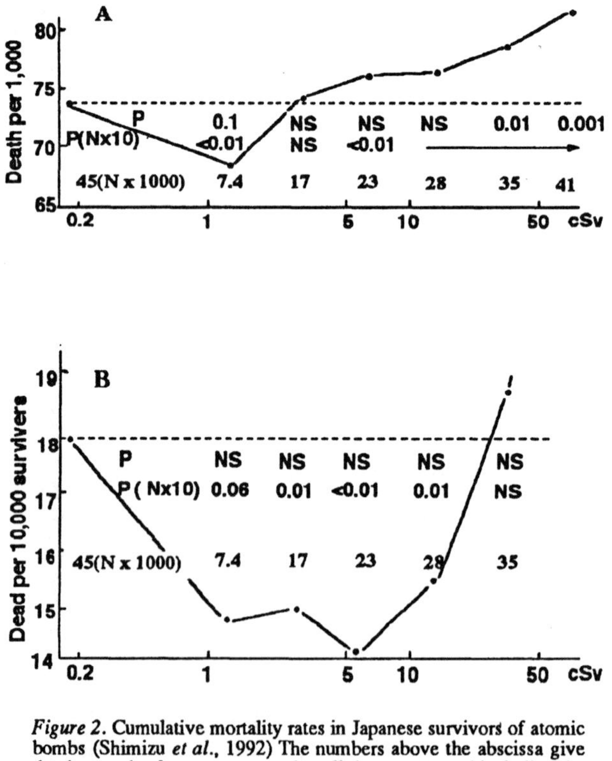

Figure 2. Cumulative mortality rates in Japanese survivors of atomic bombs (Shimizu *et al.*, 1992) The numbers above the abscissa give the thousands of persons exposed to all doses up to and including the dose indicated: A: cumulative total cancer mortality rate and B: cumulative leukemia mortality rates.

When the concentrations of radon and progeny were increased to over one million times ambient levels, lung tumors were induced in animals. This is not cause to change our perception of the results in figure 2. Low levels of radon and progeny do not cause lung cancer mortality ; they may decrease it.

Chronic exposure to low dose irradiation is exemplified by nuclear workers. A summary of recent studies in over 7,000,000 person-years of exposed nuclear workers is given in Table 1. In most studies there was a significantly lower cancer mortality rate in exposed workers than in unexposed workers in the same plants, $p<0.05$-0.001. Only in the smallest study was the difference not significant. It is suggested that this study had inadequate correction for age.

Leukemia is the classic radiation-induced cancer. The leukemia mortality rates of 130,000 exposed workers was less than that of unexposed workers, $p<0.01$ (Gilbert *et al.*, 1989; Abbatt *et al.*, 1983; Gribbin *et al.*, 1993; Kendall *et al.*, 1992).

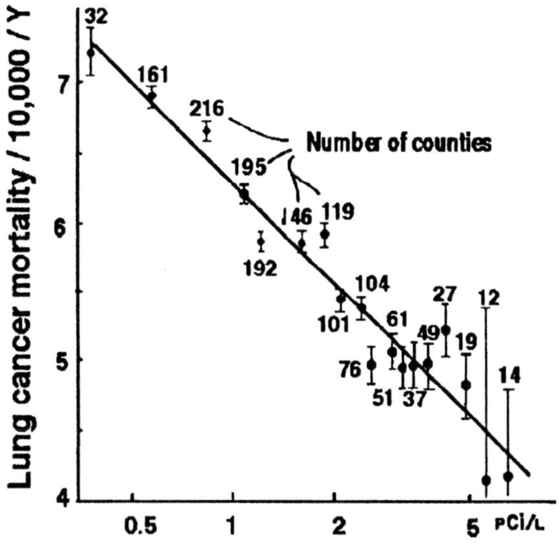

Figure 3. Lung cancer mortality rates in males compared with radon concentration in United States homes (Cohen, 1993). Each mean ± one standard deviation includes the number of counties represented. The dashed line indicates the invalid, but much used, linear model of the data.

Presumably, low dose irradiation activates the immune system to destroy the altered stem cells of leukemia. This concept is supported by many studies showing increased activity of the components of the immune system (Sugahara *et al.*, 1992).

TABLE 1. Total cancer mortality in nuclear workers

Plant	Worker		% Cancer	
	Control	Exposed[a]	Mortality[b]	
US Shipyards[c]	32,510	38,220	67*	Matanoski
US Weapons	20,619	15,817	21**	Gilbert
Canada Energy	21,000	4,000	59**	Abbatt
Canada Energy	4,000	4,000	95*	Gribbin
Brit.Weapons	24,500	70,900	3**	Kendall

* $P < 0.05$ ** $P < 0.001$

[a] Lifetime exposures were 1-20 cSv.

[b] % = Exposed/internal Control X 100.

[c] Total death rate ; total cancer mortality data were not found.

5. Discussion of Radiation Hormesis in Cancer Mortality

Most evidence showed a significant decrease in cancer mortality rates of lightly exposed humans. This was true for acute exposures, survivors of atomic bombs in Japan and observers of atmospheric nuclear explosions in the United States, Britain and Canada. It was equally true for chronically exposed workers when compared with eiter unexposed workers in the same plants or with the general population. The obvious conclusion is : low doses of ionizing radiation do not cause cancer. Experimental animals show the same result (Luckey 1990, 1993). Thus, it is highly probable that low dose irradiation actually lowers cancer mortality rates in humans.

Radiation hormesis is used with success in the clinic (Yamaha *et al.*, 1992). The best results appear following irradiation of the lower body with 10-15 cGy twice weekly. This procedure is particularly useful in treating non-Hodgkin's lymphoma patients. Obviously, this opens the way for more clinical research.

The evidence provides a new and scientifically valid concept; *cancer mortality rates are lowered by exposure to low doses of ionizing radiation.* This invalidates the zero thesis that *all* radiation is harmful. The use of linear, no threshold models should cease.

Since most studies had internal controls, the results were not due to "a healthy worker effect". The "healthy worker effect" can not explain the decreased cancer mortality in nuclear workers. Most of the studies compared exposed workers with unexposed workers ; both a) were in the same plants, b) passed the same entrance examinations and c) had comparable medical care. Secondly, there is evidence that the exposed workers have a longer average lifespan. Since cancer mortality increases dramatically with age, the exposed workers should have had a higher cancer mortality rate than the controls. They did not. Thus, the best explanation for the decreased cancer mortality following low dose irradiation is radiation hormesis. These results negate the "healthy worker effect" and show that the use of linear, no threshold models for the effects of low dose irradiation is counterproductive.

6. Compare and Contrast Hormesis and Homeopathy

How can comparisons and contrasts between radiation hormesis and homeopathy be made ? One represents science ; the other the art of medicine. Science does not intrude on the delicate nuances between physician and patient. Science attempts to simplify and separate elements of complex phenomena. Medicine accepts the complex, ever changing condition of each patient as the progress of disease and multiple treatments acrue. Science demands strict controls and knowledge from multiple sources. Medicine involves individual faith, hope, beliefs and prayers in addition to all that science can contribute.

TABLE 2. Comparison and contrast of hormesis and homeopathy

A. Hormesis and Homeopathy are comparable	
	Large doses are harmful
	Dose rate is important
	Low doses are beneficial
	Unusual compounds may be active
	Reaction initiated within hours
B. Homeopathy characteristics (not typical of hormesis)	
	Only special compounds are active
	Very minute doses are active
	First dose sensitizes the subject
	Medical use is primary
	Psychosomatic factor is great
	Primary effects are short-lived
C. Hormesis characteristics (minimal in homeopathy)	
	Non specific agents are active
	Medium doses are active
	Most dose rates are active
	Animal and human react
	Statistically valid data
	Many parameters are affected*
	Immune competence involved
	Benefits sick and well
	Effective for months/years
	Public health concern

* *Fecundity, development, lifespan*

Comparisons and contrasts of hormesis and homeopathy are listed in Table 2. Shared characteristics are listed in part A. Characteristics of homeopathy which do not typify hormesis are listed in part B. And hormesis characteristics not found in homeopathy are listed in part C. The major difference is that hormesis is a general phenomenon while homeopathy involves one, or very few, specific compounds for each malady. Many agents evoke hormesis. And each agent activates many systems ; hormetins stimulate reproduction, growth, neuromuscular develoment, mental acuity and memory, defense from infection and cancer mortality, and average lifespan. Radiation in cancer mortality is a lifetime effect ; most homeopathic treatments involve only a few day or weeks.

7. Radiation Hormesis is Essential

Finally, radiation hormesis is in a special class of hormetics. A major mechanism of action of ionizing radiation is proposed. It is probably an essential agent. Thus, radiation hormesis is comparable to those essential nutrients which are not present in adequate quantities in certain environments. The hormesis curve for ionizing radiation is comparable with those for vitamin A or selenium. An excess is harmful. Small amounts are needed for essential physiologic functions. Supplementation is usual for populations living in a partial deficiency. This is done for vitamin A and selenium. Irradiation supplementation promises increased quality of life and a new plane of health for people in the 21st century.

New concepts raise new questions. What is an optimal level of radon, alpha, beta, gamma and X rays for babies, children, teens, young adults, pregnant females, old people and sick people ? How can the art of medicine best use low dose irradiation ? Is ionizing radiation essential for optimal physiologic functions ? For life ? Should whole populations be supplemented ? How ? Should control be with health physics professionals, the medical profession, public health officials, or politicians?

8. References

Abbatt, J.D., Hamilton, T.R. and Weeks, J.L. (1983) Epidemiological studies in three corporations covering the nuclear fuel cycle. Biological Effects of Low-Level Radiation, *Proc.IAEA-STI/Publ646*, International Atomic Energy Agency, Vienna, 351-361.

Bogoljubov, W.M. (1988) Clinical aspects of radon therapy in the U.S. S. R. Z. *Phys. Med. Balneol. Med. Klimatol.* 17,59-61.

Cohen, B.L.(1993) Test of the linear-no threshold theory of radiation carcinogenesis, *ICSE Conference*, Paris.

Darby, S.C., Kendall, G.M., Fell, T.P. *et al.*(1988) A summary of mortality and incidence of cancer in men from the United Kingdom who participated in the United Kingdom atmospheric weapons tests and experimental programs, *Br. Med. J.* **296**, 332-340.

Gilbert, E.S., Fry, S.A., Wiggs, L.D. *et al.* 1989) Analysis of combined mortality data on workers at the Hanfort Site, Oak Ridge National Laboratory, and Rocky Flats Nuclear Weapons Plant, *Radiat. Res.* **120**, 19.

Gribbin, M.A., Weeks, J.L., and Howe, G.R.(1993) Cancer mortality (1956-1985) among male employees of Atomic Energy Limited with respect to occupational exposure to low-linear-energy-transfert ionizing radiation, *Radiat Res* **3**, 375.

Haynes, R.M.(1988) The distribution of domestic radon concentrations and lung cancer mortality in England and Wales, *Rad. Prot. Dosim.* 93-4.

Kendall, G.M., Muirhead, C.R., MacGibbon, B.H. *et al.* (1992) Mortality and ocupational exposure to radiation ; first analysis of the National Registry for Radiation Workers, *Brit Med J* ,**304**, 220.

Luckey, T.D.(1959) Antibiotics in Nutrition, in H.S.Goldberg (ed), *Antibiotics, Their Chemistry and Non-Medical Uses*, D. Van Nostrand Publisher, Princeton, pp174-321.

Luckey, T.D. (1990) *Hormesis with Ionizing Radiation*, CRC Press, Boca Raton Publisher, In Japanese Soft Science Inc., Tokyo.

Luckey, T.D. (1993) *Radiation Hormesis*. CRC Press, Boca Raton Publisher, 1991. In Japanese, Soft Science, Inc., Tokyo.

Luckey, T.D., Gordon, H.A., Wagner, M. and Reyniers, J.A. (1956) Growth of germfree birds fed antibiotics, *Antibiot. Chemother.* **6**, 36-40.

Matanoski, G.M. (1985) Health Effects of Low Level Radiation in Shipyard workers, *DOE Final Report*. E.1.99 DOE/EV/10095-T 1 and 2. DOE, Washington.

Moore, P.R., Evenson, A., Luckey, T.D., McCoy, E., Elvehjem, C.A. and Hart, E.B. (1946) The use of sulfasuxadine, streptothrycin and streptomycin in nutritional studies with the chick, *J. Biol. Chem.* **165**, 437-441.

Nageli, U. (1893) Ueber oligodynamische Erscheinungen in Lebenden Zellen. Gesellsch, *f.b. Naturwissenschaft* **33**, 1-4.

Raman, E., Dulberg, C.S. and Spasoff, R.A. (1987) Mortality among Canadian military personel exposed to low-dose radiation, *Can. Med. Assoc. J.* **136**, 1951-55.

Richet, C. (1906) De l'action de doses minuscules de substance sur la Fermentation Lactique - troisième mémoire - Périodes d'accélération et de ralentissement, *Arch. Intern. Physiologie* **4**, 18-50.

Robinette, C.D., Jablon, S. and Preston, T.L. (1985) Studies of Participants in Nuclear Tests, *National Research Council Final Report*. DOE/EVO1577, Washington.

Shimizu, Y., Kato, H., Schull, W.J. and Mabuchi, K. (1992) Dose response analysis among atomic-bomb survivors exposed to low-level radiation, in T. Sugahara, L. Sagan and T. Aoyama (eds.) *Low Dose Irradiation and Biological Defense Mechanisms*, Exerpta Medica Publisher, Amsterdam, pp. 71-74.

Southam, C.M. and Ehrlich, J. (1943) Effects of extract of western red-ceder heartwood on certain wood decaying fungi in culture, *Phytopathol.* **33**, 517-524.

Sugahara, T., Sagan, L.A. and Aoyama, T.(1992) *Low Dose Irradiation and Biological Defense Mechanisms*, Excerpta Medica Publisher, Amsterdam.

Yamada, S., Nemoto, K., Ogawa, Y., Yakatou, Y., Hosi, A. and Sakamoto; K.(1992) Anti-tumor effect of low dose total (or half) body irradiation and changes of the functional subset of peripheral blood lymphocytes in nonHodgkin's lymphoma patients after TBI (HBI),in T. Sugahara, L. Sagan and T. Aoyama (eds.) *Low Dose Irradiation and Biological Defense Mechanisms*, Exerpta Medica Publisher, Amsterdam, pp. 113-116.

STIMULATION OF SELF-RECOVERY BY LOW DOSES OF ARSENITE IN ARSENITE-INTOXICATED CELLS *

Self-recovery by low doses of arsenite

F.A.C. WIEGANT, J.H. OVELGÖNNE, J.E.M. SOUREN, AND R. VAN WIJK
Department of Molecular Cellular Biology, Utrecht University, PO Box 80.056, 3508 TB Utrecht, The Netherlands

1. Introduction

The aim of our studies is to understand the stimulation of self-recovery processes at the cellular level by compounds which are applied according to the similia-principle.

In order to induce self-recovery, cells are exposed to mildly damaging conditions. In our research a non-lethal concentration of sodium arsenite was used which induces the so-called heat shock- or stress-response (Nover, 1991; Welch, 1992). This response includes the synthesis of a specific class of proteins, the heat shock proteins (hsps) and the development of tolerance. It has been shown that self-recovery can be evaluated in terms of the development of tolerance and the enhanced production of hsps (Van Wijk and Wiegant, 1994). Hsps facilitate repair and refolding of stress-damaged proteins and also prevent irreversible denaturation and aggregation of cellular proteins (for review see Hightower, 1991; Welch, 1992; Parsell and Lindquist, 1994). In this latter function they may act as 'protector proteins', which may be one of the underlying molecular mechanisms by which cells acquire tolerance.

With respect to the aim of our studies, it is of interest to determine whether, after an exposure to a damaging condition, cells can be stimulated in their development of tolerance and the production of hsps, by stressor doses that are not effective in undamaged (control) cells. Surprisingly, hardly any studies have been performed to determine the effect on the synthesis of hsps or the development of tolerance of an exposure of cells to these so-called step-down protocols; a transient incubation with a high dose of a compound, which is followed by the application of a low dose of the same or a similar compound.

In the present paper, experiments are described which were performed to study the effect of low doses of arsenite on the synthesis of hsps and on the development of tolerance in arsenite-sensitized cell cultures. In first instance, however, the conditions

* Communication presented at the 7th GIRI Meeting, November 1993, Montpellier, France.

M. Bastide (ed.), Signals and Images, 41–51.

for a cell model are summarized as far as it is required for the study of self-recovery and its stimulation with low doses.

2. Optimal Conditions to Study the Stimulation of the Recovery of Damaged Cells

A variety of factors influence the effect of stressors and the process of hsp-activation. For the experiments described in this paper, mild stressor conditions are used in order to prevent the occurrence of too much irreversible damage. Cell populations should be obtained that are sensitized and are able to express self-recovery (as evaluated by the synthesis of hsps and the development of tolerance).

2.1. THE USE OF CELLS SHOWING SUB-OPTIMAL SELF-RECOVERY

In order to be able to stimulate recovery processes, a (sub-)population of cells is required which shows sub-optimal self-recovery. Using mild stressor conditions, a variability exists in reaction pattern and consequently in the speed of tolerance development in the different cells of the population. There will be a subpopulation of cells showing a fast recovery, whereas another subpopulation of cells will show a slow recovery and hence a slow development of tolerance. Our studies are focused on the sub-population of cells showing a slow recovery. This sub-population has been threatened in such a way that they are damaged but are relatively weakly activated in their production of hsps. The replenishment of the pool of protector proteins, i.e. the new production of these hsps, is hampered causing the occurrence of a sensitive state. The subpopulation of cells being in this state has the highest probability of becoming irreversibly damaged and of dying. Our research is aimed on stimulating this part of the cell population which shows a sub-optimal self-recovery, by applying low doses of the stressor. The hypothesis is formulated that a low doses is effective when cells are damage-sensitized. The effect on self-recovery will be evaluated by studying the mean reaction pattern of hsp synthesis and tolerance development as measured in the total cell population.

2.2. SELF-RECOVERY AND THE SIMILIA PRINCIPLE

According to the similia-principle, self-recovery may be stimulated by low doses of a specific substance that is capable of inducing similar symptoms in a healthy system. The stimulation of self-recovery should not occur with low doses of substances that induce different symptoms in the healthy system.

In essence, the isopathic-treatment is an example of the similia-principle in its most elementary form. The same substance which has caused a dis-balance or a pathological condition will support the recovery from this dis-balance (provided that an adequate doses is given). The application of this principle in clinical practice is difficult since in many cases the cause of pathology is unknown. Then a substance should be chosen which is able to induce symptoms in a healthy organism optimally resembling the symptoms expressed by the organism in the pathological state. With respect to the treatment of patients, one is entering here into the specialism of homeopathy. Homeopathic physicians claim self-recovery processes to be stimulated when an adequate homeopathic remedy is administered to the patient.

At the cellular level, it can be expected that - in an isopathic approach - self-recovery can be stimulated with a small doses of the substance which was responsible for the disturbance of the system. This stimulation will show itself in an increased activation of the expression of the cellular gene program for protective proteins, leading to a stimulated supplementation of protective proteins and in an increased resistance for the disturbing agent.

2.3. NUTRITIONAL CONDITIONS

When the effect of low doses are studied, the medium is changed various times during the experiments, resulting in the possibility of a stimulation of the self-recovery when nutritional components are introduced. However, when the medium is refreshed several times before and during the experiment the possibility that self-recovery is stimulated by other factors than the application of low doses of the stressor, can be excluded. For this reason, Reuber H35 hepatoma cells were grown in complete media, given fresh media 12 h before the experiment and incubated in a stressor-containing fresh medium again. After the application of the stressor, cells are rinsed extensively with phosphate buffered saline (PBS) and incubated again in fresh medium containing the low doses.

Summarising, to be able to demonstrate the stimulation of the recovery process and to avoid a cell population with an increasing number of cells with irreversible damage it is essential to expose cells to mild damaging conditions. Furthermore, the metabolic capacity of the cells to perform self-recovery processes should be optimal in order to prevent changes to be due to the nutritional condition of the cells. For this reason, the cell culture is provided with optimal nutritional conditions. The model developed can then be used to test the similia-principle and to determine the specificity of this principle.

3. Do Low Doses of Arsenite Stimulate Self-recovery in Arsenite-Intoxicated Cells?

This study has been focused on the regulation of cellular sensitivity to arsenite-intoxication. There is evidence for the existence of a state of tolerance to arsenite (Lee and Hahn, 1988; Lee *et al.* 1989) and recently the phenomenon of an arsenite-induced transition from enhanced (sensitization) to reduced (tolerance) sensitivity to arsenite has been demonstrated (Wiegant *et al.*, 1993). Previously, arsenite has been shown to be an effective activator of hsp gene expression in a variety of mammalian cell types (Johnston *et al.*, 1980; Levinson *et al.*, 1980; Li, 1983; Wiegant *et al.*, 1987). We have investigated whether a treatment of H35 rat hepatoma cells with arsenite sensitizes cells to arsenite. It can be argued that when a threat is applied which is just strong enough to stimulate cells to start the replenishment of hsps, that this replenishment can be stimulated by a lower doses of the stressor, provided that it is given directly following the first threat. For this reason we incubated arsenite-treated cells with low doses of arsenite, according to the step-down protocol, which could enhance the synthesis of the levels of hsp68-mRNA as well as the induction of the major hsps. In addition, the effect of low concentrations of arsenite on the induced development of arsenite-tolerance has been studied in arsenite-sensitized cells. According to previous definitions, 'tolerance' is described as a stress-induced cellular resistance to a lethal stress treatment, whereas 'sensitization' is defined as an increase in cellular susceptibility for a stress treatment (Wiegant *et al.*, 1993).

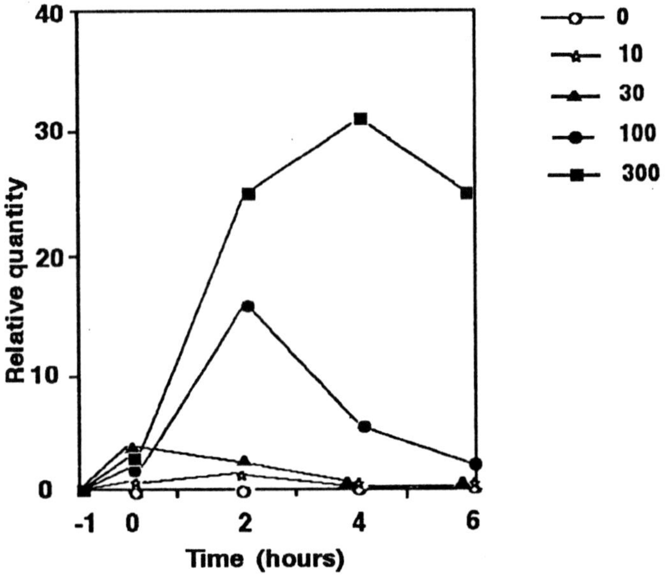

Figure 1. The induction of hsp68-mRNA after an exposure to arsenite (10, 30, 100 or 300 μM) for 1 h in H35 cells. Arsenite treatment was followed by a recovery in normal culture medium for up to 6 h. RNA samples were prepared at t = -1 hour (untreated cells), directly after the treatment (t=0) and after a recovery of 2, 4 and 6 h. Hsp68-mRNA is shown as a relative quantity (means of 2 duplicates).

3.1. LEVELS OF HSP68-mRNA

In first instance, the effect of the step-down (arsenite) protocol on the level of hsp68-mRNA was studied. H35 hepatoma cells were exposed to various concentrations of arsenite (10, 30, 100 or 300 μM) for 1 h. Samples were taken immediately after the exposure to arsenite and at 2 h intervals up to 8 h during an incubation in an arsenite-free medium. Increased hsp68-mRNA levels could be detected immediately after arsenite treatment, the increase continuing during the incubation in arsenite-free medium. An exposure to 10 μM for 1 h did not have any effect on the induction of hsp68-mRNA. As shown in figure 1, the magnitude and the kinetics of the induction depended on the arsenite concentration. In general the response was stronger at the higher arsenite concentrations. Hsp68-mRNA levels reached the highest levels later and also decayed more slowly when the concentration of arsenite was raised.

To study the effect on the stimulation of hsp68-mRNA levels in arsenite-pretreated cells of an exposure to low concentrations of arsenite, exposures of 1 h to 100 μM were chosen as the sensitizing treatments. As shown above, an exposure to 100 μM arsenite is considered to be a mild treatment, appropriate for the study of the stimulatory effect of a low doses.

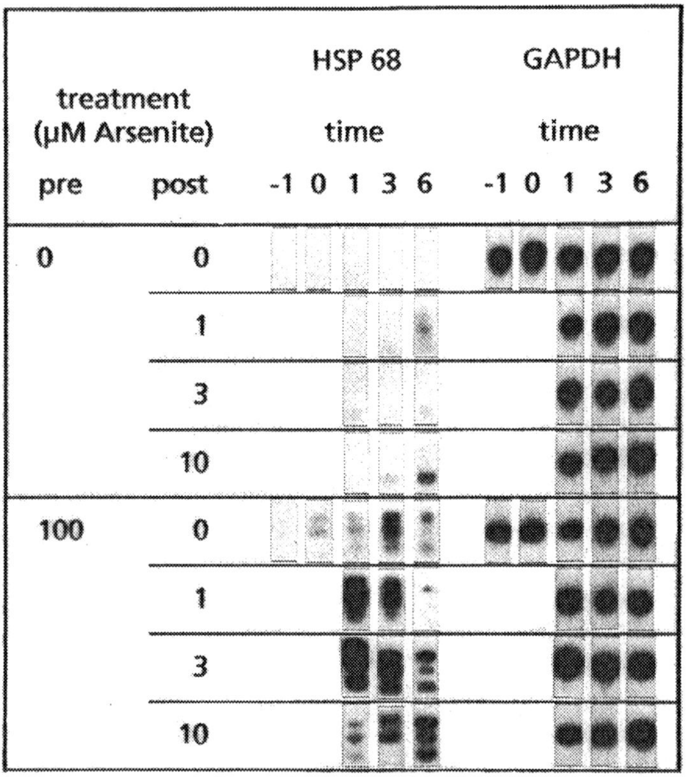

Figure 2. The influence of a step-down arsenite treatment on the induction of hsp68-mRNA. H35 hepatoma cells were pretreated with 0 or 100 µM arsenite from -1 to 0 h, thoroughly washed and subsequently incubated in 0, 1, 3 or 10 µM arsenite. At the indicated times, RNA samples were prepared and processed by northern blotting and autoradiography. Autoradiograms are shown from a representative experiment made after hybridization with a radioactive hsp68 probe and a GAPDH probe.

In order to determine the occurrence of sensitization we investigated which concentration had no effect when added alone but caused a stimulation when applied as a post-treatment. For this reason, a 1 h treatment with 0 or 100 µM arsenite was either or not followed by a continuous exposure to low concentrations of arsenite (1, 3 or 10 µM) for up to 6 h. Figure 2 and 3 illustrate the kinetics of the induction of hsp68-mRNA. A continuous exposure to concentrations of arsenite up to 10µM did not induce hsp68 messengers in cultures that had not been pretreated. After an exposure of 6 h to 10 µM only a minor spot is observed as shown in figure 2. After a 1 h pretreatment with 100 µM arsenite, a continuous treatment in the range up to 10 µM caused the hsp68-mRNA levels to increase. Lower concentrations of arsenite (1, 3 µM) tended to accelerate the response, whereas higher concentrations (10µM) delayed the response (figure 3). As the post-treatment alone did not cause any induction, it can be concluded that the cells were sensitized by the pretreatment.

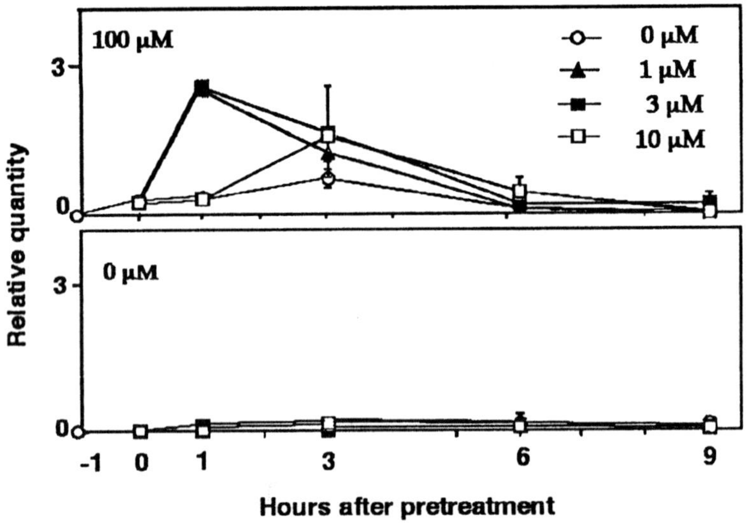

Figure 3. The influence of a step-down arsenite treatment on the induction of hsp68-mRNA. Pretreatment with 0 or 100 μM arsenite from -1 to 0 h, was followed by a subsequent incubation in 0, 1, 3 or 10 μM arsenite. The means and standard deviation are shown after quantification of five experiments.

In order to study the synthesis of hsp68 as well as of other hsps, the effect of the step-down (arsenite) protocol was studied on the induction of specific heat shock proteins.

3.2. HEAT SHOCK PROTEINS

A 1 h treatment with 0 or 100 μM arsenite was either or not followed by a continuous exposure to low concentrations of arsenite (1, 3 or 10 μM). At 3 h intervals samples were incubated with [35S]-methionine for 1 h, andsubsequently prepared for one-dimensional gel-electrophoresis. In figure 4 an example is given of a 1 hour treatment with 100 μM followed by either 0, 1, 3 or 10 μM arsenite. The relative rate of synthesis of each hsp was quantified and the time course of the different hsps, expressed relative to actin synthesis, was determined and presented in figure 5. A treatment with 100 μM, followed by an incubation in an arsenite-free medium shows a transient increase in the synthesis of hsp60, 68, 70, 84 and 100. This condition is considered to be a 'mild' treatment. The increase in relative synthesis of most hsps reaches a maximum at 3 h after the pretreatment.

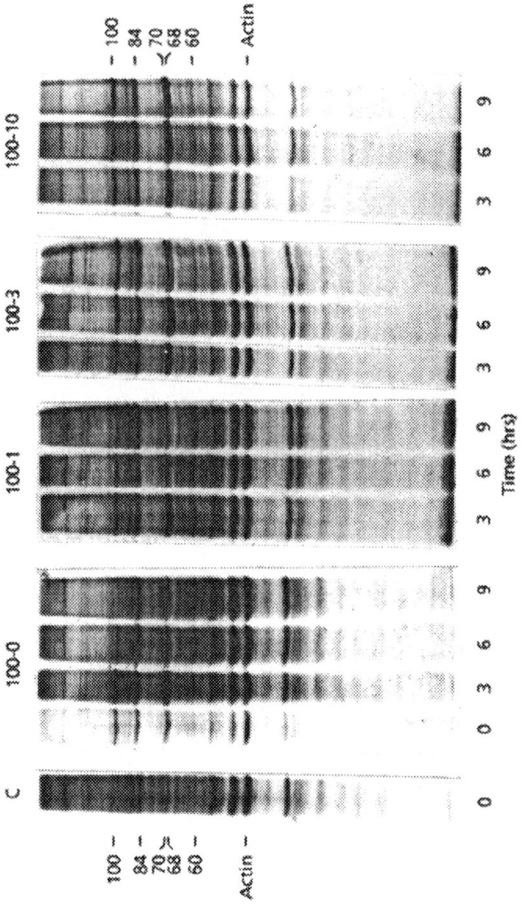

Figure 4. Induction of hsp-synthesis in H35 cells, pre-treated with 100 μM arsenite during 1 h was followed by a continuous treatment with 0, 1, 3 or 10 μM arsenite during 9 h. At various moments after pretreatment (0, 3, 6 or 9 h), cultures were pulse-labelled with [35S]-methionine for 1 h and prepared for 1-D PAGE. The same amount of protein was applied to each lane. (C= control cells).

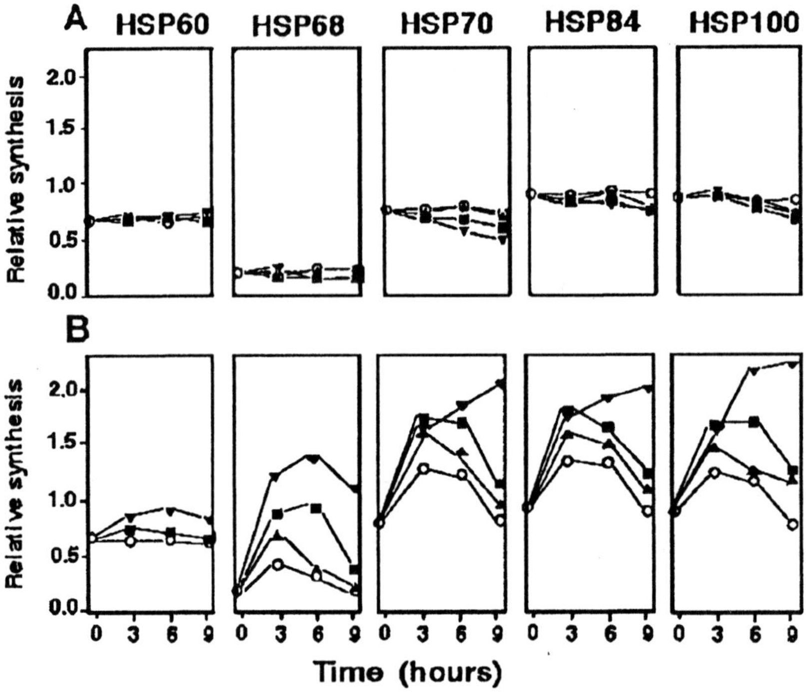

Figure 5. The relative synthesis of the major hsps (hsp60, 68, 70, 84 and 100) after a treatment with 0 μM (A) or 100 μM (B) arsenite for 1 h, subsequently followed by an incubation in 0 μM (circle), 1 μM (triangle-up), 3 μM (square) or 10 μM (triangle-down) arsenite during the indicated time (h). The data are expressed as the ratio of the optical density between the indicated hsp and actin.

When cells pretreated with 100 μM are incubated in media containing a low concentration of arsenite (1, 3 or 10 μM), an enhanced and prolonged induction of several hsps can be observed. Incubation of cultures in low concentrations of arsenite only, did not induce the synthesis of hsps. As the treatment with 3 or 10 μM arsenite alone (without pretreatment) did not cause any induction of hsps, it is concluded that a step-down arsenite treatment enhances the relative synthesis of hsps. Apparently, the sensitizing effect of an arsenite pretreatment can be determined in terms of an enhanced synthesis of specific hsps.

3.3. SURVIVAL CHARACTERISTICS AND DEVELOPMENT OF TOLERANCE IN ARSENITE-PRETREATED CELLS

It has been demonstrated that hsps facilitate repair and refolding of stress-damaged proteins and also prevent irreversible denaturation and aggregation of cellular proteins (for review see Hightower, 1991; Welch, 1992; Parsell and Lindquist, 1994). In this latter function they may act as 'protector proteins'.This may be one of the underlying molecular mechanisms by which cells acquire tolerance; i.e. tolerance to an otherwise lethal stress after exposure to a sublethal stress which is sufficient to induce stress protein synthesis.

We investigated whether the induction of arsenite-tolerance in arsenite-pretreated H35 cell cultures could be modulated by a post-incubation with low concentrations of arsenite. First, the H35 cell population was incubated for 1 h with 100 μM arsenite. Subsequently the culture was incubated with a lower concentration of arsenite (up to 10 μM) for 4 h. Then the sensitivity or the degree of tolerance was tested by establishing the survival after an exposure for 2 h to 300 μM arsenite. In Figure 6A it is shown that when cells have been pre-treated with 100 μM arsenite and post-incubated at 1, 3 or 10 μM, a decreased sensitivity is observed in comparison with cells that have been post-incubated in the absence of a low doses of arsenite. Apparently, pre-treated cells are able to develop a higher level of tolerance in the presence of low concentrations of arsenite. In cultures without pretreatment the presence of low concentrations of arsenite for up to 4 h did not induce tolerance (data not shown).

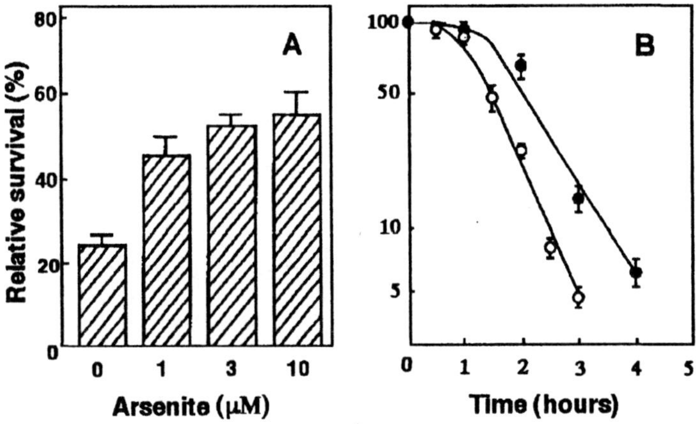

Figures 6.
6A: The percentage survival after a step-down arsenite treatment, indicating the level of tolerance obtained after a pretreatment with 100 μM arsenite, followed by a 4 h post-treatment with either 0, 1, 3 or 10 μM arsenite and finally a test treatment with 300 μM arsenite for 2 h.
6B: Survival curves of H35 cells subjected to a step-down arsenite treatment. Cell cultures were incubated during 1 h in 100 μM arsenite, followed by a post-treatment in either 0 μM (open circle) or 10 μM (closed circle) arsenite for 4 h, subsequently followed by a test-treatment with 300 μM during a period of up to 4 h. Data were corrected for the effect of pretreatment and presented as the surviving fraction of pre-treated cells.

In order to investigate whether the rate of cell killing and/or the capacity to withstand a stress treatment was affected (represented by D0 and Dq respectively); a pretreatment with 100 μM arsenite was followed by an incubation with either 0 or 10 μM for 4 h. After this incubation a test-treatment with 300 μM over a period of up to 4 h was given. As can be observed in figure 6B the survival curve of the cell culture which is post-treated with 10 μM, shows a slope which is less steep than found in cultures which were not post-treated and starts to decline at a later moment in time. Consequently, a larger D0-value is found (51.4 min in comparison with 34.7 min in post-treated and non-posttreated cultures respectively) as well as a larger Dq value (89.0 min in post-treated in comparison with 65.2 min in non-post-treated cultures), indicating that cells die at a lower rate when post-treated with 10 μM. The larger Dq value indicates that cells start to die later in time when post-treated with 10 μM, which means that the cell population has obtained a larger capacity to withstand a stress treatment.

4. Discussion

From the experimental data presented in this paper, it can be concluded that an exposure of Reuber H35 rat hepatoma cells to arsenite concentrations of 100 μM for 1 hour, rapidly changes the sensitivity to a second arsenite treatment with respect to the induction of the heat shock proteins and the development of tolerance.

The present experiments demonstrate that after a step-down arsenite treatment and under appropriate conditions, an increase of the level of hsp68-mRNA and an increased synthesis of arsenite-induced heat-shock proteins could be observed when compared with the rate of hsp68-mRNA and of hsp synthesis in cultures which received arsenite pretreatment only. An incubation of cell cultures in low concentrations of arsenite only (without arsenite-pretreatment), did not result in any detectable increase of the hsp68-mRNA level nor in a detectable induction of hsp-synthesis. It was also observed that under arsenite step-down conditions, the development of tolerance was enhanced in comparison with the level of tolerance obtained by a single treatment with arsenite.

These results are in agreement with our previous experiments using heat as the stressor condition. It has been shown that during step-down heating under appropriate conditions, the relative rate of hsp-synthesis as well as of hsp68-mRNA was increased to a greater extent in comparison with the values observed in cultures to which only the heat pretreatment was applied (Schamhart *et al.*, 1992; Van Wijk *et al.*, 1994). In addition, an enhanced development of tolerance during step-down heating conditions was described (Delpino *et al.*, 1992; Van Wijk *et al.*, 1994).

In conclusion, a stimulation of the self-recovery has been demonstrated upon the application of the isopathic principle (the similia principle in its most elementary form). For the next step in our research program, the model presented here can be used for the study of the specificity of the similia-principle. The results of the experiments aimed to identify the specificity of the similia-principle are described elsewhere in this volume.

Acknowledgement
The financial support of the HomInt organisation is gratefully acknowledged.

5. References

Delpino, A., Gentile, F.P., Di Modugno, F., Benassi, M., Mileo, A.M. and Mattei, E. (1992) Thermosensitization, heat shock protein synthesis and development of thermotolerance in M-14 human tumor cells subjected to step-down heating, *Radiat.Environm.Biophys.* **31**, 323-332.

Hightower, L.E. (1991) Heat shock, stress proteins, chaperones and proteotoxicity, *Cell* **66**, 191-197.

Johnston, D., Oppermann, H., Jackson, J. and Levinson, W. (1980) Induction of four proteins in chicken embryo cells by sodium arsenite, *J.Biol.Chem.* **255**, 6975-6980.

Lee, K.J. and Hahn, G.M. (1988) Abnormal proteins as the trigger for the induction of stress responses: heat, diamide and sodium arsenite, *J.Cell.Physiol.* **136**, 411-420.

Lee, T.C., Wei, M.L., Chang, W. J., Ho, I.C., Lo, J.F., Jan, K.Y. and Huang, H. (1989) Elevation of gluthatione levels and gluthatione s-transferase activity in arsenic-resistant chinese hamster ovary cells, *In Vitro Cell.Dev.Biol.*. **25**, 442-4480.

Levinson, W., Oppermann, H. and Jackson, J. (1980) Transition series metals and sulfhydryl reagents induce the synthesis of four proteins in eukaryotic cells, *Biochim.Biophys.Acta* **606**, 170-180.

Li, G.C. (1983) Induction of thermotolerance and enhanced heat shock protein synthesis in chinese hamster fibroblasts by sodium arsenite and by ethanol, *J.Cell.Physiol.* **115**, 116-122.

Nover, L. (1991) *Heat shock response*, CRC Press Publisher, Boca Raton.

Ovelgönne, J.H., Souren, J.E.M., Van Rijn, J., Van Wijk, R. and Wiegant, F.A.C. (1995) Enhancement of the stress response by low concentrations of arsenite in arsenite-pretreated H35 hepatoma cells, *Toxicol.Appl.Pharmacol.* (in press)

Parsell D.A. and Lindquist S. (1994) Heat shock proteins and stress tolerance, in *The biology of heat shock proteins and molecular chaperones*, R.I Morimoto, A. Tissieres and C. Georgopoulos (eds.) CSHL Press Publisher, New York, pp.457-494.

Schamhart, D.H. J., Zoutewelle, G., Van Aken, H. and Van Wijk, R. (1992) Effects on the expression of heat shock proteins by step-down heating and hypothermia in rat hepatoma cells with a different degree of heat sensitivity, *Int.J.Hyperthermia* **8**, 701-716.

Van Rijn, J., Van den Berg, J., Wiegant, F.A.C. and Van Wijk, R. (1994) Time-temperature relationships for step-down heating in normal and thermotolerant cells, *Int.J.Hyperthermia* **10**, 643-652.

Van Wijk, R., Ovelgönne, J.H., De Koning, E., Jaarsveld, K., Van Rijn, J. and Wiegant, F.A.C. (1994) Mild step-down heating causes increased transcription levels of hsp68 and hsp84 mRNA and enhances thermotolerance development in Reuber H35 hepatoma cells, *Int.J.Hyperthermia* **10**, 115-125.

Van Wijk, R. and Wiegant, F.A.C. (1994) *Cultured mammalian cells in homeopathy research; the similia principle in self-recovery*, Utrecht University Publisher, The Netherlands .

Welch, W.J. (1992) Mammalian stress response: cell physiology, structure/function of stress proteins, and implications for medicine and disease, *Physiol.Rev.* **72**, 1063-1081.

Wiegant, F.A.C., Van Bergen, E, En Henegouwen, P.M.P., Van Dongen, G. and Linnemans, W.A.M. (1987) Stress-induced thermotolerance of the cytoskeleton of mouse neuroblastoma N2A cells and rat Reuber H35 cells, *Cancer Res.* **47**, 1674-1680.

Wiegant, F.A.C., Souren, J.E.M., Van Rijn, J. and Van Wijk, R. (1993) Arsenite-induced sensitization and self-tolerance of Reuber H35 hepatoma cells, *Cell Biol.Toxicol.* **9**, 49-59.

STIMULATION OF CELLULAR SELF-RECOVERY BY APPLICATION OF THE SIMILIA-PRINCIPLE *

Self-recovery and similia-principle

R. VAN WIJK AND F.A.C. WIEGANT
Department of Molecular Cell Biology, Utrecht University
P.O Box 80.056, 3508 TB Utrecht, The Netherlands

1. Introduction

The stimulation of self-recovery by applying the similia-principle is considered to be the essence of homeopathy. The similia-principle indicates that a substance which causes certain symptoms in a healthy organism, may be applied in a curative sense to a diseased organism with identical symptoms.

Previously, it has been shown that the recovery of cells that were damaged by a certain stressor, are stimulated by the application of a low dose of the same stressor. In this respect, the stimulatory effect of low doses could be observed after a heat shock (Van Wijk *et al.*, 1994) as well as after a treatment with arsenite (Ovelgönne *et al.*,1995a; Wiegant *et al.*,-this volume-) or with cadmium (Wiegant *et al.*, in prep).

As parameters for stimulation of self-recovery the development of tolerance is used. An important line of research deals with the molecular aspects underlying the process of recovery. In this respect a particular set of proteins, indicated as heat shock proteins (hsps) or stress proteins are supposed to play a crucial role (Welch 1992; Parsell and Lindquist 1994). This set of proteins includes different protein families with approximate molecular masses (in kD) of 28 (hsp28), 60-65 (hsp60), 68-73 (hsp70), 84-90 (hsp84) and 100-110 (hsp100). The evidence for their role in protection comes from studies of hsp synthesis under differential conditions of environmental stress (Li and Laszlo 1985; Nover 1991) as well as by transfection of hsp-genes (Landry *et al.*, 1989; Li *et al.*, 1991; Angelidis *et al.*, 1991; Jaatela *et al.*, 1992; Parsell and Lindquist 1994).

A further study of the similia-principle requires additional experiments aimed at the specificity of this low-dose stimulation of tolerance development and of hsp induction. In this paper we focus on the specificity of low doses of stressors with respect to the enhancement of hsp induction.

With respect to the specificity of the similia-principle, our working hypothesis is that the most effective low-dose stimulation of hsp-induction in damaged cells is obtained with the same stressor or a stressor which has an identical pattern of induction of the different stress proteins. With less similarity between the damaged condition and

* Communication presented at the 8th GIRI Meeting, December 1994, Jerusalem, Israel.

M. Bastide (ed.), Signals and Images, 53–61.

the low-dose stimulating condition, the low-dose stimulation will be less or even absent.

If for instance arsenite is applied as the disturbing agent, it can be hypothesized that stimulation of sub-optimal hsp induction as part of self-recovery will only occur by the application of a small dose of arsenite and not, for instance, by a small dose of cadmium, unless cadmium shows the induction of a similar pattern of hsps.

The present study, including arsenite, heat shock and cadmium as damaging conditions, and the same three conditions at a low dose applied afterwards, deals with the relationship between degree of similarity and efficiency of low dose stimulation. In this way the specificity of the similia-principle can be verified.

2. Stressor-specificity with Respect to Induction of Gene Expression

In our study different stressors were used in order to identify a possible stressor-specific induction in the pattern of heat shock proteins. If a non-specific expression is observed, i.e. when all stressors induce the same set of hsps to an identical degree, then it would be impossible to study the specificity of the similia-principle.

2.1. STRESSOR-SPECIFIC INDUCTION OF HSPS

The induction of hsps by a variety of stressors including heat shock, arsenite and cadmium has been studied extensively and recently reviewed by Nover (1991) and Welch (1992). In a number of these studies the effects of two or more stressors was compared (Li 1983; Anderson *et al*.,. 1987; Pipkin *et al*.,. 1987; Lee *et al*.,1989; Aoki *et al*.,. 1990; Deaton *et al*., 1990; Edwards *et al*., 1990; Lai *et al*., 1993). The shortcoming of most of these studies is that the pattern of induced proteins was not studied over a range of concentrations nor were they studied under conditions that lead to similar consequences on the cellular physiology. The limited conditions in most studies might be the cause of severe biases. A reliable method to compare the effects stressors have on the stimulation of hsp synthesis, is to establish their concentration dependency and then compare the hsp patterns in these concentration ranges. To this end, Reuber H35 rat hepatoma cells were exposed to different concentrations of various environmental stressors (including heat shock, arsenite and cadmium). For a careful comparison of these stressor conditions, the stressors were all applied during one experiment in order to exclude slight differences in culturing conditions. The possible stressor-specific effect on the induction of hsps was studied as well as the effects of these stressors on the inhibition and recovery of protein synthesis and on cell survival (Wiegant *et al*., 1994).

A comparison of the stressor-specific induction of the major hsps was made under conditions of identical impact on cellular physiology; i.e. those stressor conditions were chosen that resulted in approximately equal percentages of cell death (iso-survival dose: 40-50%). These iso-survival doses were: 1 h at 42°C; 2 h of 200 μM arsenite and 2 h of 100 μM cadmium. The pattern of hsp synthesis of stressed cells consists of the hsp28, hsp60, hsp68, hsp70, hsp84 and hsp100 (figure 1). Cells exposed to cadmium show the highest relative synthesis of hsp28 if compared with cell cultures treated with heat shocks or arsenite. No induction of hsp60 is observed after cadmium treatment, whereas a large induction of hsp60 is observed after heat shocks and arsenite. Hsp68 and hsp70 are both induced by heat shocks as well as arsenite and to a lesser extent by cadmium. More or less the same pattern is observed with respect to hsp84 and hsp100. (Wiegant *et al.*, 1994 for a more detailed analysis).

Each hsp induced after the treatment with a stressor is supposed to have a characteristic function with respect to the recovery of specific cellular damage induced by this stressor. Although not all the effects of the various stressors have been identified at the molecular level, it can be assumed that cellular damage in general requires a differential activity of hsps and that specific cellular damage requires the presence of specific hsps. In this respect, it can be proposed that the differential induction of the group of hsps induced by a specific stressor is exactly attuned to the precise need of the cell and that it can be taken as an indicator for the type of cellular damage which is under repair.

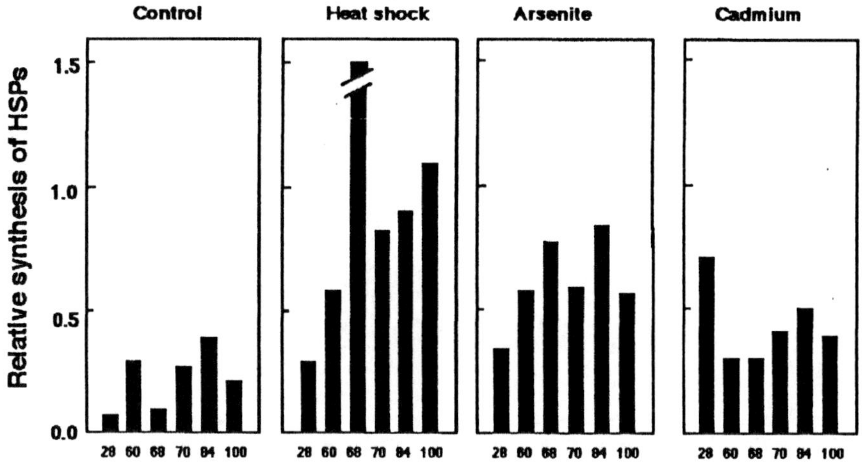

Figure 1. The effect of three different stressors (heat shock at 42°C for 60 minutes, 200 μM arsenite for 2h and 100 μM cadmium for 2h) on the synthesis of the major hsps is shown (a-hsp28; b-hsp60; c-hsp68; d:-hsp70; e-hsp84; f-hsp100). In the control situation several hsps are synthesized at a low level. Synthesis of the major stress proteins is expressed relative to the synthesis of actin.

2.2. STRESSOR SPECIFIC INDUCTION OF HSP mRNA

In order to determine whether the stressor-specific enhanced expression of HSP-genes is caused by a differential gene expression, experiments were performed in which both the levels of specific hsps and of hsp mRNAs were determined in parallel cultures after the application of a heat shock, arsenite or cadmium. The stressor-specificity of the synthesis of hsps was confirmed by hsp mRNA data (Ovelgönne *et al.*, 1995b). The inability of cadmium to induce hsp60 appeared to correlate with an absence of the induction of hsp60 mRNA at the various concentrations studied.

3. Cross-sensitization

In previous papers, the low-dose stimulation has been studied in an isopathic approach. Thus, a cell population that previously had been damaged and sensitized by a high dose of a stressor was used to study the effect of a low dose of the same stressor on stimulation of self-recovery and hsp induction. (Van Wijk *et al.*, 1994; Ovelgönne *et al.*, 1995a; Wiegant *et al.*, -this volume). The application of a mild heat treatment or a low dose of arsenite immediately after a heat shock or an intoxication with arsenite respectively, stimulated the synthesis of hsps as well as the development of tolerance.

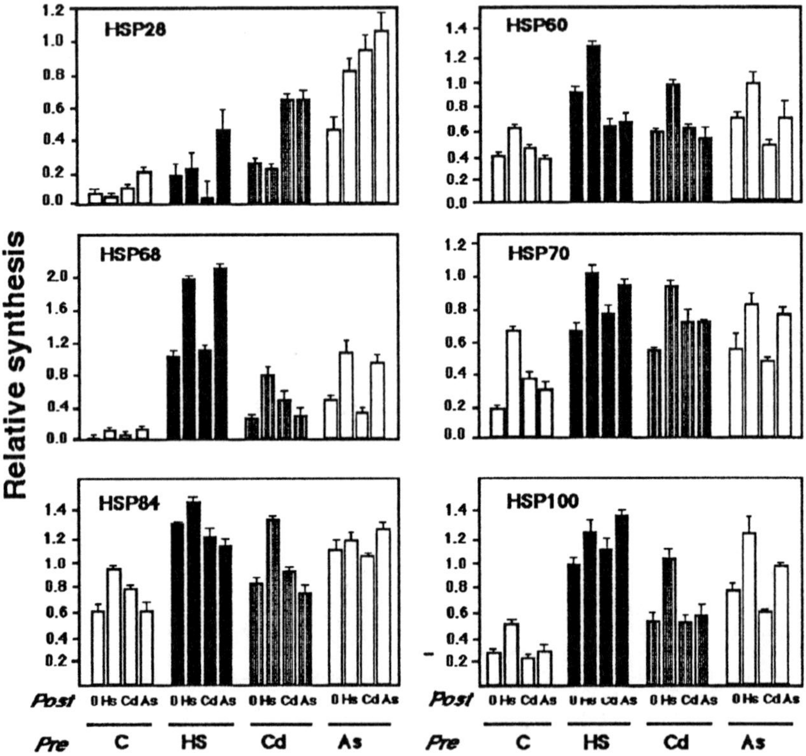

Figure 2. Cross-sensitization. The effect of a low doses of a stressor on the synthesis of the major hsps, which was applied immediately following a pretreatment with a high doses of either a heat shock (HS: 42°C for 30 minutes), cadmium (Cd: 100 μM for 1 h) or arsenite (As: 100 μM for 1 h). The incorporation of [35S]-labelled amino acids into the major hsps over an 8-h period was expressed relative to the total incorporation into cellular protein. The results for each hsp are grouped per effect of the high-doses stressor. Of each group of 4 histograms, the first bar shows the induced synthesis of the indicated hsp in control conditions (0) or by a pretreatment with the high-doses stressor (HS, As or Cd), whereas the other three show the effect on the synthesis of hsps of the post-incubation in a low doses of heat shock (hs: 8h in 40.5°C), cadmium (cd: 8h 10 μM) or arsenite (as: 8h 10 μM). Results are the mean of three experiments.

Cells were highly sensitized; i.e. they responded to relatively low doses, whereas these low doses of stressors were without any effect in control cells during the time period studied. It was argued that in this way the similia-principle can be studied in its most elementary form.

In the present study it became of interest to determine whether hsp-induction can also be stimulated by low doses of other (non)-related stressors. Are sensitized cells also stimulated by low doses of non-related stressors.

In this respect it was hypothesized that the specificity of the similia-principle reveals itself in the fact that for the stimulation of hsps by low doses identical or strongly similar conditions must be applied. If condition 'A' is applied as the disturbing agent, the stimulation of a suboptimal hsp-induction will only result from the application of a small dose of condition 'A' and not by a small dose of condition 'B' or 'C', unless B or C has a similar effect on the cellular physiology as condition 'A' has.

If, however, the stimulation of self-recovery by a low dose is an a-specific phenomenon, then a low dose of stressor 'B' or 'C' can also be expected to be able to stimulate the hsp-induction. Such a phenomenon, indicated as cross-sensitization, has been studied with Reuber H35 cells exposed to heat shock, arsenite or cadmium.

A number of experiments were performed in which a pretreatment was given with either a heat shock (30 min. at 41.5°C), sodium arsenite (100 μM for 1h) or cadmium chloride (100 μM for 1 h). After the pretreatment the cultures were exposed to low, mostly subliminal conditions of heat (40°C or 40.5°C), arsenite (3 or 10 μM), cadmium (3 or 10 μM) or by an incubation at control conditions (37°C) for a period of up to 8h. In this period the stimulating effect on the synthesis of various hsps was studied. The results are shown in figure 2.

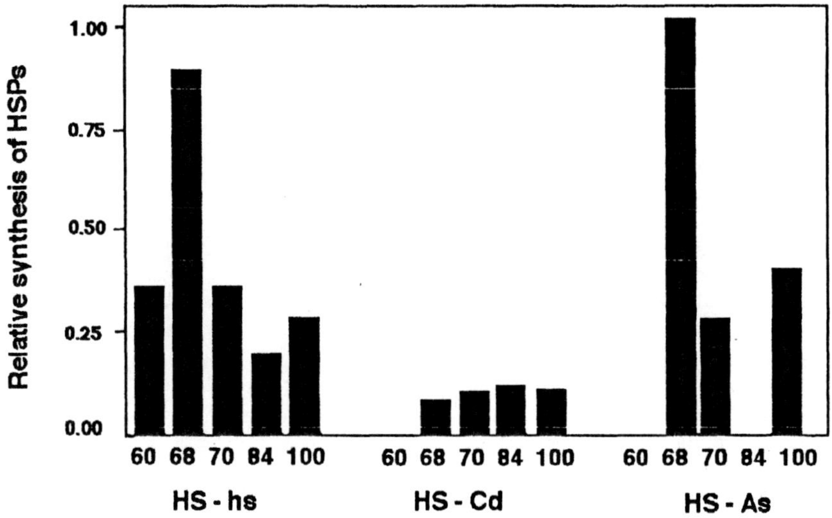

Figure 3. The effect of a low doses of a stressor (heat shock at 40.5°(hs); 10 μM cadmium (cd); or 10 μM arsenite (as), immediately applied after a high doses of heat shock (HS: 42°C for 30 min), on the extra synthesis of the major hsps over a period of 8h. The effect of the HS has been subtracted in order to show the stimulatory effect of the low doses only.

The first step in the analysis of the results of these experiments was to determine which of the species of hsps was stimulated by the low-dose iso-condition.

- A small increase in temperature after a more severe heat shock enhances the induced synthesis of all studied hsps, but mainly hsps 60, 68, 70 and 100 and to a lesser extent hsp84. The extra synthesis of the selected hsps, induced by low dose stimulation after heat shock, is also presented in figure 3 (left panel).

- Of the hsps induced by a treatment of arsenite, mainly the subgroup of hsps, consisting of hsp28, hsp68, hsp70 and hsp100, was further stimulated by arsenite at a subliminal concentration (figure 2).

- Of the hsps induced by a pretreatment with cadmium, a subsequently added subliminal concentration of cadmium, stimulated the synthesis of a subgroup of hsps which consisted of hsp28, hsp68 and hsp70 (figure 2). These subgroups of hsps thus seem to be indicative for a particular role in the stimulation of self-recovery.

The second step in the analysis involved the stimulation of synthesis of the selected hsps by subliminal concentrations of other stressors than the ones used for the pretreatment.

- If heat shock was used as the disturbing pretreatment, hardly any extra stimulation of hsp-synthesis is observed upon the subsequent application of a low dose of cadmium (figure 2 and figure 3 -middle panel-). In contrast when a low dose of arsenite is applied to cells which were previously heat shocked, a large extra synthesis of hsp68, 70 and 100 occurs. Interestingly no extra enhancement at all of hsp60 and 84 is observed (figure 2 and figure 3 -right panel-). This demonstrates a rather specific effect of low doses of the used stressors on heat-treated cell cultures.

- If arsenite was the first disturbing agent, the synthesis of hsp28, hsp68, hsp70 and hsp100 was also stimulated when a mild heat treatment was applied. Application of a subliminal cadmium concentration only stimulated hsp28, whereas no effect of cadmium was found on the synthesis of hsp 68, hsp70 and hsp100 (figure 2).

- If cadmium was used as the disturbing agent for the pretreatment, stimulation of hsp28, hsp68, hsp70 occurred after a successive application of a subliminal cadmium concentration. The synthesis of these hsp protein species was also stimulated by a very mild heat shock when applied as a second treatment. The application of a subliminal arsenite concentration stimulated only hsp28 and hsp70.

This analysis of the experiments on cross-sensitization is very elementary and takes no account of the quantitative differences between the stimulatory effect of small doses of the various stressors.

Nevertheless, from the data it can be observed that even this elementary analysis demonstrates a specificity in the synergistic action between a pretreatment and a subliminal condition. A schematic presentation of the qualitative similarity between the different conditions and the cellular capacity to enhance synthesis of specific hsps upon low dose stimulation is given in Table 1.

Based on a comparison of the hsp-induction pattern of the three stressors, it was concluded that the heat shock- and the arsenite pattern resemble each other more than the heat shock- and the cadmium-pattern. This higher degree of similarity is now also reflected in a higher degree of extra synthesis of hsps upon low dose stimulation and thus confirms the aspect of specificity of the similia-principle.

TABLE 1: Cross-sensitization

	First stressor Cadmium	Heat shock	Arsenite
Second stressor			
Heat shock	++++	++++	++++
Arsenite	+++	++++	++
Cadmium	-	+	++++

The range -, +, ++, +++, ++++ is indicative for the degree of extra synthesis of hsps by a low dose of the second stressor.

4. Discussion

The aim of our research program was to study the process of self-recovery and its stimulation by low doses of compounds applied to diseased biological systems according to the similia-principle. For the study of the similia-principle it is for the researcher of importance to be able to concentrate on the most elementary level where changes manifests themselves. Therefore, we have proposed the following question in our research program:

Which molecular processes and interactions are of importance in the process of cellular self-recovery and which molecular parameters should be used to evaluate (stimulation of) cellular self-recovery ?

In our study we have focused on the synthesis of hsps as a measure of cellular self-recovery. This choice is based on our own experimental results (Van Wijk and Wiegant, 1994) as well as on and data from the literature (for recent review see: Nover, 1991; Welch, 1992; Parsell and Lindquist, 1994). The pattern of synthesis of the individual hsps can be considered as the pattern of symptoms specifically linked to the recovery process of the cell after damage.

Our comparative study of hsp induction in H35 cells by three different stressors: -heat shock, cadmium and arsenite- have shown that the heat shock- and arsenite-induced patterns of synthesis of the major hsps resemble each other more than the heat shock- and cadmium-induced patterns of hsp-synthesis. Furthermore, the possible existence of a stressor-specific attunement to the cellular requirements of the induced hsps, has been tested in experiments in which a pretreatment was given with a heat shock, cadmium or arsenite, then followed by incubation in stressor-free conditions for 16h. This provided for a period of time in which all hsps could reach their maximal level. Subsequently, the effect on the induction of hsps of a second application of a stressor of the same dose (heat shock, arsenite or cadmium) was studied in a cross-tolerance design. The synthesis of hsps was followed over a period of 8h after application of the second stressor. A more detailed description of the results will be described elsewhere (Wiegant *et al.*, 1995). From the results however, three aspects of interest for the present discussion can be mentioned:

1. In general, a reduced induction of hsps is observed when a stressor is applied during a tolerant period. Therefore a certain degree of attunement to the induced cellular damage appears to be present in the effect of all stressors used in this study. The degree of attunement depends on the similarity between the first and second stressor with respect to their effect on the synthesis of hsps.

2. The repeated application of the same stressor (heat shock-heat shock, arsenite-arsenite, cadmium-cadmium) provides the highest degree of attunement and therefore the highest degree of tolerance, since in these conditions the lowest induction of all hsps occurs after the application of the second stressor.

3. The application of different stressors (heat shock-arsenite, heat shock-cadmium, arsenite-heat shock, arsenite-cadmium, cadmium-heat shock, cadmium-arsenite) results in a larger induction of the different hsps. Apparently the transient insensitivity as observed in self-tolerance has not developed to a similar degree. The attunement of the various hsps to the specific damage inflicted by the first stressor does not seem to include the damage inflicted by the second stressor since the induction of (other) hsps is required. However, it is interesting to note that of the different combinations of stressors, cross-tolerance between heat shock and arsenite develops to a higher degree than the cross-tolerance between heat shock and cadmium or cadmium and arsenite. This suggests that the pattern of induced hsps by a heat shock is better attuned to the damage inflicted by arsenite than to the damage caused by cadmium. It can therefore be concluded that of the three stressors studied, heat shock and arsenite shows the largest similarity in effect.

With respect to the substantiation of the similia-principle - the following expectation has been formulated:

The specificity of the similia-principle will reveal itself in the fact that for the stimulation of self-recovery always identical conditions or conditions with a reasonable degree of similarity must be applied however in a smaller concentration or in a lighter form. If condition 'A' is applied as a disturbing agent, stimulation of suboptimal self-recovery will only happen by applying a small dose of 'A'. By applying a small dose of condition 'B', when 'B' provokes symptoms that are highly different from 'A' stimulation will not happen. Finally stimulation of suboptimal self-recovery to a lesser degree will happen with a small dose of condition 'C', when 'C' provokes symptoms partly similar to 'A'.

In this paper we have provided evidence in which the specificity of the similia-principle is substantiated. Further studies applying detailed quantitative analysis of hsp patterns will be presented elsewhere. For the interest of homeopathy it is of importance to ask the question:

5. Do Studies on the Similia-principle at the Cellular Level Have any Significance?

It has been argued that the similia-principle has been formulated for the organismal level and therefore can not be verified or falsified by research on the cellular level alone.

There is, of course, a clear difference between research on the similia-principle as described here and the clinical practice. The first difference is the systemic level or level of organization which is studied or treated (cells versus organisms). Secondly, in our experiments cells are brought into a pathological condition, whereas patients who come to a doctor already are in this condition. Thirdly, cells are grown in cultures and treated as a group. Their behaviour and response with respect to the process of self-recovery is evaluated as the mean of the total population. There is of course a variability in self-recovery. Thus, not only cells showing a fast and spontaneous recovery, but also cells showing a slow and hampered self-recovery are evaluated. In contrast to research with patients, a single case approach or a time-series design for studies on the behaviour of individual cells is not possible or is extremely complex.

However, we expect the assumption of the similia-principle as being a general biological phenomenon to gain weight by the study at the cellular level of regulatory processes underlying this principle. The demonstration of the stimulation of self-recovery by compounds according to the similia-principle will substantiate the fundamental basis of homeopathy and will improve the acceptance of this basic point of view in bio-medicine.

Acknowledgement

The financial support of the HomInt organisation is gratefully acknowledged.

6. References

Angelidis, C.E., Lazaridis, I. and Pagoulatos, N. (1991) Constitutive expression of heat-shock protein 70 in mammalian cells confers thermoresistance, *Eur.J.Biochem.* **199**, 35-39.

Hightower, L.E. (1991) Heat shock, stress proteins, chaperones and proteotoxicity, *Cell* **66**, 191-197.

Jäättelä, M., Wissing, D., Bauer, P.A. and Li, G.C. (1992) Major heat shock protein hsp70 protects tumor cells from tumor necrosis factor cytotoxicity, *EMBO J.* **11**, 3507-3512.

Landry, J., Chretien, P., Lambert, H., Hickey, E. and Weber, L.A. (1989) Heat shock resistance conferred by expression of the human HSP27 gene in rodent cells, *J.Cell Biol.* **109**, 7-15.

Li, G.C. and Laszlo, A. (1985) Thermotolerance in mammalian cells: a possible role for heat shock proteins, in B.G. Atkinson and D.B. Walden (eds) *Changes in eukaryotic gene expression in response to environmental stress*, Academic Press Publisher, New York, pp.227-254.

Li, G.C., Li, L., Liu, Y.K., Mak, J.Y., Chen, L. and Lee, W.M.F. (1991) *Thermal response of rat fibroblasts stably transfected with the human 70-kDa heat shock protein-encoding gene*, Proc.Natl.Acad.Sci.U.S.A. **88**, 1681-1685.

Nover, L. (1991) *Heat shock response*, CRC Press, Boca Raton.

Ovelgönne, J.H., Souren, J.E.M., Van Rijn, J., Van Wijk, R. and Wiegant, F.A.C. (1995) Enhancement of the stress response by low concentrations of arsenite in arsenite-pretreated H35 hepatoma cells, *Toxicol.Appl.Pharmacol.* (in press).

Ovelgönne, J.H., Souren, J.E.M., Wiegant, F.A.C. and Van Wijk, R. (1995), Relationship between cadmium-induced cell killing, inhibition of total protein synthesis and increased expression of heat shock genes, *Toxicology* (in press).

Parsell, D.A. and Lindquist, S. (1994) Heat shock proteins and stress tolerance, in R.I. Morimoto, A. Tissières and E. Georgopoulos C (eds.) *The biology of heat shock proteins and molecular chaperones*, CSHL Press Publisher, New York, pp.457-494.

Van Wijk, R., Ovelgönne, J.H., de Koning, E., Jaarsveld, K., Van Rijn, J. and Wiegant, F.A.C. (1994) Mild step-down heating causes increased transcription levels of hsp68 and hsp84 mRNA and enhances thermotolerance development in Reuber H35 hepatoma cells, *Int.J.Hyperthermia* **10**, 115-125.

Van Wijk, R. and Wiegant, F.A.C. (1994) *Cultured mammalian cells in homeopathy research; the similia principle in self-recovery*, Utrecht University Publisher, Utrecht.

Welch, W.J. (1992) Mammalian stress response: cell physiology, structure/function of stress proteins, and implications for medicine and disease, *Physiol.Rev.* **72**, 1063-1081.

Wiegant, F.A.C., Souren, J.E.M., Van Rijn, J. and Van Wijk, R. (1994), Stressor-specific induction of heat shock proteins in rat hepatoma cells, *Toxicology* **94**, 143-159.

Wiegant, F.A.C., Spieker, N. and Van Wijk, R. (1995) Stressor-induced tolerance of hsp-synthesis is maximal for the same stressor only; quantisation of hsp re-induction as a measure of self- and of cross-tolerance,*Toxicology* (submitted).

Wiegant, F.A.C., Van Rijn, J. and Van Wijk, R. Enhancement of the stress response by low concentrations of cadmium in cadmium-pretreated H35 hepatoma cells, (in preparation).

PREVENTIVE EFFECTS OF A MONOCLONAL ANTI-IDIOTYPIC ANTIBODY COMPARED TO A HORMETIC MODEL ON RAT SARCOMAS *

Preventive effects of an antibody on rat sarcomas

S. FAIDERBE, J.L. CHAGNAUD, M. BASTIDE[1], P. DORFMANN[2] and M. GEFFARD
Laboratoire PIOM, ENSCPB, BP 108, Avenue Pey Berland, 33402 Talence cedex, France
[1]Laboratoire d'Immunologie, Faculté de Pharmacie, Université de Montpellier I, 15 av Charles Flahault, 34060 Montpellier cedex 1, France
[2]Laboratoires Dolisos, 71 rue Beaubourg, 75003 Paris, France

1. Introduction

Increased levels of autoAb have previously been found in patients with malignant tumors (Chagnaud *et al.*, 1992 ; Faiderbe *et al.*, 1990; 1991; 1992a). These autoAb reflect antigenic disturbances between endogenous B(a)P ligand and its cytosolic receptor(s), and disturbances in PtdIns turn over. In the same time, a chemically-induced malignant tumor model was obtained after subcutaneously (s.c.) administration of a single dose of B(a)P (0.1mg/g body weight at the top of the right thigh of SD female rats) diluted in sesame oil (Faiderbe *et al.*, 1992b). Significantly elevated levels of circulating anti-PtdIns autoAb were found in sera of B(a)P-treated rats 40 days after B(a)P administration, whereas no significant levels of anti-PtdIns autoAb were noted in control rats (oil-treated rats) (Faiderbe *et al.*, 1992b). After day 60, autoAb levels plateaued in B(a)P-treated rats, and highly malignant sarcomas appeared with 100% efficiency around day 100. When a monoclonal anti-idiotypic Ab, internal image of conjugated B(a)P, called AIB1 Ab (Chagnaud *et al.*, 1993a), was injected into SD female rats to mimic carcinogenic B(a)P effects, significant anti-PtdIns autoAb levels appeared. These circulating autoAb were found immunochemically equivalent to that found after injection of 2mg B(a)P (Chagnaud *et al.*, 1993b). These results suggested possible relationships between carcinogen ligand/receptor interactions and PtdIns turn over.

AIB1 Ab effects on rat sarcomas have been evaluated before (day 73) and after (day 102) the appearance of clinically palpable tumors (Chagnaud *et al.*, 1993c ; Chagnaud *et al.*, 1994).We have found that when AIB1 Ab was administered to SD female rats, anti-PtdIns autoAb levels were modified (Chagnaud *et al.*, 1993b, 1994). Furthermore, we have demonstrated that AIB1 Ab has the capacity to delay tumor onset, slow down growth, and to increase animal survival when administered biweekly

* Communication presented at the 7th GIRI Meeting, November 1993, Montpellier, France.

M. Bastide (ed.), Signals and Images, 63–70.

to SD female rats, at different time after tumor induction (Chagnaud *et al.*, 1993c, 1994).

The present investigation was designed to evaluate the possible vaccine effects of AIB1 Ab. Taking into account our previous results, we hypothesize that this anti-idiotypic Ab might have preventive effects when administered prior to carcinogen injection (tumor induction). To support this hypothesis, we have studied the effects of 500 µg AIB1 Ab (anti-idiotypic antibody) administration 8 days before tumor induction, compared to that of 1 µg of B(a)P (dose equivalent to that of 500 µg of AIB1 Ab sites).

The preventive effects of a treatment involving a monoclonal anti-idiotypic antibody (Ab) compared to a hormetic model were evaluated on a benzo(a)pyrene [B(a)P]-induced malignant sarcomas in Sprague-Dawley (SD) female rats. This Ab was an anti-idiotypic Ab, internal image of conjugated B(a)P, called AIB1 Ab. It was administered to female rats (500 µg of AIB1 Ab at the top of the left thigh) 8 days prior to tumor induction. Other animals received an injection of 1µg of B(a)P, dose equivalent to that of 500 µg of AIB1 Ab sites. Anti-phosphatidylinositol (PtdIns) autoantibody (autoAb) levels observed in AIB1 Ab-treated rats sera were 70% weaker than those observed in rats treated with 1 µg of B(a)P or control animals. Conversely, no differences were noted in the profil curves between the control group and the group treated with 1µg of B(a)P prior to tumor induction dose. There was a delay of 16 days in tumor onset between 75% of AIB1 Ab-treated rats and the controls and, the mean survival of rats treated with AIB1 Ab was increased compared to the control group. Conversely, there was no delay in tumor onset between rats treated with 1µg of B(a)P and controls. These results revealed that the preventive effects of 1 µg of B(a)P injected prior to tumor induction were lower than those of the monoclonal anti-idiotypic Ab, internal image of B(a)P.

3. Material and Methods

3.1. MONOCLONAL ANTIBODY

The monoclonal AIB1 Ab used in this study was IgG1 (k chain). This monoclonal anti-anti-conjugated B(a)P Ab, has been previously characterized as the internal image of conjugated B(a)P (Chagnaud *et al.*, 1993a). Producing hybridomas were propagated in the peritoneal cavity of BALB/c mice after prior injection with Freund's incomplete adjuvant (Difco, 0.5 ml/mouse). Mouse ascitic fluids were harvested 8-10 days later and monoclonal Ab partially purified by a 50% $(NH_4)_2SO_4$ precipitation followed by gel filtration (Sephadex G-200).

3.2. SARCOMA MODEL

Eleven female SD rats (Janvier, Le Genest Saint-Isle, France) were housed under controlled conditions (22°C, monitored light-dark cycles, with light from 8:00 a.m. to 8:00 p.m.), and were supplied with food (UAR, Versailles, France) and water ad libitum. After preventive treatment (see below), the 50-60 day-old female SD rats weighing 180-200 g received under anesthesia a single s.c. injection of 2 mg of B(a)P diluted in 500 µl of sesame oil, at the top of the right thigh. Blood of the treated rats was regularly sampled and tested (Figure 1).

3.3. TREATMENT

Ab solution were prepared as follows : G-200 purified fractions of Ab were diluted at 2 mg/ml in NaCl 9‰ (Meram) and stored at - 80°C. Eight days before tumor induction, female SD rats were devided in 3 groups. The first group (n=4) received at the top of the left thigh only one s.c. injection of 500 µg of AIB1 Ab. The second group (n=4) was injected with 1 µg of B(a)P (dose equivalent to that of 500 µg of AIB1 Ab sites). A control group has received an equivalent volume of solvent (n=3). Eight days later, a s.c. injection of 2 mg of B(a)P diluted in sesame oil was administered to all SD female rats at the top of the left thigh (Figure 1).

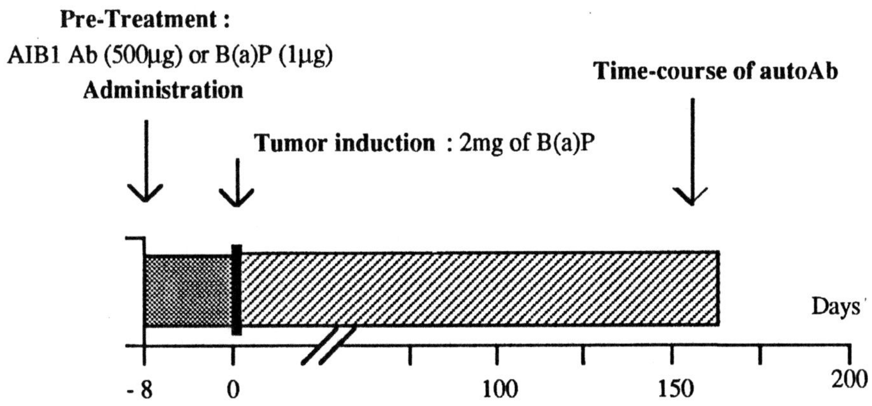

Figure 1. Description of the experimental protocol : Time sequence of the events (Pre-treatment, tumor induction, time-course of circulating anti-PtdIns autoAb).

3.4. IMMUNOENZYMATIC ASSAY

Detection of anti-PtdIns autoAb was performed with an adapted ELISA method as previously described by Faiderbe *et al.* (1992b) on rat sera diluted at 1/1000. Experimental absorbance values obtained from well-plates coated with PtdIns were corrected by subtracting blank values read from wells coated with control protein (thyroglobulin).

3.5. TUMOR GROWTH MEASUREMENT

B(a)P-induced tumors were mechanically measured with a calliper rule as described by Faiderbe *et al.* (1992b). Their volume was calculated using a standard formula : width^2 x length x 0.52, according to Ingber *et al.* (1990), and expressed as cubic centimeters.

3.6. HISTOLOGICAL ANALYSIS

Some animals were randomly chosen in each group and their tumors were removed, cut in pieces and fixed in Bouin de Hollande.

4. Results

4.1. EFFECTS OF PRE-TREATMENT ON ANTI-PHOSPHATIDYLINOSITOL AUTOANTIBODY LEVELS

Detection of anti-PtdIns autoAb was performed as previously described in Material and Methods in order to monitor the malignant transformation and effects of AIB1 Ab (500 µg) or B(a)P (1 µg) pre-treament (Figure 2).

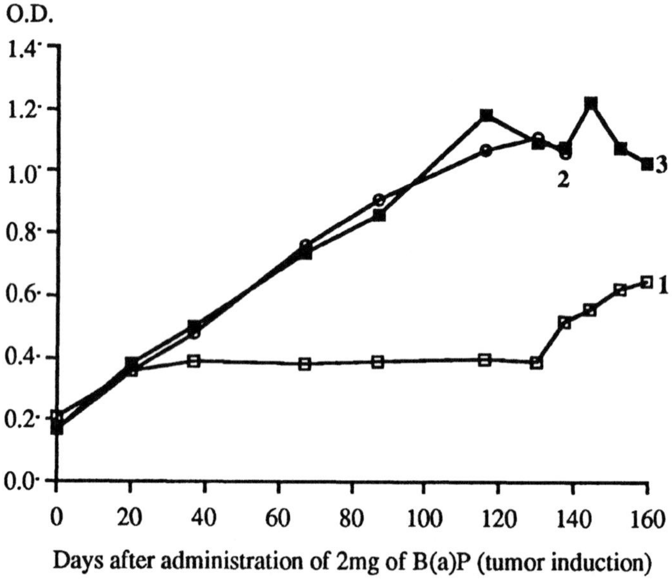

Figure 2. Effects of AIB1 Ab or B(a)P pre-treatment on circulating anti-PtdIns autoAb. Optical density (O.D.) represents the binding of rat autoAb (1/1000) to PtdIns coating on well-plates. Curves show mean O.D. values with standard error : curve 1 (p), AIB1 Ab pre-treatment (500 µg) 8 days before tumor induction (2mg of B(a)P) (n=4) ; curve 2 (m), B(a)P pre-treatment (1 µg) 8 days before tumor induction (n=4); curve 3 (n), solvent-treatment (control group, n=3).

When AIB1 Ab was administered 8 days before carcinogen injection (2mg of B(a)P, tumor induction), anti-PtdIns autoAb levels were around 70% weaker (curve 1, Figure 2) than those found in rats treated with 1µg of B(a)P (curve 2, Figure 2), or control animals (curve 3, Figure 2). Anti-PtdIns autoAb remained at a low level (70% of control group) until day 140. From that day onwards, autoAb level increased (50% of control group level), and seemed to reach those of control group. No differences were noted in the profil curves between the control group (curve 3) and the group treated with 1μg of B(a)P (curve 2) prior to the tumor-induction dose.

4.2. EFFECTS OF PRE-TREATMENT ON TUMOR GROWTH EVOLUTION

When tumors appeared, they were mechanically measured as previously described in Material and Methods. Figure 3 shows the increase of tumor growth for the three groups treated with AIB1 Ab (500 μg), B(a)P (1 μg), or solvent (control group).

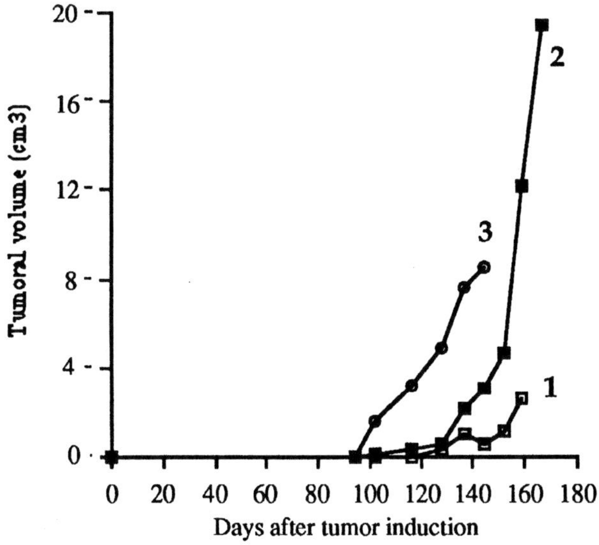

Figure 3. Time-course of tumor growth in rats pre-treated with: curve 1 (p), 500 μg of AIB1Ab (n = 4) ; curve 2 (n), 1 μg of B(a)P (n = 4) ; curve 3 (m), solvent (n =3). The tumor volume was calculated with the formula width^2 x length x 0.52.

Ten days after detection of clinically palpable tumors (0.065 cm3), control rat tumor volumes rapidly increased (around day 100, as previously described (Faiderbe et al., 1992b) (Figure 3, curve 3). The animals did not survive beyond day 137.
For SD female rats treated with AIB1 Ab, there was a delay of 16 days in tumor onset between 75% of AIB1 Ab-treated rats and the control group (Figure 3, curves 1 and 3). Moreover, tumor growth volume increased more slowly than those of control group (Figure 3, curves 1 and 3). 75% of animals died between day 140 and day 168. Moreover, 25% of the animals of this group were still alive 168 days after tumor induction, bearing little clinically palpable tumors (0.065 cm3).

For SD female rats treated with 1μg of B(a)P prior to tumor induction, there was no delay in tumor onset between these rats and the animals of the control group (Figure 3, curves 2 and 3). Moreover, tumor growth evolution was slowed down for around 30 days (until day-130). From that day onwards, tumor volumes rapidly increased and became equivalent to that of controls. All the animals treated with 1μg of B(a)P died around day 160.

4.3 EFFECTS OF PRE-TREATMENT ON ANIMAL SURVIVAL

Effects of AIB1 and B(a)P pre-treatment on animal survival have been evaluated.

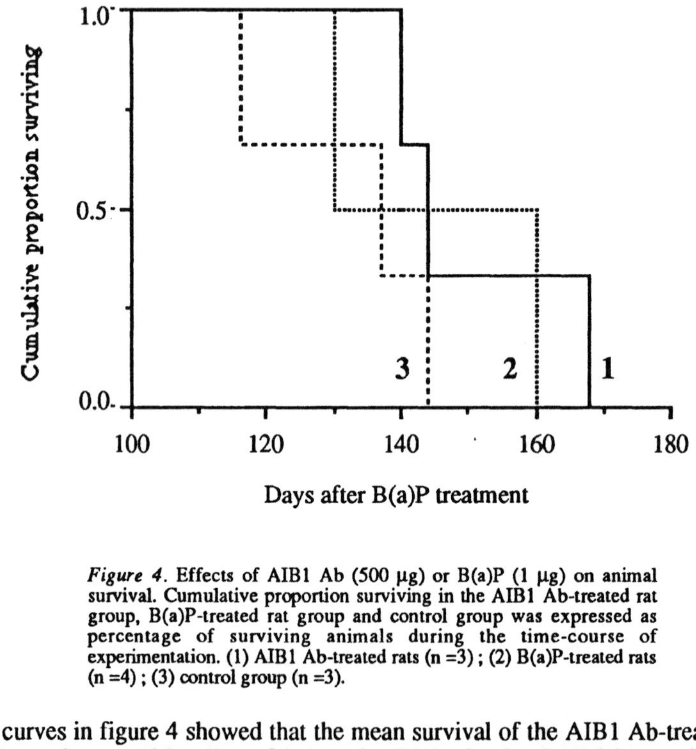

Figure 4. Effects of AIB1 Ab (500 μg) or B(a)P (1 μg) on animal survival. Cumulative proportion surviving in the AIB1 Ab-treated rat group, B(a)P-treated rat group and control group was expressed as percentage of surviving animals during the time-course of experimentation. (1) AIB1 Ab-treated rats (n =3) ; (2) B(a)P-treated rats (n =4) ; (3) control group (n =3).

Survival curves in figure 4 showed that the mean survival of the AIB1 Ab-treated rats (curve 1) was increased by about 34 days for 75% of animals (25% are still alive) relative to the control group (curve 3). For rats treated with 1μg of B(a)P, the survival was increased by about 20 days relative to the control group (curve 2).

4.4 HISTOLOGICAL ANALYSIS

Histological analysis of the tumors showed the presence of highly malignant sarcomas with fusiform cells for control rats as previously described by Faiderbe *et al.*, (1992 b) and sarcomas with cells presenting no morphological changes whatever the Ab (500 μg of AIB1 Ab) or carcinogen (1 μg of B(a)P) pre-treatment used.

5. Discussion

Using our tumor model induced by a single dose of 2 mg of B(a)P, increased anti-PtdIns autoAb titers were found in all B(a)P-treated SD female rats. All the animals developed a highly malignant sarcoma at the carcinogen injection site (top of the right thigh) around 100 days after administration of 2 mg of B(a)P diluted in sesame oil (Faiderbe *et al.*, 1992b). Tumor incidence and/or growth were easily monitored prior to death. This model was thus reliable for the evaluation of Ab treatment effects. To examine the possible vaccine effects of AIB1 Ab treatment, a single administration of AIB1 Ab before tumor induction was assayed, compared to administration of a single dose of B(a)P (Figure 1). When 500 µg of AIB1 Ab was injected 8 days before tumor induction (injection of 2 mg of B(a)P), anti-PtdIns autoAb levels were 70% weaker in relation of control group (Figure 2, curves 1 and 3). This low level remained stable until day 160 where it slowly increased. Moreover, tumor growth slowed down in relation to that of control group. Animal survival was increased by about 34 days relative to the controls (Figure 4, curve 1). The effects of AIB1 Ab injected prior to tumor induction, confirm the interaction between metabolic pathways involving B(a)P ligand/ cytosolic receptor binding, and the PtdIns as both second messenger and membran component. Comparatively similar effects were observed when AIB1 Ab was administered after tumoral induction (Chagnaud *et al.*, 1993c, 1994).

Anti-PtdIns autoAb levels decreased around 80% in relation to control group when AIB1 Ab was administered biweekly 73 days after carcinogen injection. This decrease was only of 30% when AIB1 Ab was administered biweekly after tumor appearance. For SD female rats treated from day 73 a significant delay of 30 days in tumor onset and a significant mean survival of 35 days were observed. For curative studies on SD female rats bearing little tumors (0.065 cm^3) a mean survival of 32 days was observed. Moreover, in 25% of the SD female rats pre-treated with AIB1 Ab tumor growth evolution was not observed even a long time after tumor induction (10 months over).

Whatever the period of AIB1 Ab treatment (preventive or curative) a slowing down in tumoral growth was observed and AIB1 Ab-treated rats survived longer than solvent treated rats. But what is notewhorthy in preventive treatment is that a single injection was necessary to obtain results equivalent to those obtain with repetitive injections in animals treated at differents times after tumor induction.

These results confirmed the capacity of a monoclonal anti-idiotypic Ab, internal image of a ligand involved in tumoral processes, to slow down tumor onset and growth, and to increase animal survival.

Conversely, for SD female rats treated with 1µg of B(a)P, autoAb levels were not modified relative to the control group (Figure 2, curves 2 and 3) and there was no delay in tumor appearance between these rats and the control group (Figure 3, curves 2 and 3). The tumor growth was slowed down for around 30 days before getting equivalent to that of controls (Figure 3, curves 2 and 3). It seemed that the preventive effects of 1 µg of B(a)P injected prior to tumor induction were lower than those of 500 µg of AIB1 Ab.

This preliminary study on preventive effects opens new perspectives. Using larger experimental groups and statistical studies the results will be further convincing.

Acknowledgements : The present work was supported by grants from Dolisos Laboratories (Paris) and IDRPHT (Talence).

6. References

Chagnaud, J.L., Faiderbe, S., Gosset, I. and Geffard, M. (1993a) Characterization of monoclonal anti-idiotypic antibody, Internal image of conjugated benzo(a)pyrene, in P. Guarrigues and M. Lamotte (eds), *Polycyclic Aromatic Compounds : Synthesis, properties, analytical measurements, occurrence and biological effects*, Gordon and Breach Science Publishers, pp 663-672.

Chagnaud, J.L., Faiderbe, S., Gosset, I. and Geffard, M. (1993b) Comparison of in vivo administration of benzo(a)pyrene and monoclonal anti-idiotypic antibody, internal image of benzo(a)pyrene to Sprague-Dawley female rats, in P. Guarrigues and M. Lamotte (eds), *Polycyclic Aromatic Compounds : Synthesis, properties, analytical measurements, occurrence and biological effects*, Gordon and Breach Science Publishers, pp 1119-1126.

Chagnaud, J.L., Faiderbe, S. and Geffard, M. (1993c) Effet d'un anticorps monoclonal anti-idiotypique, image interne du benzo(a)pyrène sur des sarcomes de rats, *C.R. Acad. Sci. Paris* **316**, 1266-1269.

Chagnaud, J.L., Faiderbe, S. and Geffard, M. (1994) Curative effects on rat sarcomas obtained after a treatment combining two monoclonal antibodies, *Anticancer Drugs* **5**, 361-366.

Faiderbe , S., Chagnaud, J.L., Wafflart, J. and Geffard, M. (1990) Autoanticorps dirigés contre un lipide membranaire dans les sérums de malades porteurs de tumeurs malignes, *C.R? Acad. Sci. Paris* **310**, 49-52.

Faiderbe , S., Chagnaud, J.L., Diolez-Bojda, F., Kerdelhue, B. and Geffard, M. (1991) Increase of auto anti-phosphatidylinositol antibodies in plasma of female rats during the appearance of DMBA-induced malignant mammary tumors, *Cancer Letters* **57**, 15-19.

Faiderbe , S., Chagnaud, J.L., Wafflart, J. Charrier M.C., Peyron, M.A. and Geffard, M. (1991) Autoantibodies directed against lipid membrane components in sera of patients with malignant tumors, *Cancer Detection and Prevention* **15**, 199-203.

Faiderbe , S., Chagnaud, J.L., Wafflart, J. and Geffard, M. (1992a) Anti-phosphoinositide autoantibodies in sera of cancer patients : Isotypic and immunochemical characterization, *Cancer Letters* **66**, 35-41.

Faiderbe, S., Chagnaud, J.L. and Geffard, M. (1992b) Identification and characterization of an autoimmune response during the time-course of benzo(a)pyrene-induced malignant tumors in female Sprague-Dawley rats, *Cancer Res.* **52** , 2862-2865.

Ingber, D., Fujika, T., Kishimoto, S., *et al.* (1990) Synthetic analogues of fumagilin that inhibit angiogenesis and suppress tumor growth, *Nature* **348**, 555-557.

EVIDENCE AND MECHANISTIC APPROACH OF THE PROTECTIVE EFFECTS OF HEAVY METAL HIGH DILUTIONS IN RODENTS AND RENAL CELL CULTURES *

Protective effect of metal high dilutions

A.DELBANCUT, M.P. BAROUILLET AND J. CAMBAR
Groupe d'Etude de Physiologie et Physiopathologie Rénales. Faculté de Pharmacie - 3, place de la Victoire - 33000, Bordeaux (France).

1. Introduction

Chronic exposure to heavy metals (e.g cadmium or lead), which are major environmental toxicants, can be associated with a variety of pathologies such as renal, hepatic or nervous damages (Mery and Fillastre, 1983; Humes and Weinberg, 1986). For decades, several authors have attempted to reduce *in vivo* metal toxicity using chelating agents or other metals as competing agents. Another method successfully applied consisted of pre- or post-treatment using either low concentrations (Terhar *et al..*, 1965; Ito and Sawauchi, 1966; Yoshikawa, 1973) or high dilutions (Fisher *et al..*, 1987; Cazin *et al..*, 1987; Herkovitz and Perez-Coll, 1991) of the same toxic inducing agent. Critical rewiews on the effectiveness of serially dilute agitated compounds in experimental toxicology have been recently published (Linde *et al.*, 1994; Cambar *et al.*, 1994).

For twelve years, investigations have been carried out in our laboratory, aimed at underscoring and ascertaining the effectiveness of very high dilutions of heavy metals in experimental toxicology. Our previous *in vivo* experiments performed in rodents had clearly pointed out a fair protective effect of mercury high dilution in mercury-induced intoxication (Cal *et al.*, 1986, 1988; Larue *et al.*, 1985).

The recent development of cell cultures permitted to largely reduce animal experimentation and introduced a convincing alternative method for a better approach of molecular mechanisms at cellular level. To date, renal cell cultures can be considered as a fruitful tool for studying the nephrotoxic potency of numerous substances (Bruggeman *et al.*, 1989; Lagroye *et al.*, 1995; Toutain *et al.*, 1995).

More recently in our laboratory, renal cell cultures were pretreated with very highly dilute cadmium or cisplatin prior to incubation with high cytotoxic concentrations of these metals. As a result, the high dilution pretreatment prevented the cytotoxic effect induced by high concentrations of cadmium or cisplatin; furthermore, these *in vitro* data could be correlated with our previously reported *in vivo* results.

---

* Communications presented at the 7th GIRI Meeting, November 1993, Montpellier, France, and at the 8th GIRI Meeting, December 1994, Jerusalem, israel.

M. Bastide (ed.), Signals and Images, 71–82.

Consistently, a recent book reported similar *in vitro* results when other mammalian cell lines were post-treated with low doses of heavy metals (Van Wijk and Wiegant, 1994).

2. Personal Results

1.1 IN VIVO PROTECTIVE EFFECT OF HIGHLY DILUTE HEAVY METAL AS ASSESSED BY DEATH RATE

Female Swiss mice weighing 18-20 g were housed by 15 in a room where the lighting conditions (light/dark 12:12 with light on from 08.00 to 20.00) remained constant all year long. Food and water were available *ad libitum.* Two series of experiments were carried out to test the influence of highly dilute mercury pretreatment on $HgCl_2$-induced mortality in mice and also to determine whether this effect is reproducible at different months in the year.

Low concentrations of mercury (10^{-18} to 10^{-30} M) were prepared by Dolisos laboratories (Paris, France) from a mother solution (1g Hg / ml in distilled water) by serial centesimal dilutions 1:99 and the solutions were succussed between the successive dilution steps.

After a standardization period of 10 days, the mice were daily injected i.p. throughout a 7-day period as follows: 0.5 ml of the 10^{-18} M and 10^{-30} M $HgCl_2$ dilutions (pretreated groups) or agitated (succussed) distilled water (control groups). On the 7th day of treatment, one hour after the last preventive injection, the animals were given a single i.p. dose of 5, 6 or 7 mg Hg/kg of $HgCl_2$.

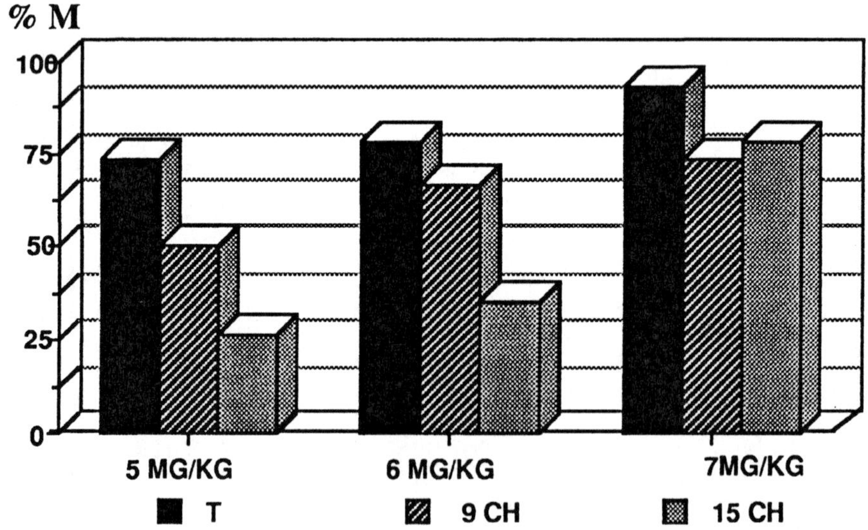

Figure 1. Mercury high dilution (10^{-18} and 10^{-30} M) protective effect against mercury-induced acute toxicity

Two hundred sixty five mice were pretreated with 10^{-30} M mercury according to the previous protocol with only 5 mg Hg/kg of $HgCl_2$ at 5 different months in the year (october, december, february, april and june).

In all experiments, the mortality was recorded daily during ten days after intoxication to assess the kinetics of the mortality rate.

Statistical analysis was performed using both the frequency comparison test and the Cosinor method.

The results of these experiments clearly show that the pretreatment with mercury high dilutions can induce an important reduction of the letal toxicity of $HgCl_2$, especially when the amount of $HgCl_2$ administered is as close as possible to the letal dose LD50 value.

Indeed, as outlined in figure 1, when the mice received 5 mg/kg, 73.4% of the animals died in the control group, against 50% and 26.7% ($p < 0,01$) only in the samples pretreated with 10^{-18} or 10^{-30} M mercury respectively. Similarly, when the mice were given 6 mg Hg/kg, 78.5% of the controls died while 66.7% of the 10^{-18} M mercury-pretreated and only 35.7% of the 10^{-30} M mercury-pretreated mice were lost ($p < 0,01$). This protective effect disappeared when the highest challenging dose of Hg/kg (7 mg) was administered, since the mortality rates were 93.3 for the controls and 73.4 and 78.5% (NS) for the pretreated groups.

The experiments carried out all year long showed a constant replication of the protective effect brought about by 10^{-30} M mercury against 5 mg/kg induced-mortality.

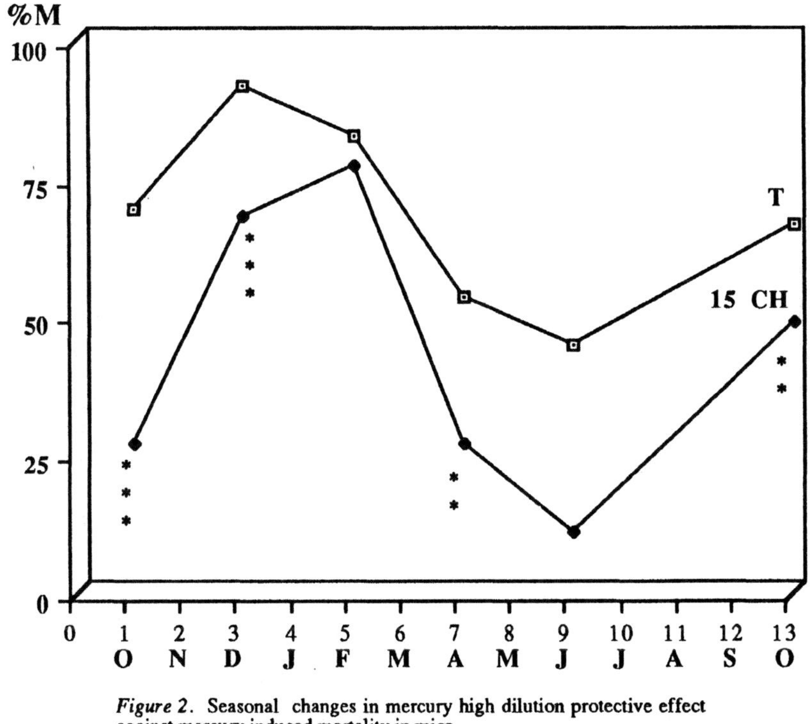

Figure 2. Seasonal changes in mercury high dilution protective effect against mercury induced mortality in mice.

Figure 2 shows the difference in mortality rates between pretreated and control groups was 66.6% in October, 26.1% in December, 11.1% in February, 49.9% in April and 75% in June. Moreover, the effectiveness of the mercury high dilution against $HgCl_2$-induced mortality exhibited a circannual rhythm ($p < 0.01$). The Cosinor analysis located the maximum of this protective effect in August.

1.2. IN VITRO PROTECTIVE EFFECT OF HEAVY METAL HIGH DILUTIONS ASSESSED BY THE CYTOTOXICITY RATE IN RENAL CELL CULTURES.

The protective effect of heavy metal high dilutions was assessed using an *in vitro* model applied to renal cell cultures. Cadmium or the vehicle was hundredfold serially diluted in the buffer (culture medium) and agitated for 30 s. We use very low cadmium concentrations (10^{-10} to 10^{-40}M).

Cultured cells issued from proximal tubular cell line (LLCPK 1) were incubated for 120 h with either dilute cadmium (pretreated cell group) or dilute culture medium (control cell group) and then intoxicated by massive amounts of cadmium. Cell viability was assessed by the neutral red test ; letality rate was compared to non-intoxicated cells. During the whole procedure (pretreatment and intoxication), calf fetal serum (CFS) concentration was only 2 %, in order to prevent the chelation of the metal by proteins of the CFS. A protective index of very high dilution was determined by comparing the death ratio of the pretreated cell group with the control one. The first experiment is summarized in figure 3.

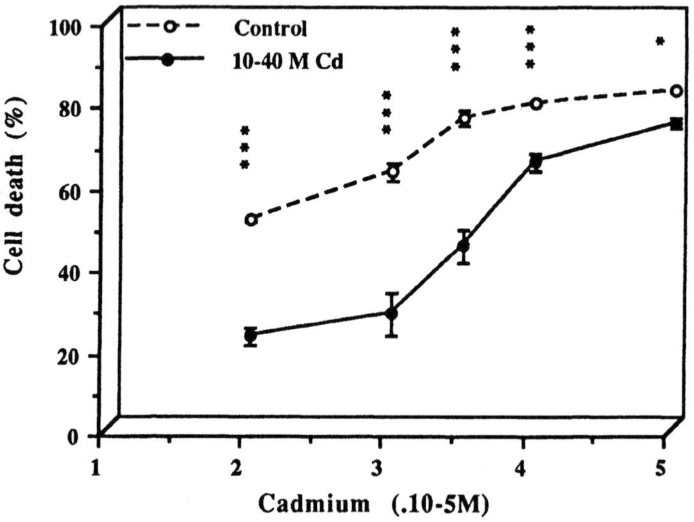

Figure 3. Protective effect of Cd 10^{-40} M dilution against Cd intoxication in tubular cultured cells.

Figure 3 shows LLCPK 1 cell mortality towards ponderal cadmium 24 h after intoxication, in a case where the cell pretreatment with either highly diluted solutions of cadmium (pretreated cell group) or diluted culture medium solutions (control cell group) lasted 120 h.

The difference between both cell groups was highly significant, fairly illustrating the protective effect of the 10^{-40} M cadmium dilution: cell mortality in the pretreated group was very low as compared with the untreated sample; 2.10^{-5} M cadmium induced 51 % of mortality for controls against 22% only for pretreated cells. This protective effect of highly diluted cadmium occurred irrespective of the toxic dose of cadmium used. However, the pretreatment appeared less effective and statistically not significant when associated with the highest toxic concentration (5.10^{-5} M). The statistical analysis was performed by the ANOVA global analysis and t Student test. Similar results have been found at five different periods in the year, showing a good reproducibility of the described protective effect.

To conclude, the protective effect of very high dilutions of a metal against toxic doses of the same metal seems clearly demonstrated in proximal tubular cell cultures. Nevertheless, the pretreatment must be applied for a 120 h period. Indeed, we established that no protective effect could be observed for a 48 h pretreatment (data not shown).

In the same way, a protective effect has been obtained with very low concentrations of cisplatin applied during 120 h against intoxication with ponderal doses of cisplatin. Whatever the cisplatin dose used for intoxication, the death rate was higher in the control than in the pretreated cell group. Hence, for the 4.10^{-5} M toxic dose, the percentage of dead cells decreased from 60 % for untreated cells down to 40 % for pretreated cells. Nevertheless, the index of protection was lower for cisplatin than for cadmium.

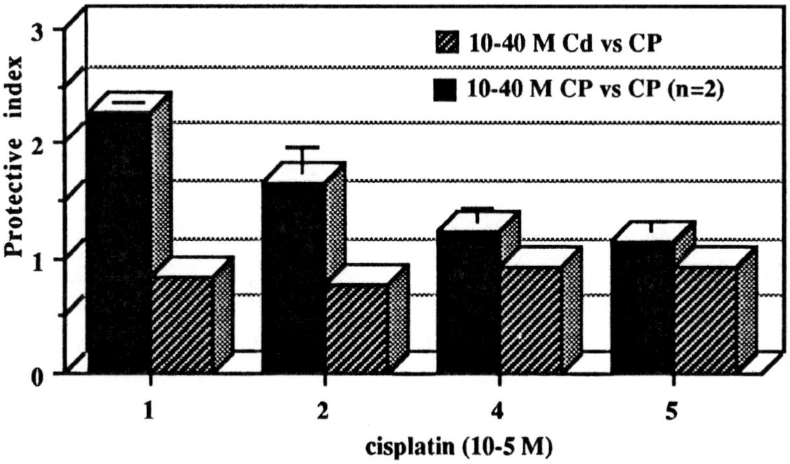

Figure 4. Dilution height dependent effect against Cd intoxication

Therefore, the protective effectiveness seems to depend on the toxic dose used, the pretreatment duration and, as we can see in the figure 4, also on the height of the pretreatment dilution. For a 120 h period, we had incubated the cells with serial centesimal dilution of cadmium from 10^{-10} to 10^{-40} M. Then, cells were intoxicated with our usual ponderal range of cadmium.

For those four dilutions (10^{-10}, 10^{-20}, 10^{-30}, 10^{-40} M), a protective index was determined. It is defined by the ratio of control cell death rate to pretreated cell one. If this index is equal to 1, there is lack of protective effect; if it is higher than 1, then it expresses a favourable effect and consequently a protective one. On the contrary, an index below 1 means an increase of cell death induced by the pretreatment itself.

As outlined in the figure 4, the protective effect of the highest dilutions is neither efficient for high toxic doses of cadmium nor for the lower ones. The index of protection is dilution dependent and it increases with the dilution from 10^{-10} to 10^{-40} M with a maximum for 10^{-40} M of cadmium.

The same data was obtained with another metal like cisplatin. The protection index was also dilution height dependent. However, the effectiveness of this protection was less important for the cisplatin than for cadmium. So, the most important index for cisplatin was 1.68 (2.10^{-5} M cisplatin) against 4.70 for cadmium (2×10^{-5} M cadmium).

The protective effect of highly diluted metal towards metal cytotoxicity revealed to be specific of the metal used for the dilution. We compared the effect of pretreatment with 10^{-40} M of cisplatin towards an intoxication with cisplatin (cisplatin versus cisplatin) and pretreatment with cisplatin towards an intoxication with cadmium (cisplatin versus cadmium). The data are shown in figure 5.

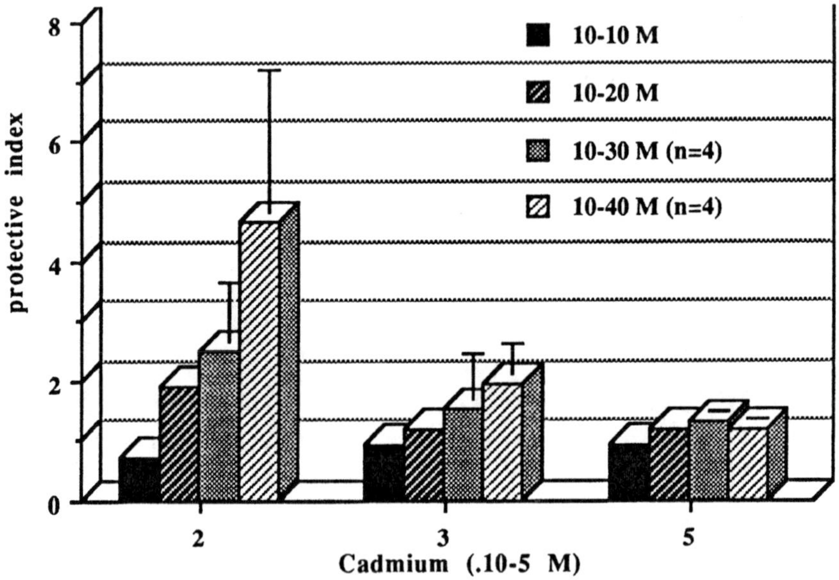

Figure 5. Protection index of cisplatin and cadmium high dilution (10^{-40}M) in intoxicated tubular cultured cells by the same metals.

The protection index was less than 1 for each cadmium versus cisplatin experiment, which means total lack of protection. We even noted a significant decrease in death rate for the pretreated cells as compared with their control counterparts (CP vs CP): 5.10^{-5} M cisplatin killed 63.1 % of cisplatin pretreated cells when it killed 68.4 % of cadmium pretreated cells.

The effectiveness of the pretreatment requires the use of the same metal for high dilution and intoxication; indeed the protection induced by the pretreatment with cadmium or cisplatin dilution is significant only when the intoxication is caused by cadmium or cisplatin respectively; the principle of similarity, the basis of Homeopathy, is confirmed.

After bringing to the four experimental proofs of the effectiveness of the pretreatment by highly diluted metals in reducing cell death associated with heavy metal intoxication, we decided to study the intracellular structures, especially the organisation of cytoskeleton proteins, marker of metabolic and functional state of the cell.

By an indirect immunofluorescence technique , it is possible to visualize actin and vimentin networks respectively diffused and clearcut in the cytoplasm.

We compared the aspect and the organization of the network of both proteins in pretreated and control cells towards an intoxication by ponderal high cadmium doses (0,125, 0,25, 0,5, 2 x 10^{-5} M). Whatever the network studied, it seemed that the pretreatment by high dilutions is able to decrease cadmium-induced cytoskeleton protein desorganization. Indeed, for the same intoxication concentration, the network aspect of actin and vimetin appeared more normal in pretreated cells than in controls. The vimentin network seemed yet clearcut, well-organized and developed even in 24 h intoxicated cells by 0,25 x 10^{-5} M of cadmium (figure 6).

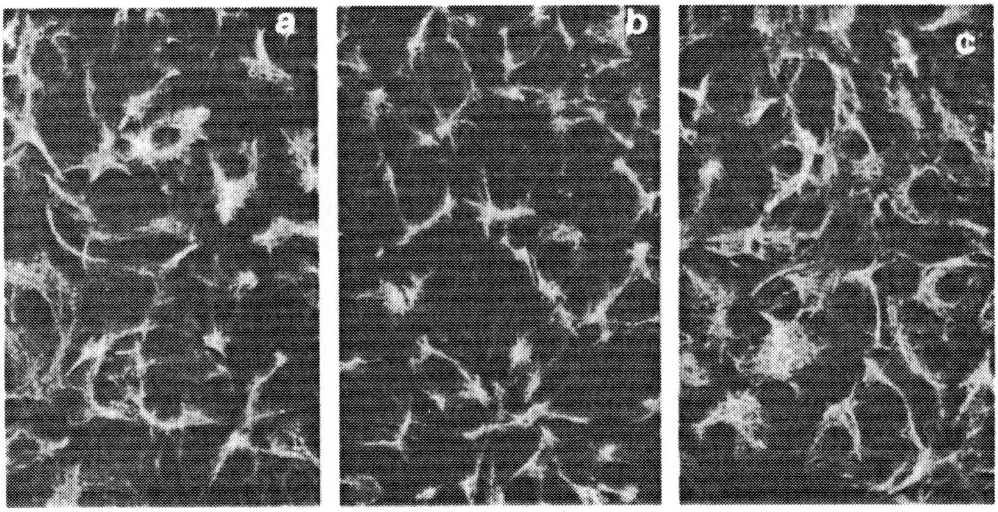

a b c

Figure 6 . Vimentin cytoskeleton protein organization in tubular cultured cells (a: control non intoxicated cells, b: intoxicated non pre-treated cells, c: intoxicated pre-treated cells.)

3. Mechanistic Approach of Heavy Metal High Dilution Protection

3.1.INFLUENCE OF VERY HIGH DILUTIONS OF CADMIUM ON INTRACELLULAR CADMIUM PENETRATION

Flame spectrophotometry allows a precise determination of cadmium concentration. The penetration rate of cadmium in the cell can be deduced from the difference between the cadmium present in the cells and the total cadmium present in the mixture and in the supernatant.

At the end of the pretreatment period (120 h), the cells were intoxicated by introducing ponderal concentrations of cadmium from cadmium standard solutions directly in the cell medium.

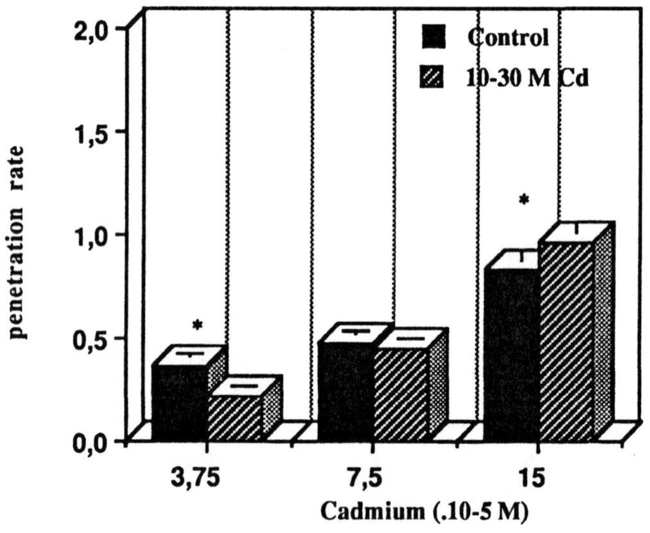

Figure 7. Cd penetration ratio in control intoxicated and intoxicated pre-treated tubular cultured cells.

As we can see in figure 7, cadmium penetration in the pretreated cell group was 35 % lower than in the control group; for the lowest intoxication concentration, at 3.75 mg Cd, the penetration rate decreased from 0.36 to 0.22. For this concentration, the very high dilution seemed to reduce cadmium penetration and therefore lowered intracellular cadmium concentration which induced less dramatic cell damages.

Nonetheless, except from this concentration, no significant difference between the pretreated and control groups could be found. Furtheremore, these results are confirmed by atomic absoption , a more sensitive method for cadmium dosage.

3.2. INFLUENCE OF VERY HIGH DILUTIONS OF CADMIUM ON METALLOTHIONEIN SYNTHESIS

As previously reported by different authors, cadmium is a well-known metallothionein (MT) synthesis stimulating agent. Therefore, we wondered whether MT could be involved in the observed protective effect of highly diluted metal in heavy metal intoxication. MT was detected and quantified by radioimmunoassay (RIA) techniques by Drs LAMBOEUF and CARRERA from the laboratory of " Signalisation et Differenciation des Macrophages" (Unité I..N.S.E.R.M. du C.H.U de Rangueil à Toulouse).

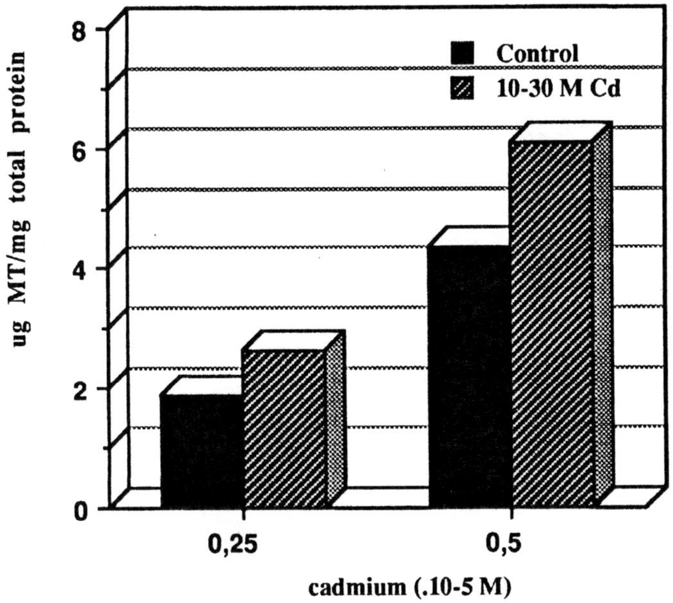

Figure 8 . Metallothionein synthesis in intoxicated control and intoxicated pretreated tubular cultured cells.

Figure 8 shows clearly that cadmium induces metallothionein synthesis in renal tubular cell culture, and that the pretreatment with very high dilutions of cadmium enhances metallothionein biosynthesis after a 24 h cadmium intoxication. The 120 h pretreatment with 10^{-30} M dilution increases significantly the concentration of metallothionein as compared with the control group.

After 24 hour incubation with 0.25 x 10^{-5} M of cadmium, the concentration of MT decreases from 1.93 mg MT / mg of protein for control cells to 2.67 mg MT / mg protein for pretreated cells. In the same way, 0.5 x 10^{-5} M cadmium induces the synthesis of 4.0 mg of MT / mg protein in the control cell group against 6.03 mg of MT/ mg protein in the pretreated cell group.

So, it seems that very high dilutions of cadmium enhance metallothionein biosynthesis consecutively to a ponderal intoxication with cadmium; this MT biosynthesis seems to remain dilution-dependent. Indeed, data not shown here, revealed that a solution of 10^{-10} M cadmium still induces metallothionein sythesis but less than cadmium 10^{-20} and 10^{-30} M solutions. The pretreatment with 10^{-20} M favours MT synthesis but not enough for protecting against an intoxication with 0.5 x 10^{-5} M cadmium.

In fact, it seems that only the pretreatment with cadmium 10^{-30} M dilution has a protective effect, which could explain its ability to enhance MT biosynthesis consecutively to ponderal intoxication of cadmium.

4. Discussion and Conclusion

The presentation of the results obtained in our laboratory for more than 12 years has been divided into two chronological and methodological parts, considering previous *in vivo* and recent *in vitro* experiments.

We have shown, *in vivo* in rodents, that high dilutions of mercury can significantly reduce the death rate induced by toxic doses of the same metal. Thus, the death rate of mice intoxicated with high doses of mercury was markedly reduced by a 7 day pretreatment with 10^{-30} M concentrations of that metal. For example, the death rate following a single injection of 5 mg/kg $HgCl_2$ was 73.4% in control, 50% for those pretreated with 10^{-18} M and only 26.7% for those pretreated with 10^{-30} M.

We found the same profile of response with our *in vitro* model in which 10^{-30} and 10^{-40} M dilutions provided a fair protection against cadmium induced cytotoxicity in renal tubular cell cultures. Similar results have recently been described for high dilutions of thymulin (Bastide *et al.*, 1987), bursin (Youbicier-Simo *et al.*., 1993) and silica (Oberbaum *et al.*., 1992).

The classical neutral red test permits a standardization of cytotoxicity evaluation procedures using renal cell cultures intoxicated by toxic cadmium solutions. Our results are reproducible (5/5 for 10^{-30} M and 4/4 for 10^{-40} M). This *in vitro* model using cell cultures eliminates numerous sources of variability, especially chronobiological ones, present in *in vivo* tests. Our results are confirmed by similar *in vitro* experiments in other cultured mammalian cells intoxicated by cadmium and protected by low dilutions of the same metal (Van Wijk and Wiegant , 1994).

The protective effect of very high cadmium dilutions against toxic doses of this metal seems clearly demonstrated. Similar protection against cisplatin intoxication is also demonstrated for ultra low doses of cisplatin. This protection seems specific since cadmium dilutions do not protect against cisplatin toxicity despite the fact that cisplatin is also a tubulotoxic metal.

Recent experiments carried out in our laboratory have shown that MT synthesis was increased in Cd intoxicated cells after exposure to very high Cd dilutions (10^{-20}, or 10^{-30} and 10^{-40} M) for 120 hours. Similar results have been reported by Pellegrini *et al.* (1994) with low doses of cadmium (1 μM to 0.1 nM) in an immature human T cell line. A 24 h pretreatment with non toxic concentrations of cadmium confered protection against subsequent cytotoxic doses and was able to induce metallothionein - IIa gene expression. An increase of MT synthesis has been already observed with toxic or subtoxic concentrations of heavy metals such as zinc, cisplatin or cadmium. These results are in fair accordance with other ones from our laboratory showing that such a treatment decreases the metal storage in the cell by, for example, changes in membrane permeability.

These *in vitro* data support our *in vivo* results suggesting that very high dilutions of metals may be useful in decreasing or even eliminating the toxic effects of nephrotic metals in patients chronically exposed to such products. Such a protective procedure was already well accepted for pretreatments with milligram or microgram doses in experimental toxicology (Flora *et al.*, 1982a,1982b; Stacey and Klaassen, 1981; Magos *et al.*, 1979; Yamane and Koizumi, 1982).

These data demonstrate not only the relevance of *in vitro* models in this area of toxicology but also confirm the specific protective effects of high dilutions of cadmium or cisplatin against subsequent cytotoxic doses of the same metals.

Aknowmedgements

We thank Dr P. Dorfman for assistance with statistical analysis and helpful and kind discussion. These studies were supported by a grant of Laboratoires Dolisos with a CIFRE contract with the French Ministery of Research and Technology.

5. References

Bastide, M., Daurat, V., Doucet-Jaboeuf, M., Pelegrin, A. and Dorfman P. (1987) Immunomodulatory activity of very low doses of thymulin in mice, *Int. J. Immmunotherapy* **3**, 191-200.

Bruggeman, I.M., Mertens, J.J., Temmink, J.H., Lans, M.C., Vosr,.M. and Van Bladeren, P.J. (1989) Use of monolayers of primary rat kidney cortex cells for nephrotoxic studies, *Toxicol. Vitro* **3**, 261-269.

Cal, J.C., Larue, F., Guillemain, J. and Cambar, J. (1986) Chronobiological approach of protective effect of *Mercurius corrosivus* against mercury induced nephrotoxicity, *Ann. Rev. Chronopharmacol.* **3**, 99-103.

Cal, J.C., Larue, F., Guillemain, J. and Cambar, J (1988) Chronobiological approach of mercury induced toxicity and the protective effect of high dilutions of mercury against mercury-induced nephrotoxicity, in A. Guillouzo (ed.) *Liver Cells and Drugs*, John Libbey Eurotext Publisher, Paris, pp 481-485.

Cambar, J., Delbancut, A. and Dorfman, P (1994) Applications des très hautes dilutions en toxicologie, *Cahiers de Biothérapie*, **125**, 27-34.

Cazin, J.C., Cazin, M., Gaborit, J.L., Chaoui, A., Boiron, J., Belon, P., Cherruault, Y., and Papayanotou, C.(1987) A study of the effect of decimal and centesimal dilutions of arsenic on the retention and mobilization of arsenic in the rat, *Human Toxicol.* **6**, 315-320.

Fisher, P., House, I., Belon, P. and Turner, P. (1987) The influence of the homeopathic remedy *Plumbum met* on the excretion kinetics of lead in rats, *Human Toxicol.* **6**, 321-324.

Flora, S.J.S., Behari, J.R., Ashquin, M. and Tandon, S.K. (1982a) Time dependent protective effect of selenium against cadmium-induced nephrotoxicity and hepatoxicity, *Chem. Biol. Inter.*, **42**, 345-351.

Flora, S.J.S ., Jain, V.K., Berahi, J.R. and Tandon, S.K. (1982b) Protective effect of trace metals in lead intoxication, *Toxicol. Letters* **13**, 51-56.

Herkovits, J. and Perez-Coll, C. (1991) Potentized microdoses of cadmium reduce the lethal effect of this heavy metal in amphibian embryos, *Berlin J. Res. Homeo.* **1**, 93-96.

Humes, H.D. and Weinberg, J. M. (1986) Toxic nephropathy, in B.M. Brenner and F.C. Rector (eds) *The Kidney*, Saunders Publisher, Philadelphia, pp 1491-2008.

Ito, T. and Sawauchi, K. (1966) Inhibitory effects on cadmium-induced testicular damage by pretreatment with smaller cadmium dose, *Okajima. Fol. Anat. Jap.*, **42**, 107-117.

Lagroye, I., L'Azou, B., Lakhdar, B., Plande, J. and Cambar, J. (1995) Nephrotoxicité in vitro: modèles glomérulaires, in M. Adolphe and A. Guillouzo (eds) *Toxicologie cellulaire in vitro* , Editions INSERM Publisher, Paris, pp 121-150.

Larue, F., Cal, J.C., Dorian, C., Guillemain, J. and Cambar, J. (1987) Influence du prétraitement de dilutions infinitésimales de *Mercurius corrosivus* sur la mortalité induite par le chlorure mercurique, *Néphrologie*, **6**, 86.

Linde, K., Jonas, W.B., Melchart, D., Worku, F., Wagner, H. and Eitel F. (1994). Critical review and meta-analysis of serial agitated dilutions in experimental toxicology, *Human Exp.Tox.* **5**, pp 481-492.

Magos, L., Webb, M. and Buttler, M. (1979) The effect of Cd pretreatment on the nephrotoxic action and kidney uptake of mercury in male and female rats, *Brit J Pathol*, **55**, 589-594.

Mery, J.P and Fillastre, J.P. (1983) Reins et médicaments, in J. Hamburger, J.P. Grunfeld (eds.), *Nephrologie*, Flammarion, Paris, pp 1193-1207.

Oberbaum, M., Markovits, R., Weissman, Z., Klinkevits, A. and Bentwich, Z. (1992) Wound healing by homeopathic silica dilutions in mice, *Harefuah*, **123**, 79-82.

Pellegrini, O., Davenas, E., El Azzouzi, B., Jurgens, P., Benneniste, J., Manuel, Y. and Thomas, Y. (1994) Stress proteins in human lymphocytes. Induction of metallothionein and heat shock protein genes by cadmium in an immature T cell line, *Cellular Pharmacology*, **1**, 91-97.

Stacey, M.A. and Klaasen, C.D. (1981) Interaction of metal ions with cadmium-induced cellular toxicity, *J. Toxicol. Environ. Health* **7**, 149.

Terhar, C.J., Vis, E., Roudabush, R.L. and Fasset, D.W. (1965) Protective effects of low doses of cadmium chloride against subsequent high oral doses in the rat, *Toxicol. Appl Pharm.* **7**, 500.

Toutain, H.J.- Néphrotoxicité in vitro: modèles de cultures de cellules tubulaires rénales (1995). in M. Adolphe and A. Guillouzo (eds) *Toxicologie cellulaire in vitro* , Editions INSERM Publisher, Paris, pp 151-189.

Van Wijk, R. and Wiegant, F.A.C. (1994) *Cultured mammalian cells in homeopathy research; The similia principle in self-recovery.* Utrecht University Publisher, Utrecht.

Yamane, Y. and Koizumi, T. (1982) Protective effect of Mo on acute toxicology of HgCl2, *Toxicol. Pharmacol.* **15**, 14-221.

Yoshikawa, H. (1973) Preventive effect of pretreatment with cadmium on acute cadmium poisoning in rats *Industrial health (Japan)* **11**, 13-119.

Youbicier-Simo, B.J., Boudard, F., Mekaouche, M., Bastide, M. and Baylé, J.D. (1993) Effects of embryonic bursectomy and in ovo administration of highly diluted bursin on adrenocorticotropic and immune responses of chickens, *Int. J. Imunotherapy* **9**, 169-180.

CHAPTER II

IN VITRO AND IN VIVO MODELS

This point raises a question: is the pharmacology of successed high dilution a classical one related to stimulus in biological systems (Bonne) or are these high dilutions informative structures different from the molecules and able to be received (and treated) directly by the organism ? The physical study by Demangeat *et al.* opens the door to potentially a modification of successed high dilutions (succussion or potentization is obtained by a strong vertical manual or mechanical agitation of the dilution for at least 30 seconds). This appears as a physical modification

M. Bastide (ed.), Signals and Images, 83–84.

of the agitated dilutions whatever the real phenomenon which occurs. All the pharmacological papers presented here (except the Bellavite *et al.* results) were performed with non-molecular succussed high dilutions (beyond the Avogadro number), either diluted endognenous molecules as hormones (Youbicier-Simo *et al.* and Endler *et al.*) or homeopathic remedies (Cristea *et al.* and Oberbaum *et al.*). High dilutions of antigens were also tested in immune response (Weissman *et al.*). The Bellavite paper is more related to the mechanistic models reviewed by Bonne. One may think that some low dose effects can be interpreted in the framework of classical science. However, as soon as the work is performed with non-molecular succussed solutions, either it must be considered as nonsense in the context of the molecular dogma, or "something" is present in the solutions which can be received and interpreted by the organism. The highly diluted hormones are significantly active in frogs or chickens as soon as they bring a lacking and important piece of information (thyroxine is important for tadpole metamorphosis and bursin is crucial for the bursectomized chick embryos). The "message" can be transmitted through the glass or by a compact disk to tadpoles (Endler). We are a step further in the "information" hypothesis. The use of homeopathic remedies are more difficult to perform considering the results of Cristea and Oberbaum. The "receivers" (live organ, living organism) have to interprete the transmitted information, either in an Arndt-Schultz law in the Belladonna Cristea model (related to hormesis by inversion of effects) or according to the true homeopathic "similia-law" based on the framework of the symptoms (Chamomilla or Silicea). The reproducibility of the results depends on the quality of the interpretation of the receiver due to its "sensitivity".

These experimental models are well documented, scientifically accepted by academic journals and will help us to understand the next chapter.

M.B.

FROM STIMULUS TO MESSAGE IN BIOLOGICAL SYSTEMS, AN ILLUSTRATED SURVEY *

A review

C. BONNE
Université Montpellier I, Faculté de Pharmacie
34060 Montpellier, France

1. Introduction

For a century, a major part of physiology and pharmacology has been devoted to the investigation of the mechanisms which transform a stimulus into a message. All the present notions are still based on the revolutionary concept of Paul Ehrlich: *"Corpora nun agunt nisi fixata"*. Substances cannot act without binding, and, as far as the physical stimuli are concerned, one can consider that the stimulus modifies the binding of pre-existing biochemical components. The aim of this survey, is to illustrate the various mechanisms which transform external stimuli into cellular messages.

A stimulus is an agent that produces a response in a living organism. One stimulus has no significance *per se* but acquires one particular meaning, becoming a message, when applied to one particular biological system. Stimuli can be of very different natures; they can be physical, chemical, static or dynamic. Ultimately, the absence of stimulus can be a stimulus by itself if we consider the environment of the living organisms.

2. Physical Stimuli

As an illustration of this paradox, darkness is a stimulus as far as the retinal photoreceptors are concerned, since sensory cells are very active in the absence of light in terms of metabolism and synaptic transmission ! Actually, under this condition, rhodopsin, the photosensitive pigment located in the outer segment disks of the cell, does not interact with cation channels of the membrane. These channels are maintained in the open confirguration by cyclic GMP and consequently, the membrane is depolarized and the neurotransmitter, glutamate, is continuously released (Figure 1).

Obviously, light is also a stimulus for these cells. When a photon interacts with rhodopsine, the conformation of retinol, a component of the pigment, changes from eleven-cis to all trans.

This event allows rhodopsin to bind another protein (Gt), a G-Protein also called transducin, which activates an enzyme (PDE) able to metabolize cGMP. Without

---

* Plenary lecture presented at the 8th GIRI Meeting, December 1994, Jerusalem, Israel.

M. Bastide (ed.), Signals and Images, 85–93.

cGMP, cation channels are closed, the cell becomes hyperpolarized and the synaptic transmission is stopped (Pugh and Lamb, 1993).

This is a good illustration of a stimulus transformation into a message, through the activation of a receptor, a transduction and an effector system.

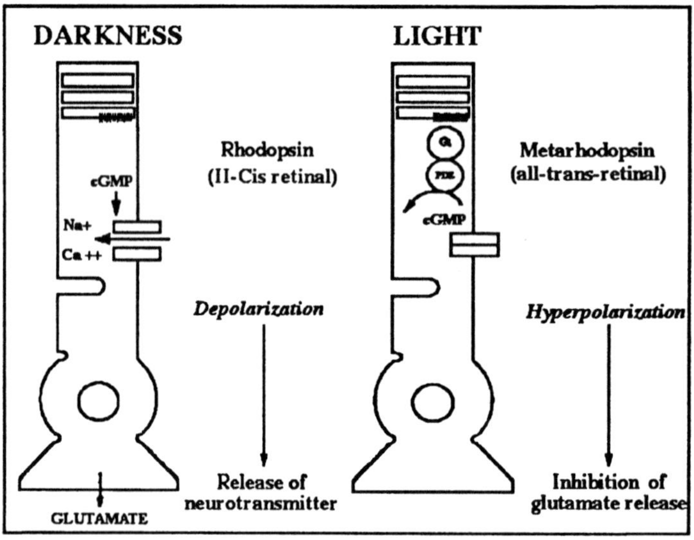

Figure 1. Physical stimuli, darkness and light, modulate the release of the photoreceptor neurotransmitter.

Another example of physical stimulus is mechanical force. Mechanosensitive membrane ion channels provide a means of transducing membrane deformation or stretch into an electrical signal. These mechanoreceptors are ubiquitous, they are expressed in a wide variety of cell types including both sensory and non-sensory cells.They are implicated in such diverse functions as cell volume regulation, stretch-activated reflexes in vascular cells, but also in cell growth and embryogenesis.

Among the physical stimuli, a particular one, magnetic field, has to be considered with interest but caution. There is a lot of contradictory data reported in the medical and scientific litterature, which concern the effect of magnetic fields on biological systems.

To the present knowledge, it seems however quite clear that biochemical parameters can be modified by exposure to 50 Hz magnetic fields at very low flux densities (Loscher and Mevissen, 1994; Sandyk, 1992). For example, animal studies have shown that magnetic field exposure reduces the nocturnal plasma level of melatonin, a hormon secreted by the pineal gland. Another example is given by racing pigeons which are believed to use a magnetic map for homing. This question raises a great deal of controversies but, if some reports are very impressive, no magneto-receptor has been yet identified in terms of molecular entity.

3. Chemical Stimuli and the Key/Lock Image

Molecules are the key information vectors in biological systems. They are the extracellular and intracellular stimuli which regulate cell functions, metabolism as well as genetic expression. As previously quoted, molecules become informative by binding to their receptors. The classical concept of ligand/receptor interaction is illustrated by the key/lock image, where the dimensional characteristics of the ligand are complementary to the structure of the binding site. Recent developments of molecular biology, in particular experiments based on site-directed mutagenesis have established the role of specific residues of the receptor proteins and clearly identified the binding pocket of an increasing number of hormon and neurotransmitter specific receptors.

In fact, specificity of binding sites for ligands is only relative (fortunately for the pharmacist, otherwise there were no possible drugs !). The figure 2 shows the functional structure of a corticosteroid hormon receptor, which is a soluble protein of the cytoplasm. The glucocorticoid receptor has now been cloned and we know that the primary structure consists of approximately 800 amino-acid residues. Using deletion mutagenesis, several distinct domains within the structure have been defined.

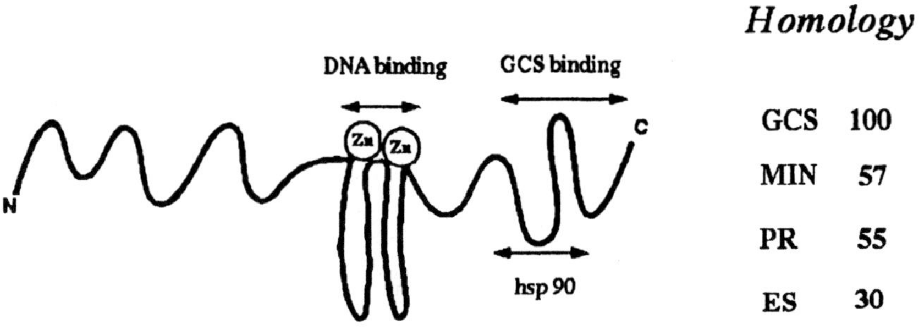

Figure 2. Structure and homology of steroid hormon receptors.

The steroid-binding domain is at the C-terminal end of the molecule. The other domains are implicated in interactions with other structures such as DNA. As far as binding specificity is concerned, it is interesting to note that there is a high percentage of homology of the binding domain in the various steroid hormon receptors (Barnes and Adcock; 1993). This can explain that some synthetic steroids present a low specificity for one particular receptor, can bind to several classes of hormon receptors and consequently have dual hormonal activities (e.g. the anti-aldosterone drug, spironolactone, presents anti-androgenic side effects). These data recall to pharmacologists that the lock is not really a safety one !

As everybody knows, a majority of informative molecules interact with their receptor according to the action of mass law. It is important to remark that Kd, the dissociation constants at equilibrium, are always in the range of physiological concentrations of ligands, such as hormons, neurotransmitters and other informative molecules (Kelly and Beaulieu, 1989). That is to say that the message can be quantitatively modulated as a function of stimulus concentration. Kd values are most frequently in the range of micro- to nanomolar, but can also be lower, for instance at picomolar for several growth factors. In fact, possible lower values could occur but there is a technical limit for their measurement even with radioligands. Moreover, in some cases, chemicals can be detected at extremely diluted concentrations by receptors of relatively low affinity. This is the case for odorous molecules which are recognized by receptors of olfactory cells due to prior dissolving and accumulation in nasal mucus. This process concentrates the ligand up to the Kd value of the receptor.

If a lot of chemical stimulus/receptor interactions are governed by this sample low, some other receptor activation processes are more complex : cooperativity of binding has been reported for various ligands either positive or negative. In these cases, the binding kinetic parameters depend on the concentration of the unbound ligand. In a similar way, it can be noted that the Kd of some receptors for their ligand is modulated by the interaction of other molecules with allosteric binding sites. For example, the sedative benzodiazepins bind to the GABA-A receptor on a specific site and increase the affinity of the receptor for this inhibitory neurotransmitter. (This mechanism explains why benzodiazepins are not active in GABA-depleted organisms).

4. Stimuli and the 4th D

Another aspect of the stimulus/receptor interaction had to be considered from the point of view of the 4th D, i.e. time. Indeed, the transformation of a stimulus into a message depends on the previous state of the reception machinery which is not a frozen apparatus but a plastic one.

One illustration of the dynamic nature of the conversion of a stimulus into message is given by chemotaxis. This is a general phenomenon in biology which is responsible, for example, for sexual attraction of butterflies by pheromons, or for the directional migration of leukocytes toward an inflammation site. This mechanism requires recognition of a molecular gradient, in other words, recognition of a molecular concentration as a function of time. Such a phenomenon implicates a dynamic participation of the receptive system to give significance to the stimulus.

This notion can also be illustrated by tachyphylaxis : a first stimulus can inactivate the receptor/transduction machinery which become desensitized to a second stimulus. For example, the inflammatory mediator, leukotriene B_4, down-regulates its own receptor by producing a low-affinity phosphorylated form of this protein.

5. Transduction from Receptor to Effector

It is fascinating to discover both the great homogeneity and the large variety of the receptor-tranduction systems in biology. If we consider only the membrane receptors, we can distinguish four general systems. Ionotropic receptors which are linked to ion channels and metabotroptic receptors which are linked to enzymes. They can also be classified according to their dependence or independence on G-Proteins (Figure 3).

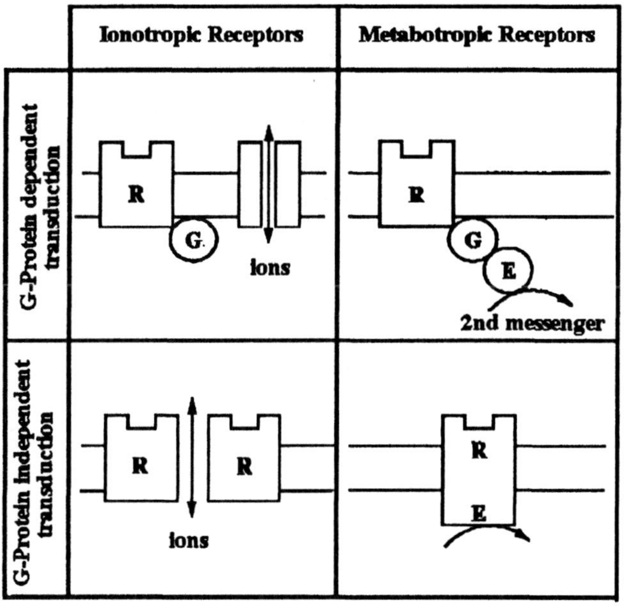

Figure 3 : Membrane receptor transduction systems.

5.1. METABOTROPIC RECEPTORS

Recently, Alfred Gilman and Martin Rodbell have received the Nobel Price for their discovery of the G-Proteins (Hanoune, 1994). G-Proteins are a superfamily of proteins which are formed by 3 subunits : the hydrophylic α and β and the lipophylic γ subunits.

These proteins are called G-Proteins because they bind GDP or GTP and hydrolyse GTP. Finally these protein complexes provide a link between activated receptors and effector proteins. There are several G-Proteins which can either activate or inhibit various enzymes and ion channels.All the receptors linked to G-Proteins present the same general structure with seven helix transmembrane domains. This is the case for β adrenoreceptors, the muscarinic acetylcholine cerebral receptor, the numerous olfactory receptors recently identified in the nasal mucosa as well as rhodopsin.

So, a chemical stimulus and a physical stimulus as light, use a similar machinery to be converted into a specific message in a particular cell. Metabotropic receptors, when stimulated by their ligand, trigger the formation of a metabolite called second messenger. The first discovered second messenger is cyclic adenosine monophosphate (cAMP). This messenger becomes an intracellular stimulus capable to activate a protein kinase then a cascade of enzymes which lead to cellular responses.

This transduction system is also a very efficient amplifier : it can be calculated that one molecule of β_1-Receptor agonist is able to trigger the synthesis of 2000 molecules of cAMP/min. Other enzymes than adenylate cyclase are linked to receptors via G-Proteins. An ubiquitous one is phospholipase C which produces two second messengers, inositol triphosphate and diacylglycerol. The first one triggers the release of calcium from intracellular stores then the activation of Ca^{++}-dependent enzymes. The second one, diacylglycerol, activates protein kinase C. This transduction mechanism is shared by a lot of chemical stimuli including hormones and neurotransmitters.

Since the last few years, it appears that receptor coupling is frequently multifunctional. For example, epinephrine, via the activation of α_2 receptors in platelets for instance, can simultaneously inhibit adenylate cyclase and stimulate phospholipase C, involving the different subunits of a G-Protein. At last, we have previously shown (Figure 1) that second messengers can directly operate ion channels, in addition to protein kinases.

5.2. IONOTROPIC RECEPTORS

In this class of receptors which operate ion-channels two sub-classes are distinguished according to the involvement of G-Proteins in the tranduction mechanism. This difference can be illustrated with two examples : i) the cardiac M_2 muscarinic receptor is a G-Protein-coupled receptor, the activation of which directly triggers K^+-channel opening, responsible for cell hyperpolarization and bradycardia ; ii) The skeletal muscle nicotinic receptor which is an hetero-oligomeric protein complex which binds acetylcholine and becomes permeable to Na^+/K^+, leading to depolarization and muscle contraction.

5.3. RECEPTOR-ENZYMES AND ENZYMES AS RECEPTORS

The fourth membrane receptor class quoted in Figure 3, is a class of transmembrane proteins which present both a receptor site and an enzymatic moity. This kind of receptors is the model of growth factor receptors, which exhibit protein kinase activity. To this mechanism which basically regulates cell growth, we can also link those of integrins, transmembrane proteins which allow cell adhesion and participate in information transfer between cells in contact. These integrins do not present enzymatic activity but directly activate some intracellular protein kinases.

In addition to receptor-enzymes, one can also consider the case where an enzyme is a target for an informative molecule. An example is given by nitric oxide (Feldman *et al.*, 1993). This very simple molecule is known to be an ubiquitous informative entity with many biological effects, responsible for vasodilation for instance, when released from vascular endothelium. This labile mediator can cross the membranes and reach the cytoplasmic enzyme of smooth muscle cells, the guanylate cyclase. This heme enzyme is then activated and cyclic GMP concentration increases leading to muscle relaxation.

5.4. CYTOSOLIC RECEPTORS

When the chemical stimuli are able to enter the cell, for instance due to their lipophilicity, they can have specific receptor sites within the cell. This is the case for steroid hormons.

Their receptors are cytosolic proteins associated with other molecules such as heat shock proteins (HSP). When the steroid binds this complex, HSP dissociate and the receptor-hormon complex is then translocated to the nucleus where it modulates mRNA transcription through interaction with responsive elements. In the case of glucocorticoid hormones, about one hundred genes are either repressed or activated.

6. Integration of Primary Messages

In the last part of this survey, I would like to move on from this quite simplistic description of biological cybernetics to a more integrated view of the transformation of stimuli into messages. Even at the cell level the response to a stimulus depends on the state of the cell and on the concomitant stimuli which are integrated by a system of an enormous combinatory complexity. The combinatory complexity can be illustrated by the communication networks summarized in figure 4.

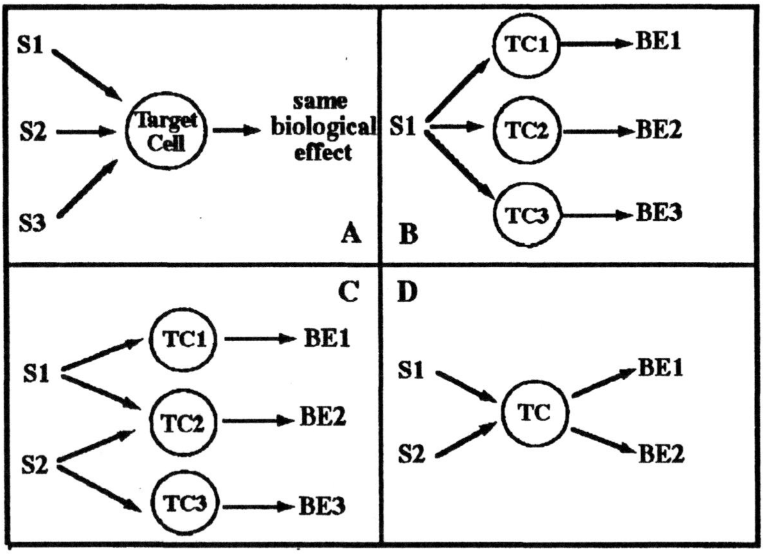

Figure 4 : Communication networks. In A : Several stimuli acting on a same target cell, through the same transduction mechanism, can induce the same response. This is the case of glucagon, calcitonine and PTH on electrolytic transport in the kidney. In B : this example shows that receptor multiplicity allows a single molecule to be used in a great variety of functions in different circuiteries or development stages. In other words, one particular chemical stimulus can lead to various messages according to the target cell. In C : there is a combination of A + B. In D : at last, two informative molecules acting on the same target cell, but through different transduction mechanisms, can induce different responses of the same effector. An example in the kidney again : PTH, via adenylate cyclase activation, inhibits the Na^{+}/H^{+} antiport, when angiotensin, via phospholipase C activation, stimulates this antiport.

7. High Level Integration

In addition to antagonism and synergy, permissive action and all-or-none action from thresholds are also mechanisms involved in the integration of messages. One example can be given with the "gate control" of pain (Figure 5).

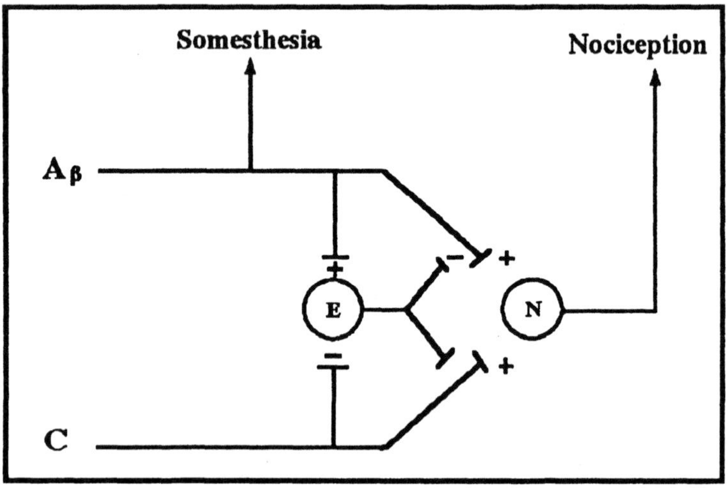

Figure 5 : The gate control theory

This theory has been proposed by Melzack and Wall in the sixties, to explain how thermal, mechanical or chemical stimuli are, fortunately, not always interpreted as painful. This neuronal circuitery is present in the posterior roots of the spinal cord : Aβ and C fibers coming from the skin for instance, stimulate the neuron N implicated in nociception, but this stimulation cannot occur when the peripheral stimulus is weak because enkephalinergic interneurons (E) stimulated by sensory somesthetic fibers Aβ inhibit nociceptive transmission. This is only when the stimulus is strong that the nociceptor C lowers the efficacy of this inhibitory control. From such an example, we can imagine the complexity of the circuitery of the central nervous system which is a prodigious machine to transform stimuli into significant messages.

The last example could illustrate the highest level of stimulus integration. A positron emission tomography (PET) of the brain of a British subject showed a small activity in the occipital and temporal cortex when a visual stimulus was a series of letters without any significance in the language of William Shakespeare. By contrast the

PET picture showed a bright illumination of these area when the stimulus was an English word ! It appears that only the meaningful stimulus is able to activate neuron metabolism in visual and temporal cortical areas, because neuronal networks have been selected by education (Edelman, 1992). This is again an illustration that a stimulus has no significance by itself, and that a message depends on the reception machinery.

8. References

Barnes, P.J. and Adcock, I. (1993) Anti-inflammatory actions of steroids : molecular mechanisms, *Trend Pharmacol. Sci.* **14**, 436-444.
Edelman, G.M. (1992). *Bright air, Brilliant Fire : on the matter of mind.* Basic Books Publisher, New York.
Feldman, P.L., Griffith, O.W., and Stuehr, D.J. (1993) The surprising life of nitric oxide, *C&EN Dec.*, 26-38.
Hanoune, J. (1994) Les proteines G, relais de transmission du signal entre récepteur et second messager cellulaire, *Prix Nobel de médecine, Médecine/Science* **10**, 1183-1184.
Kelly, P. and Baulieu, EE (1989) *Hormones*, Hermann Publisher, Paris .
Loscher, W. and Mevissen, M. (1994) Animal studies on the role of 50/60- Hertz magnetic fields in carcinogenesis, a review, *Life Sciences* **54**, 1531-1543.
Pugh, E.N. and Lamb, T.D. (1993) Amplification and kinetics of the activation steps in phototransduction, *Biochim. Biophys Acta.* **1141**, 111-149.
Sandyk, R.(1992) The influence of the pineal gland on migraine and cluster headaches and effects of treatment with picoTesla magnetic fields. A review, *International Journal of Neuroscience* , **67**, 145-171.

MODIFICATION OF 4 MHz N.M.R. WATER PROTON RELAXATION TIMES IN VERY HIGH DILUTED AQUEOUS SOLUTIONS *

NMR Relaxation in Very High Diluted Aqueous Solutions

J.L. DEMANGEAT[1], P. GRIES[1] AND B. POITEVIN[2]
[1] *Nuclear Medicine Department, General Hospital, Haguenau, France*
[2] *French Association for Homeopathy Research, Vendat, France*

1. Introduction

The NMR properties of water in the vicinity of molecules in solution differ from those of water in the pure state. Water molecules interact with the solute via hydrogen bonds and/or electrostatic forces related to their high dipolar momentum. At room temperature, the correlation times t of rotational and translational movements, of the order of 10^{-11} to 10^{-12} s in pure water may increase to 10^{-8} or even 10^{-6} in the hydration layer of macromolecules (Packer, 1977; Fung, 1977; Mathur-DeVré, 1979), giving rise to a drastic reduction in T1 and T2 water proton NMR relaxation times from 2-3 s to a few hundred milliseconds. Similarly, lowering the temperature of pure water promotes hydrogen bonding and affects relaxation in the same way as the introduction of a solute, leading to water correlation times as high as 10-5 s in ice (Glasel, 1972). Due to the wide 10-5-10-11 s range in correlation times observed in biological or non biological systems, NMR relaxation in the 1-100 MHz domain is one of the potent techniques capable of demonstrating modifications in the mobility and degree of organization of water molecules in solutions or tissues, although it is not very precise in determining the absolute structure. NMR medical imaging is a clear strong successful application of this principle.

Experimentally, protons of H_2O possess a magnetic spin (I=1/2) which align along an external magnetic field Bo; in the mechanical description of NMR phenomenon, excitation of the solution by an electromagnetic wave in the radiofrequency field causes absorption of energy by protons and their displacement from initial orientation; maximal absorption occurs when the excitation frequency equals the Larmor free precession of the protons (42.57 MHz/T) in the applied magnetic field Bo (Tesla), achieving thus the *Nuclear Magnetic Resonance* conditions. After release of the radiofrequency excitation, protons realign along the Bo direction (*Relaxation*) at a rate which can be described by a time constant observed in the direction of Bo (T1=longitudinal or spin-lattice). A second time constant (T2=transverse or spin-spin) characterizes the disappearance of the NMR signal in the plane perpendicular to Bo. Relaxation is a complex process resulting from the dipolar magnetic interaction between

---

* Communication presented at the 8th GIRI Meeting, December 1994, Jerusalem, Israel.

M. Bastide (ed.), Signals and Images, 95–110.

adjacent intra and intermolecular protons, the molecular movements of rotation and translation, the exchange of protons between neighbouring molecules and the presence of paramagnetic substances (Mn^{++}, Cu^{++}, Fe^{++}, Fe^{+++}, free radicals, molecular O_2...). These mechanisms are moreover strongly influenced by temperature. For an isotropic molecular movement that can be described by a unique correlation time τ (this is the case of liquid water where rotation and translation are strongly coupled):

$$1/T1 = C\ \{2\tau /(1 + \omega_0^2\tau^2) + 8\tau /(1 + 4\ \omega_0^2\tau^2)\} \quad (1)$$
$$1/T2 = C\ \{3\tau + 5\tau /(1 + \omega_0^2\tau^2) + 2\tau /(1 + 4\ \omega_0^2\tau^2)\} \quad (2)$$

where ω_0 is the Larmor frequency, i.e. 4 MHz in our study. For movements which are rapid compared to Larmor frequency, T1 and T2 are equal, long and independent of the frequency. When movement is limited, both T1 and T2 decrease, the latter more rapidly, and become frequency dependent: rapid movements influence T1 and T2 but those of low frequency influence primarily T2. T2 can become much lower than T1, since any process influencing T1 necessarily influences T2 while the reverse is not true. So, the T1/T2 ratio is a more sensitive parameter than T1 and T2 separately.

If NMR is known as a powerful tool to study dynamics of water in biological and non biological systems, what about aqueous high dilutions such as those used in homeopathic practice? The obvious advantage is given by the abundance of water generating quasi-maximal and constant signal/noise ratio in opposite to techniques designed to detect the solute and which become ineffective at high dilution. But at least two main difficulties arise: *first* relaxation is very slow in quasi-pure water, thus particularly sensitive to temperature and contaminants, especially the paramagnetic ones among them dissolved atmospheric oxygen is the most common. *Second*, very minor modifications are expected, hardly distinguishable from random experimental fluctuations. So, in this field of high dilutions, a drastic controlled protocol is required for preparation and measurements. In a preliminary part of this work we exhaustively examined several experimental chemicophysical sources of variations and their influence on NMR measurements.

2. Preliminary Experimental Considerations

2.1. NMR MEASUREMENTS

Measurements were carried out on an inframillimetric NMR imaging device especially developed in our laboratory and purposed for the study of the biomechanics of the arterial wall (Gries and Constantinesco, 1987; Constantinesco *et al*, 1989); it consisted in a 4 MHz Bruker PC104 Minispec connected to a Drusch EAF 16A 0.09406 Tesla resistive magnet with a 5.18 cm air-gap and a central field uniformity of ± 10 ppm. The radiofrequency transmitter-receiver coil was 7.5 mm in diameter and 19 mm in height. Standard Bruker 7.5 mm diameter test tubes were used, the filling level (1.5 ml) and the position in the coil of which were strictly controlled. A temperature of 1 ± 0.2 °C was achieved by circulation of cryostatic proton-free fluid (Fluorinert FC 40) and controlled by thermocouple. According to Glasel (1972) such a stability corresponds to maximal variations in relaxation times of about 12.5 ms. T1 was determined using the inversion-recovery method and a monoexponential fitting of an 8-points experimental curve; error on the computerized fitting was in the range 10-25 ms. Each sample was measured at least 10 to 15 times. T2 was determined by the Carr-Purcell-Meiboom-Gill sequence and monoexponential fitting of a 160-points curve, with a fitting error of 2-15 ms. Each sample was measured 15 to 20 times.

2.2. PRELIMINARY CONTROLS

Since dissolved O_2 in the test samples is a powerful proton relaxant, variability in relaxation times due to uncontrolled PO_2 levels resulting from preparation should be evaluated. All procedures were carried out under a laminar-flow exhaust hood set to laboratory temperature. Before use, all solvents or solutions were equilibrated to temperature under the hood over 30 min. All pipetting/sampling procedures were performed with single-use Eppendorf tips. PO_2 measurements (Corning 158 blood analyzer) were carried out within a maximum of 20 min on samples collected using a 2 ml plastic syringe without a needle, by carefully avoiding any intrusion of air. Detailed procedures for preparation of potencies are given in section 3.

2.2.1. *Influence of Barometric Pressure*

It is well-known that atmospheric oxygen dissolves proportionnally to partial pressure according to Henryís law. In water samples vortexed during 60 s in contact with air, a linear relationship between dissolved PO_2 and barometric pressure was indeed found (figure 1) but with a higher slope than expected, indicating enhancement of dissolution by the agitation process.

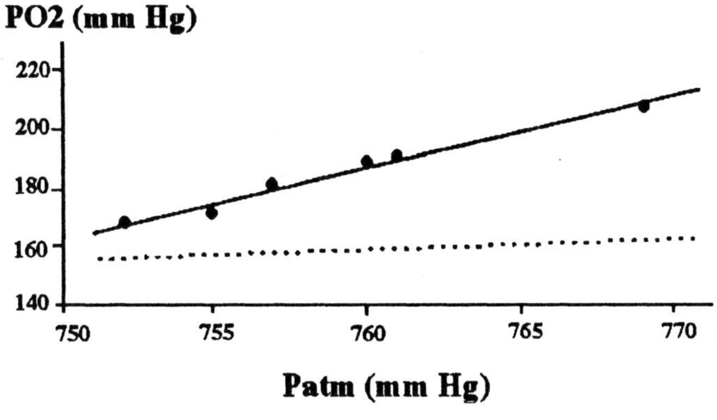

Figure 1. O_2 partial pressure dissolved in pure water samples vortexed during 60 seconds in contact with air at laboratory temperature as a function of barometric pressure at the time of preparation (samples from the same batch). Dotted line: relationship expected from Henry's law in non agitated samples (PO_2 = 0.21Patm).

The relaxant effect of dissolved O_2 on water protons is shown on figure 2, where a decrease of about 40 ms in T2 is induced by a 20 mmHg increase in barometric pressure; such fluctuations in laboratory conditions may be current. The linear relationship observed corroborates the dependence between relaxation rates (1/T) and No, the number of O_2 molecules per milliliter in pure water (Glasel, 1972):

$$1/T1 = 0.284 + 4.55 \,.\, 10^{-19} \text{No} \quad (3)$$
$$1/T2 = 0.455 + 5.25 \,.\, 10^{-19} \text{No} \quad (4)$$

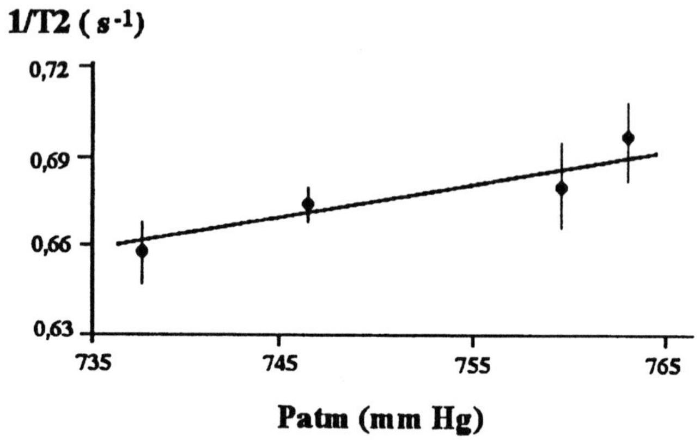

Figure 2. Relaxation rate 1/T2 of water samples as a function of barometric pressure at the time of preparation (results from 19 experiments). Samples from the same batch were equilibrated in contact with air at laboratory temperature without agitation for 15 min before NMR measurement.

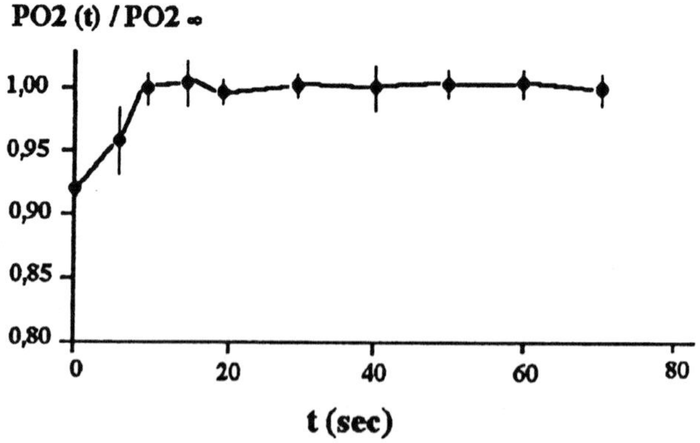

Figure 3. Influence of the vortexing duration in contact with air on atmospheric O_2 dissolution (results of 7 experiments with water, saline or 3CH Sil/Lac samples). PO_2 ∞ for each sample corresponds to the asymptote calculated from the 50, 60 and 70 s durations average.

2.2.2. *Influence of Vortexing Duration on Atmospheric Oxygen Dissolution*
O2 dissolution was proportional to duration up to 10 s; beyond this, dissolution became asymptotic with maximal deviations of 2 to 2.5% of the PO2 ∞ value (figure 3).

2.2.3. *Influence of the Dilution/Agitation Process on Atmospheric Oxygen Dissolution*
Dissolved O_2 was determined in successive centesimal dilutions/agitations over the whole range concerned in the study (figure 4). The dilution/agitation process itself, when performed over a short period of time, was unlikely to produce PO_2 variations exceeding 2.5% either for diluted substances or for solvent.

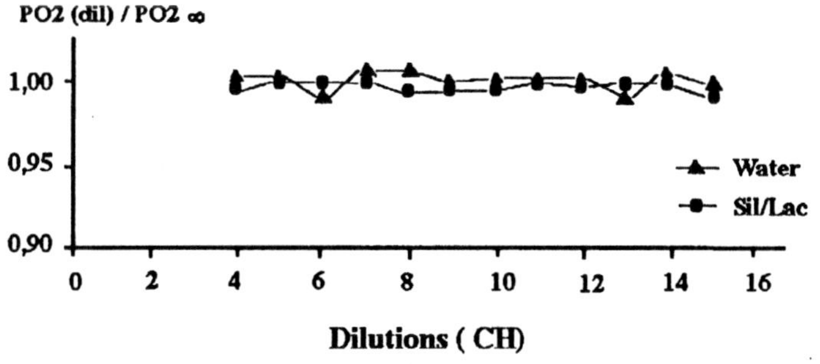

Figure 4. Influence of the centesimal dilution/agitation process in contact with air on atmospheric O_2 dissolution (constant vortexing duration of 60 s). PO_2 (dil) was standardized in each case to PO_2 ∞, the mean value of the 3 highest dilutions 13CH, 14CH and 15CH.

2.2.4. *Implications*
According to pharmacopea, potencies used in homeopathy are prepared in the presence of air; moreover it has been argued that they were ineffective when prepared under pure nitrogen (Cazin *et al*, 1991). So it would not be logical that physicists working on high dilutions solve out the problem by suppressing oxygen, as it is usually done for most NMR studies. In order to ensure a similar if not identical level of dissolved oxygen in samples, 3 conditions must be respected: *i)* preparation of all samples of a given series within the same day - *ii)* minimum 20 s vortexing duration at controlled rate for the dilution/agitation process - *iii)* closing the test tubes for delayed NMR measurements. Despite these precautions, PO_2 fluctuations from about 2 to 2.5% might be observed. For T1 and T2 values from 1.5 to 2 s and according to eqs (3) and (4), such PO_2 fluctuations can produce T1 and T2 variations of 20-25 ms and 5-12 ms respectively. The table 1 summarizes the expected fluctuations despite maximal care in our experimental protocol, which are due to oxygen, temperature and mathematical fitting. Observed standard deviations in solvents are found of that order or slightly greater, as logically expected; the maximal differences between high dilutions and solvents will hardly exceed these values, indicating very minor modifications, attesting thus the need for a drastically controlled protocol.

TABLE 1. Expected and observed experimental variations

		T1	T2
Expected fluctuations			
- Dissolved O2	2.5 %	20 -25 ms	5 - 12 ms
- Temperature	0.2 °C	12.5 ms	12.5 ms
- Monoexponential fitting		10-25 ms	2-15 ms
- Total		42.5 - 62.5 ms	19.5 - 39.5 ms
Standard deviation observed in solvents *		57 - 69 ms	30 - 32 ms
Maximal mean value differences observed between solutions or solvents*s		52 ms	49 m

* *From Table 2 data*

3. Preparation of Potencies and Particular Conditions of the Study

Two successive experiments are described. In the first one we study Silica/Lactose (Sil/Lac) saline potencies above 3CH (dilution level 106) which were previously demonstrated by Davenas *et al.* (1987) to be active at very high dilution on mouse peritoneal macrophages when administered in drinking water; the production of paf-acether, a mediator in allergy and inflammation, was significantly increased in the isolated, zymosan-stimulated macrophages of the mice. The effect was found at a dilution level as high as 10^{18} (9CH). The involvement of silica as active constituent (lactose being an excipient) was attested from controls receiving saline, lactose or a non-specific substance (Gelsemium) at the same level of dilution. Besides, in homeopathic physiciansí all-day practice the effect of Silicea (Sil/Lac) in recurring infections and chronic suppurations is one of the most regular clinical observations. In the second experiment additional Manganese/Lactose (Mn/Lac) and Histamine (Hist) aqueous dilutions were studied simultaneously as Sil/Lac in saline. Manganese was chosen as a control due to its paramagnetism, in order to evaluate a possible influence on water proton relaxation at high dilution; histamine was chosen because, as for Sil/Lac, biological effects have been reported at high dilution (Sainte-Laudy and Belon, 1993). All samples were prepared by the Laboratoires Homéopathiques de France.

3.1. FIRST EXPERIMENT (WATER - SALINE - SIL/LAC)

Silica dioxyde (insoluble) was first homogenized by three successive triturations with pure lactose than added with 0.9% NaCl saline (0.3 g in 3 ml). This initial solution was successively diluted to 1/100th strength 13 times by addition of 0.3 ml to 30 ml 0.9% NaCl (Aguettant, pH 6.0-6.9), using a pipette the plastic tip of which being discarded after each operation, under controlled agitation of the mixture in a Cenco Vortex rate 5 for 25 seconds. Subsequently only 5 centesimal dilutions were retained (3CH, 6CH, 9CH, 12CH, 15CH) corresponding to 10^{6}-10^{12}-10^{18}-10^{24}-10^{30} dilution levels and containing theoretical silica concentrations of $1.66.10^{-5}$ M to $1.66.10^{-29}$ M and lactose concentrations of $2.92.10^{-3}$ M to $2.92.10^{-27}$ M. As references, 0.9% NaCl and distilled water (Meram, pH 6.5-7.5) were submitted to identical concomitant cycles of dilution/agitation and also called centesimal dilution although the term dilution was inappropriate. All samples of a given series (Sil/Lac, water, saline) were imperatively prepared on the same day. After filtration by 0.22m Millipore, precise 1.5 ml aliquots of these dilutions were directly deposited in the NMR Bruker PC 7.5 test tubes previously cleaned with ether, copiously rinsed with bidistilled water and autoclaved. These tubes were flame-sealed and labeled with a code number for NMR measurements. Seven independent series of all samples were prepared on a monthly basis, using new solvent batches each time.

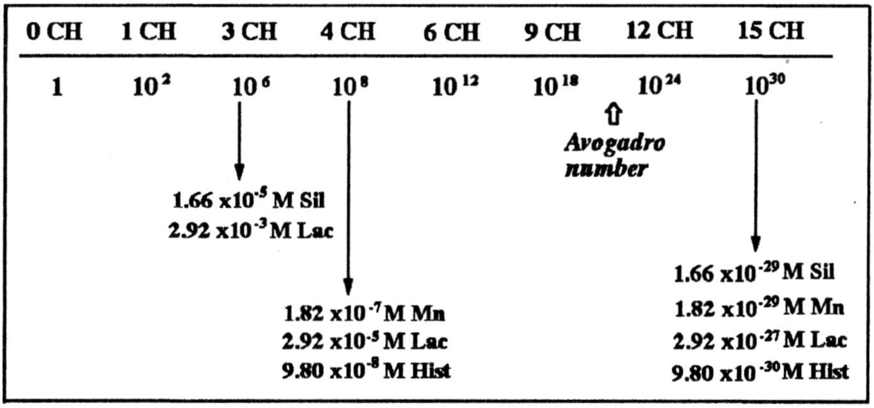

Figure 5. Correspondance between centesimal designation (CH), dilution level and molar concentrations.

3.2. SECOND EXPERIMENT (WATER - SALINE - SIL/LAC - MN/LAC - HIST)

Among four new independent series of Sil/Lac, saline and water, three series of Manganese/Lactose (Mn/Lac) and Histamine (Hist) dilutions were prepared in the same conditions except that the solvent was water instead of saline and the first potency was 4CH instead of 3CH, in order to minimize a potential effect of the magnetic (Mn) and protonated (Hist) solutes on the NMR signal. As manganese is insoluble, Mn/Lac was prepared from triturations in lactose, as for Sil/Lac. In final retained samples (4CH, 6CH, 9CH, 12CH, 15CH), manganese concentrations ranged from $1.82.10^{-7}$ M to $1.82.10^{-29}$ M and lactose concentrations from $2.92.10^{-6}$ M to $2.92.10^{-28}$ M; histamine ranged from $9.8.10^{-8}$ M to $9.8.10^{-30}$ M.

3.3. SUMMARY OF THE SPECIAL CONTROLLED CONDITIONS OF THE STUDY

. Preparation under laminar-flow exhaust hood
. 25 s vortexing at controlled rate
. Discarding pipette tip for each operation
. One-day preparation for all samples of a given series

. Several independent series using new batches of solvent
. Samples Millipore-filtered and flame-sealed
. Code-labeling for blind NMR measurements
. 0.2°C temperature stability during NMR measurements
. Code only broken at the end of the second experiment

4. Results

4.1. FIRST EXPERIMENT (WATER - SALINE - SIL/LAC)

4.1.1. *Saline Versus Water*
The relaxation times of pure water at 1°C were found slightly lower than those reported in the literature (Engel and Hertz, 1968; Cope, 1969; Glasel, 1972; James and Gillen, 1972), probably due to a higher level of dissolved O_2 resulting from the dilution/agitation process. Saline differed from water by a lower T2 (1366 ± 30 ms and 1415 ± 32 ms respectively) and a higher T1/T2 ratio (1.127 ± 0.046 and 1.082 ± 0.032 respectively); but there was no significant difference in T1.

TABLE 2. Relaxation times (ms) measured in Experiment 1.

Substance	Dil. (CH)	N	T1	T2	T1/T2
Saline	0-1-3-6-9-12-15	48	1538 ± 69	1366±30	1.127±0.046
Water	0-1-3-6-9-12-15	47	1526 ± 57	1415±32	1.082±0.032
Saline	3-6-9-12-15	34	1533 ± 67	1371±30	1.121±0.036
Sil/Lac in saline	3-6-9-12-15	34	1570±73	1367±35	1.148 ± 0.053
Saline	9-12-15	20	1520 ± 67	1375±30	1.109± 0.032
Sil/Lac in saline	9-12-15	21	1572 ± 86	1370±37	1.145 ± 0.061

Such a result could be expected as it reflects the orientation and reduction in mobility of the dipolar water molecules in the powerful electric field of Na^+ and Cl^- ions, since T2 is more sensitive than T1 to restriction of movement. It should be noted in support of our results that Engel and Hertz (1968) also reported the absence of decrease, and even a slight increase, in T1 values of inframolar NaCl solutions.

4.1.2. *Sil/Lac in Saline Versus Saline*
The Sil/Lac solution differed from its saline solvent by higher T1 (1570 ± 73 ms and 1533 ± 67 ms respectively) and higher T1/T2 values (1.148 ± 0.053 and 1.121 ± 0.036 respectively) without significant variation in T2. Moreover, differences did persist when the least diluted samples (3CH and 6CH, possibly containing subponderal amount of solute) were excluded from analysis; in the Sil/Lac dilutions thus remaining under consideration and which were still different from the solvent, there only remained theoretical concentrations lower than $1.66.10^{-17}$ M silica and $2.92.10^{-15}$ M lactose. These changes observed in Sil/Lac dilutions were unexpected and paradoxical: T1 found to be higher than in pure water appeared in favour of water destructuralization while the

T1/T2 increase suggested the reverse. T1/T2 progressively increased from water to saline and from saline to Sil/Lac.

4.1.3. *Analysis of Variance and Discriminant Analysis*
The two factors variance analysis (factor substance and factor dilution - table 3) clearly showed significant differentiation of substances. T1 recognized Sil/Lac from water and saline, T2 recognized water from Sil/Lac and saline whereas T1/T2 differentiated all substances; differences persisted with an almost identical significance level over the 10^{18} dilution (9CH). On the contrary no global significant effect was found with potency, although figure 6 suggested a tendency of enhancement of the changes in the most diluted samples (above 9CH).

Discriminant analysis is a multivariate statistical procedure that enables an optimized grouping of individuals who are described by a set of variables. This analysis provides so-called discriminant functions that are linear combinations of the variables:

$$F_i (X_j) = \Sigma_k a_k x_{jk}$$

where F_i is the ith discriminant function of the X_j individual, a_k the kth coefficient of this function and x_{jk} the kth parameter value for the X_j individual. The coefficients of the discriminant function are calculated so that the between group/within group ratio is maximum. The separation between groups can be evaluated by performing the Wilkís test, or more qualitatively by looking at the well-classified percentage. This analysis demonstrated separation of all substances with a half-plane discrimination between water and saline on one hand and between solvent and solution on the other hand ($p<0.021$, figure 7). When this analysis was performed above the 10^{18} dilution level, the discrimination was no more significant ($p = 0.16$) but the diagram looked nevertheless illustrative of separation with a 57.1 %, 85.7 % and 100 % of well-classified observations in the water, saline and Sil/Lac groups respectively (non represented).

TABLE 3. Analysis of variance in Experiment 1.

	Dil.(CH)	T1	T2	T1/T2
Factor Substance - differentiable	3-6-9-12-15	$p=0.0008$ Sil/Lac≠Water Sil/Lac ≠ Saline	$p<0.0001$ Saline≠Water Sil/Lac ≠ Water	$p<0.0001$ Sil/Lac≠Water≠Saline
Factor CH	3-6-9-12-15	ns	ns	ns
Factor Substance - differentiable	9-12-15	$p=0.0013$ Sil/Lac ≠Water Sil/Lac ≠ Saline	$p=0.0008$ Saline ≠Water Sil/Lac ≠ Water	$p<0.0001$ Sil/Lac ≠Water Sil/Lac ≠Saline
Factor CH	9-12-15	ns	ns	ns

Multifactorial analysis of variance with interaction - ns: $p>0.01$.
Differentiable substance determined by Newman-Keuls a posteriori test - ns: $p>0.01$.

4.2. SECOND EXPERIMENT (WATER - SALINE - SIL/LAC - MN/LAC - HIST)

4.2.1. *Sil/Lac in Saline Versus Saline*
No significant differences were found between Sil/Lac and saline in experiment 2 comprising fewer series (table 4); T1 values were quite identical and T2 exhibited a

slight decrease in Sil/Lac contrary to experiment 1. Nevertheless, a higher T1/T2 ratio was again observed in Sil/Lac with respect to saline which maintained above 9CH. Moreover discriminant analysis succeeded in demonstrating separation of Sil/Lac from saline and from water according to a similar pattern as in the first experiment with 100% of well-classified observations (figure 7). Above the 9CH dilution level, the substances remained separated on the diagram with a 100% amount of well-classified observations (figure 8).

TABLE 4 . Relaxation times (ms) measured in Experiment 2.

Substance	Dil. (CH)	N	T1	T2	T1/T2
Saline	3-6-9-12-15	19	1587±75	1575±45	1.008±0.050
Sil/Lac in saline	3-6-9-12-15	20	1581±78	1550±57	1.021±0.040
Saline	9-12-15	12	1595±68	1576±39	1.013±0.049
Sil/Lac in saline	9-12-15	12	1581±80	1544±54	1.024±0.036
Water	3-6-9-12-15	15	1591±75	1555±58	1.023±0.022
Hist in water	4-6-9-12-15	15	1623±36	1557±44	1.043±0.020
Mn/Lac in water	4-6-9-12-15	14	1613±26	1560±53	1.035±0.029
Water	9-12-15	9	1589±78	1561±54	1.017±0.022
Hist in water	9-12-15	9	1629±39	1548±44	1.052±0.014
Mn/Lac in water	9-12-15	9	1616±23	1570±58	1.029±0.034

Overall data concerning four series of Sil/Lac and three series of Mn/Lac and Hist.

4.2.2. *Mn/Lac and Hist Versus Water*

T1 was increased in Mn/Lac and Hist compared to water (1613 ± 26 ms and 1623 ± 36 ms respectively; water = 1591 ± 75 ms) as well as T1/T2 (1.035 ± 0.029 and 1.043 ± 0.020 respectively; water = 1.023 ± 0.022). In the analysis of variance the differences were significant at the 1% level for T1 in Hist (table 5) and maintained (p = 0.0285) above the 10^{18} dilution level where histamine concentration was lower than $9.8.10^{-18}$M. Besides, discriminant analysis succeeded in separating at the same time water, Hist and Mn/Lac with a 100% amount of well-classified observations (figure 7) despite a non significant Wilkís test, possibly due to the small number of samples. Above the 9CH dilution level, the different substances remained well-separated on the diagram (figure 8). Curiously, the analysis of variance showed a tendency of T1/T2 to differentiate the substances (p=0.062) above 9CH only.

TABLE 5. Multifactorial analysis of variance in Experiment 2.

	Dil.(CH)	T1	T2	T1/T2
Factor Substance - differentiable	3-4-6-9-12-15	p=0.0086 Hist ≠ Water Saline≠Water	p=0.056	ns
Factor CH	3-4-6-9-12-15	ns	ns	ns
Factor Substance - differentiable	9-12-15	p=0.0285 none	p=0.060	p=0.062
Factor CH	9-12-15	ns	ns	ns

Performed only on the three complete series containing Sil/Lac, Mn/Lac and Hist. - ns: p>0.05.
Differentiable substance determined by Newman-Keuls a posteriori test - ns: p>0.05.

4.3. SYNTHESIS OF BOTH EXPERIMENTS

The figure 6 presents another expression of the results enabling comparison of water and saline solutions at the same time. It appears a common feature that T1 and T1/T2 were increased in all cases in high dilutions, whatever the solvent and whatever the level of dilution. Above 9CH there was a tendency of enhancement of the differences in both T1 and T1/T2, but not established statistically. For T1 values in experiment 2 only, a discrepancy was observed, but non significant. Considering discriminant analysis it is worth noting that separation of Sil/Lac, water and saline was found as a similar graphic pattern in both experiments and that the five species Sil/Lac, Mn/Lac, Hist, saline and water could hardly be separated in a so limited number of series and measurements (figure 7). From figure 7 it was possible to distinguish three areas assigned to water (A), aqueous solutions (B) and saline solutions (C) respectively, corresponding to three increasing degrees of complexity. To our amazement, the paramagnetic solute (Mn/Lac) exhibited the smallest changes in relaxation times.

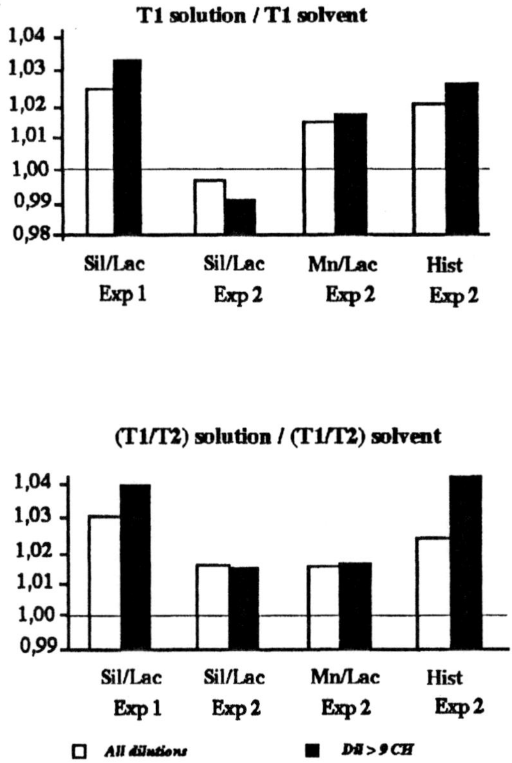

Figure 6. T1 and T1/T2 ratio in solution versus solvent (water or saline) for all concentrations and for very high diluted solutes only (above 9CH).

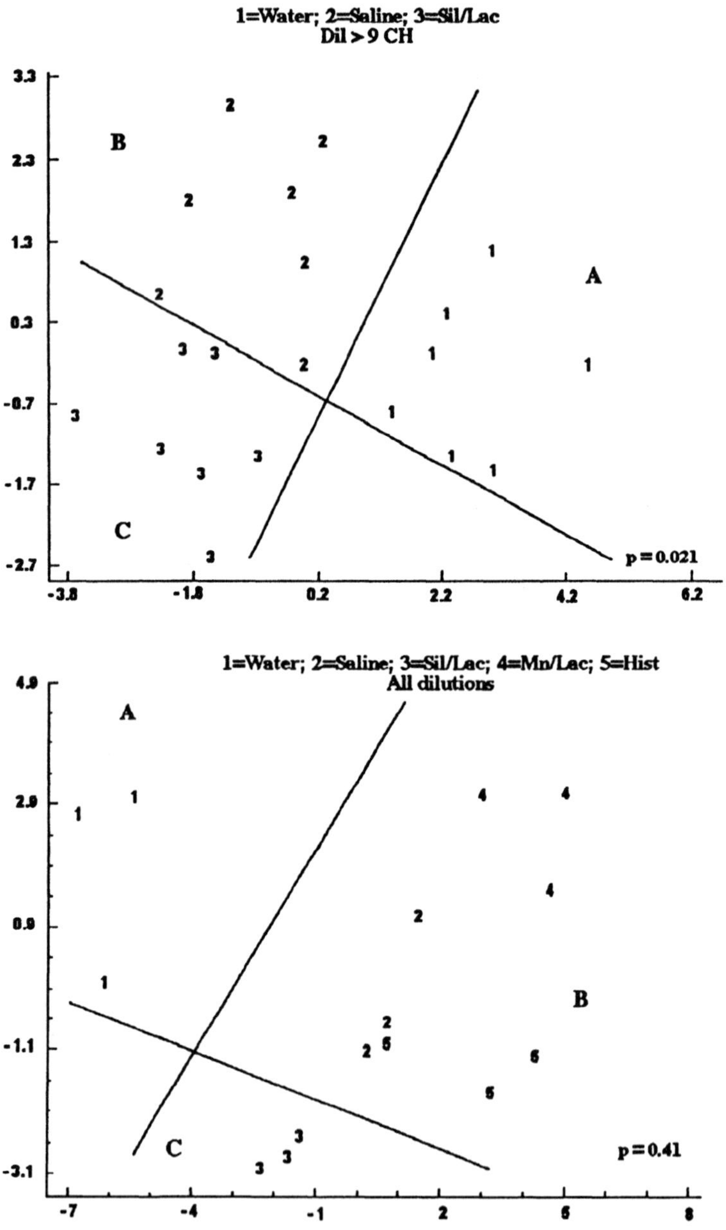

Figure 7. Discriminant analysis. X and Y axes represent 1st and 2nd discriminant functions and p, the probability from the Wilkís test. Upper diagram: Experiment 1. Lower diagram: Experiment 2. A: pure water, B: solutes in water, C: solutes in saline.

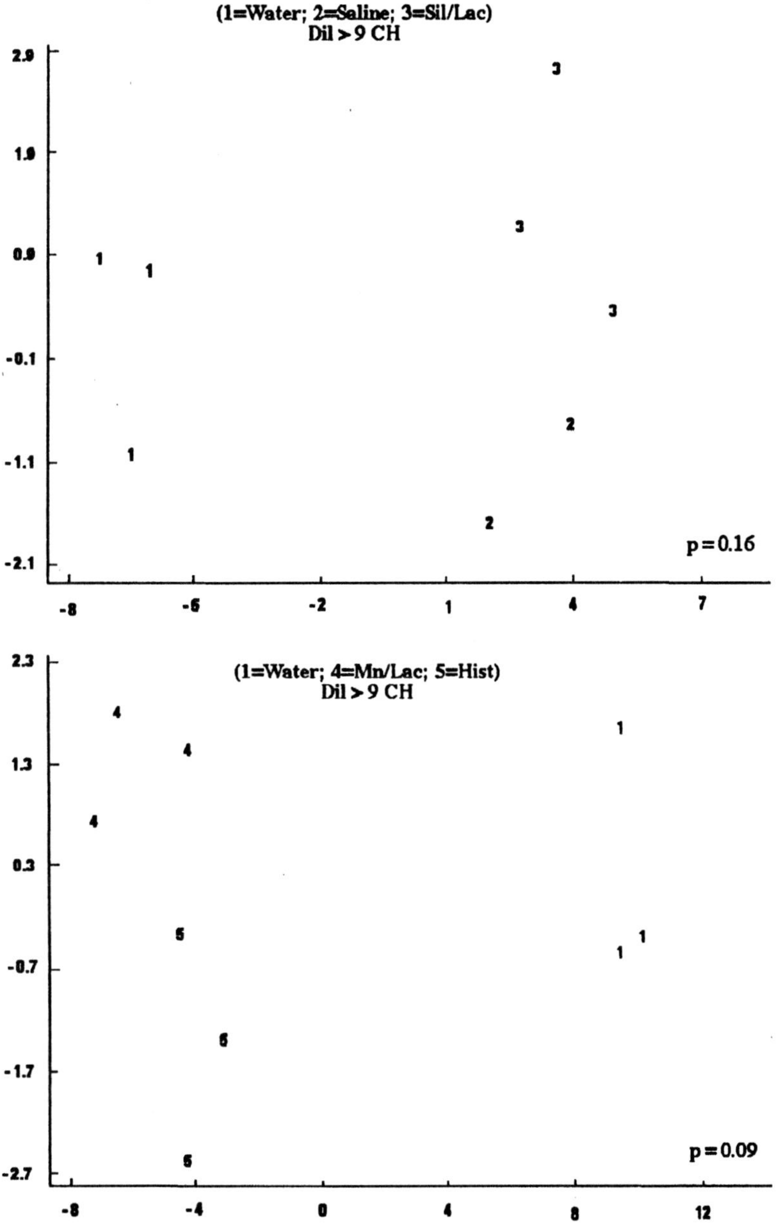

Figure 8. Discriminant analysis in Experiment 2 limited to dilutions above 9CH. In both diagrams the Wilkís test was not significant but all substances were well-classified (100%).

5. Discussion

This study demonstrated changes in 4 MHz NMR proton relaxation times in very high diluted aqueous solutions compared to solvent. Results of experiment 1 have been partially published and discussed previously (Demangeat *et al*, 1992; 1993); they have nevertheless been reported here and re-examined statistically together with a second set of new results (Experiment 2). Increases in T1 and T1/T2 ratio were observed in Sil/Lac saline solutions compared to pure saline for Sil/Lac at concentrations lower than $2.92.10^{-3}$ M lactose and $1.66.10^{-5}$ M silica. These changes were still present when the lactose and silica concentrations were lowered below $2.92.10^{-15}$ M and $1.66.10^{-17}$ M respectively; above a 10^{18} dilution level changes looked even more pronounced, but non significantly. High dilutions of Mn/Lac and Hist in water did confirm the increase in T1 and T1/T2 with respect to pure water, non significantly for Mn/Lac but significantly for Hist. Above the 10^{18} dilution level where differences were still present but with a lower degree of significance, there only remained theoretical concentrations of less than $9.8.10^{-18}$ M histamine. Of course the changes observed were minute and in the range of experimental fluctutions; but higher changes in such very high dilutions were a priori not expected. Therefore a rigorously controlled protocol was imperatively needed as well as multiplicity of independent series, multiplicity of substances and use of powerful statistical tests; so, where the classical Student's test or analysis of variance failed to prove differences, discriminant analysis succeeded in separating substances. Such unexpected and paradoxical results necessitate critical analysis of the experimental protocol, looking for possible biases.

Magnetism of Solutes. (i) Sil/Lac. The silica molecule $^{28}Si^{16}O_2$ (spin number I=0) does not yield an NMR signal. The ^{29}Si isotopic form of silica possesses a spin (I=1/2) but its contribution to the signal is negligible due to its low natural abundance (4.7%) and the distance of its resonance frequency (8.46 MHz/T) from that of the proton (42.7 MHz/T). As for ^{17}O isotope (I=5/2) of silica, which could influence relaxation via its electric quadrupolar momentum, its concentration is negligible compared to that of oxygen atoms in water (1/95000 in the first potency). Protons and ^{17}O isotopes of lactose $C_{12}H_{22}O_{11}$ could affect both signal and relaxation more than silica since the first potency (3CH) contained $2.92.10^{-3}$ M lactose; however this represents 1700-fold less protons and oxygen atoms than in water and a negligible amount in the subsequent dilutions. In fact, this first Sil/Lac potency did not differ in relaxation times from higher dilutions. *(ii) Mn/Lac.* Mn possesses a spin (I=5/2) but the influence on signal and relaxation may be neglected due to the distance of its resonance frequency (10.49 MHz/T) and to the small amount present in the first potency ($1.82.10^{-7}$ M). As for lactose, it was 1000-fold less abundant than in Sil/Lac dilutions. *(iii) Hist.* The possible magnetic influence of protons and of 14N (I=1) and 15N (I=5/2) isotopes of histamine can reasonably be neglected at concentrations as low as $9.8.10^{-8}$ M.

Paramagnetic Solutes or Contaminants. Paramagnetic contaminants in sodium chloride, silica or lactose may induce, at micromolar concentrations, up to 20 ms changes in relaxation times, that would be non negligible, if any, in our experiments. But the paramagnetic effect is essentially a reduction in relaxation times, contrary to what was observed; moreover, differences in relaxation times persisted at very high dilution levels where contaminants of solutes could no more be present. The absence of a particular behavior of the paramagnetic solute Mn/Lac (which, even, exhibited the smallest changes in relaxation) does neither argue for the responsability of paramagnetic contaminants of solutes for the observed NMR modifications. Dissolved O_2 was the main paramagnetic agent present in our experiments; the protocol used aimed to minimize the PO_2 fluctuations in the samples of a given series but we cannot formally exclude a minute variation of the oxygen dissolution in the course of the preparation, not detectable at the PO_2 determination, but which could have influenced relaxation times. A modification in the oxygen solubility within a series could result from room temperature fluctuations or from interaction with the molecules of solute, the number of which decrease with the dilution/agitation process. The relaxant effect of paramagnetism generally affects mainly T2 and particularly above the MHz domain for contaminants such as Mn^{++} or Cr^{+++} (Morgan and Nolle, 1959), resulting in a higher T1/T2 ratio. But the observed simultaneous increase in T1/T2 and T1 appears paradoxical since the increase in T1/T2 suggests a higher amount of paramagnetic agents in dilutions compared to solvent and the increase in T1 suggests the reverse. The effect of pH can neither be neglected as it influences differentially the T1 and T2 via ^{17}O quadrupolar relaxation (Meiboom, 1961) leading to a T1/T2 increase; but this parameter was not examined.

Preparation Conditions. The control samples, water and saline, underwent simultaneous and strictly identical dilution/agitation, filtering, sealing, storage and measurement. Moreover new batches of solvent were used for each series. These conditions statistically compensated the influence of *(i)* the barometric influence on O_2 dissolution *(ii)* any possible contamination from the glass test tube and especially silica which form a gel on contact with glass and alter the chemicophysical properties of water (so-called poly-water; Menès, 1970; Prigogine, 1971) *(iii)* contaminants of solvents *(iv)* filtering and sealing procedures.

Thus, there is no obvious experimental bias to explain our results. Effects of pH, of dissolved O_2 (the dissolution rate of which being enhanced) or of paramagnetic contaminants in the successive dilutions/agitations cannot be formally excluded but, if present, would probably constitute a common and non specific phenomenon, unable to explain the differentiation of substances, especially at dilution level higher than 10^{18}. In the absence of satisfactory answer, one cannot exclude a modification of the solvent structure in diluted/agitated samples; the increase in T1 suggests a decrease in correlation times. When substances such as urea, acetamide and numerous, especially monovalent, ions are dissolved in water, they break hydrogen bonds thus increasing the mobility of water molecules. This so-called negative hydration state has been described using techniques as different as diffusion, conductivity, viscosity, ultrasound absorption, and NMR T1 relaxation time (Hammes and Swann, 1967; Engel and Hertz, 1968; Brun *et al*, 1969). This phenomenon is concentration-dependent, with many substances acting as destructuralizers at low concentration (increasing T1 compared to pure water) but generating a structure at higher concentration (lowering T1). While this hypothesis of water destructuralization, suggested by T1, is not supported by the non variation of T2, it is neither refuted since T2 is an experimental less reliable parameter than T1 and since several factors acting differently on T1 and T2 may be intricated to produce the observed result.

Whatever the phenomenon described here, true modification of molecular interactions or common chemicophysical changes induced by the dilution/agitation process, the NMR relaxation, associated with a rigorous protocol and an appropriate statistics, emerges as an unsuspected sensitive technique to study water protons in very high diluted aqueous solutions, prepared according to the successive 100-fold hahnemanian principle required for homeopathy.

10. References

Brun, B., Servent, M. and Salvinien, J. (1969) Action des gros ions sur la structure de l'eau; étude par diffusion, *C.R.Acad.Sc.Paris* **269**, 1-4.

Cazin, J.C., Cazin, M., Chaoui, A. and Belon, P. (1991) Influence of several physical factors on the activity of ultra-low doses, in C. Doutremepuich (ed.), *Ultra-low doses*, Taylor and Francis Publisher, London, pp. 69-80.

Constantinesco, A., Gries, P., Vinée, P., Heckert, B. and Carlier, P. (1989) In vitro 4 MHz NMR proton relaxation times and in vitro mechanical behavior at low extension of human common carotid arterial wall, *Magn.Reson.Imaging* **7**, 173-177.

Cope, F.W. (1969) NMR evidence using D_2O for structured water in muscle and brain, *BiophysJ.* **9**, 303-319.

Davenas, E., Poitevin, B. and Benveniste, J. (1987) Effect on mouse peritoneal macrophages of orally administered very high dilutions of silica, *Eur.J.Pharmacol.* **135**, 313-319.

Demangeat, J.L., Demangeat, C., Gries, P., Poitevin, B. and Constantinesco, A. (1992) Modifications des temps de relaxation RMN ‡ 4 MHz des protons du solvant dans les très hautes dilutions salines de silice/lactose, *J.Med.Nucl.Biophys.* **16**, 135-145.

Demangeat, J.L., Gries, P., Demangeat, C. and Poitevin, P. (1993) Nachweis von NMR-Protonen-Relaxations-Zeitenver%anderungen von hochverd,nnten Wasserl^sungen bei 4 Mhz, *Nukl.Med.* **32**, P70.

Engel, G. and Hertz, H.G. (1968) On the negative hydration. A nuclear magnetic resonance study, *Ber.Bunsenges Phys.Chem.* **72**, 808-834.

Fung, B.M. (1977) Proton and deuteron relaxation of muscle water over wide ranges of resonance frequencies, *Biophys.J.* **18**, 235-239.

Glasel, J.A. (1972) Nuclear magnetic resonance studies on water and ice, in F. Franks (ed.), *Water. A comprehensive treatise. The physics and physical chemistry of water*, Plenum Press Publisher, New York, pp. 215-254.

Gries, P. and Constantinesco, A. (1987) Développement d'un imageur de laboratoire à 0.094 T à partir d'un minispectromètre RMN, *J.Biophys.Biomec.* **11**, 139-143.

Hammes, G.G. and SWAN, J.C. (1967) Influence of denaturing agents on solvent structure, *Biochemistry* **6**, 1591-1596.

James, T.L. and Gillen, K.T. (1972) NMR relaxation time and self-diffusion constant of water in hen egg white and york, *Biochim.Biophys.Acta* **286**, 10-15.

Mathur-DeVrè, R. (1979) The NMR studies of water in biological systems, *Prog.Biophys.Molec.Biol.* **35**, 103-134.

Meiboom, S. (1961) NMR study of the proton transfer in water, *J.Chem.Phys.* **34**, 375-388.

Menes, F. (1970) L'eau anormale, *J.Chimie Phys.* **67**, 2059-2064.

Morgan, L.O. and Nolle, A.W. (1959) Proton spin relaxation in aqueous solutions of paramagnetic ions, *J.Chem.Phys.* **31**, 365-368.

Packer, K.J. (1977) The dynamics of water in heterogeneous systems, *Phil.Trans.R. Soc.- Lond.* **B278**, 59-87.

Prigogine, M. and Fripiat, J.J. (1971) L'eau dite anormale: thèse nouvelle sur sa genèse et sa nature, *Bull.Soc.Chim.Fr.* **12**, 4291-4296.

Sainte-Laudy, J. and Belon, P. (1993) Inhibition of human basophil activation by high dilutions of histamine, *Agents Actions* **38**, C245-C247.

DOSE-DEPENDENCE OF THE VARIOUS FUNCTIONAL RESPONSES OF NEUTROPHILS TO FORMYLPEPTIDES*

Activation, Regulation, and Inverse Effects According to the Agonist Dose and CellCondition

P. BELLAVITE, S. CHIRUMBOLO, C. SANTONASTASO, D. BIASI*, S. LUSSIGNOLI AND G. ANDRIOLI
*Institute of Chimica Microscopia Clinica and of *Patologia Medica, University of Verona*
Laboratorio Centrale, Ospedale Policlinico, 37134 Verona, Italy

1. Introduction

The up- and down-regulation of the response to a specific stimulatory compound, according to its doses and to the sensitivity of the cells, is a widespread process in biology, immunology and endocrinology. Leukocytes are particularly suitable models for studying these changes of biological response as these cells may be easily isolated from blood and from inflammatory exudates and appear to be regulated by a large variety of mediators both *in vitro* and *in vivo* (disease states).

Neutrophil granulocytes are one of the main cell types involved in the first defense lines against infections and in the acute inflammation. Following the local or systemic generation of membrane perturbing agents and biological signals such as cytokines, complement factors and bacterial products, neutrophils undergo a series of highly regulated functional modifications. The expression and activation of specific membrane glycoproteins induce cell adherence to endothelium or to subendothelial structures, thus allowing the leukocyte extravasation and the chemotactic movement into the connective tissue.

These cells are endowed with a powerful armamentarium of enzymes and antimicrobial peptides. Moreover, upon activation, they produce a huge amount of excited oxygen species, including superoxide anion (O2-) and its derivatives, that participate in the microbial killing but also in the possible tissue damage caused by the dysfunction of neutrophil activation in a number of pathologies (Weiss, 1989; Smith, 1994). Since the duration and the magnitude of the functional responses are important factors in determining the final balance - positive or unfavourable for the host - of neutrophil activation, these phenomena are subject to a fine and complex regulation at the level of receptors, transduction mechanisms and effector enzymes.

Here we summarize a series of studies carried out in our laboratory with the aim of characterizing the sensitivity limits of normal and primed human neutrophils to the chemotactic peptide formyl-methionyl-leucyl-phenylalanine (fMLP). We studied: a)

---

* Communication presented at the 7th GIRI Meeting, November 1993, Montpellier, France.

M. Bastide (ed.), Signals and Images, 111–119.

the oxidative metabolism, assessed as the burst of O2- production, b) the adhesion to serum-coated surfaces, c) the release of lysozyme, a bactericidal enzyme, d) the actin polymerization, a rapid agonist-stimulated response of the cell cytoskeleton, e) the expression of fMLP receptors, f) the activation of two important intracellular biochemical changes that follow receptor stimulation, i.e. free Ca2+ increase and cyclic AMP (cAMP) increase.

These data are of interest both for an understanding of neutrophil physiology in the context of the inflammatory reactions and, in a wider perspective, for the understanding of the complex events involved in the changes of sensitivity and of responsiveness of the cells treated with different doses of receptor agonists. Several experiments where the phenomena of direct activation, priming, desensitization, and reverse effects have been observed are reported.

2. Dose-response Curves of Activation, Priming and Desensitization

The dose-response curves of the effect of increasing doses of fMLP on O2- production and on adhesion of human blood neutrophils are shown in figure 1.

The left panels of the figure (A and C) show the *direct* stimulatory effect of the agonist: O2- production was triggered by fMLP doses of 10 nmoles (nM) and higher, reaching a maximum at 100 nM, while adhesion required at least 30 nM fMLP and increased up to 1000 nM. The right panels of the figure (B and D) show the net effect of a *second addition* of high doses of fMLP (100 nM) to the cells that had been previously treated with the range of fMLP concentrations indicated in the X axis of the figure. It can be seen that the response of neutrophils pre-treated with doses ranging from 0.1 to 5 nM (i.e. sub-stimulatory doses) was higher than that of neutrophils not pre-treated with fMLP, both in terms of O2- production (B) and in terms of adhesion (D). Therefore, low doses of the peptide primed the cells to a subsequent higher response to the same agent, a phenomenon that we called *homologous priming* (Bellavite *et al.*, 1993a). Figures 1B and 1D also show that the response of neutrophils pre-treated with doses ranging from 10 nM to 1000 nM (i.e. stimulatory doses as shown in the left panels of the figure) had a decreased response both in terms of O2- production and in terms of adhesion. In other words, the stimulatory doses of the peptide *desensitize* the cells to a subsequent challenging with the same agent, and the extent of the desensitization is proportional to the extent of the previous stimulation. Similar dose-dependent stimulatory, priming and desensitising responses were observed by measuring the release of lysozyme.

Priming induced by low doses of fMLP is not only homologous but also heterologous, because we and others have observed that pretreatment with fMLP increases also the response to other unrelated agents, such as phorbol-myristate acetate (PMA), the active principle of croton oil. On the other hand, the desensitization induced by high doses of fMLP is exclusively homologous, because the response to PMA was not inhibited but, instead, was further augmented (Table 1).

3. Other Dose-dependent Phenomena

We then investigated the possible mechanisms underlying these low-dose (priming) and high-dose (direct activation and homologous desensitization) effects, by looking at the correlation between these phenomena and other fMLP-induced biochemical events.

DOSE DEPENDENCE OF NEUTROPHIL RESPONSE

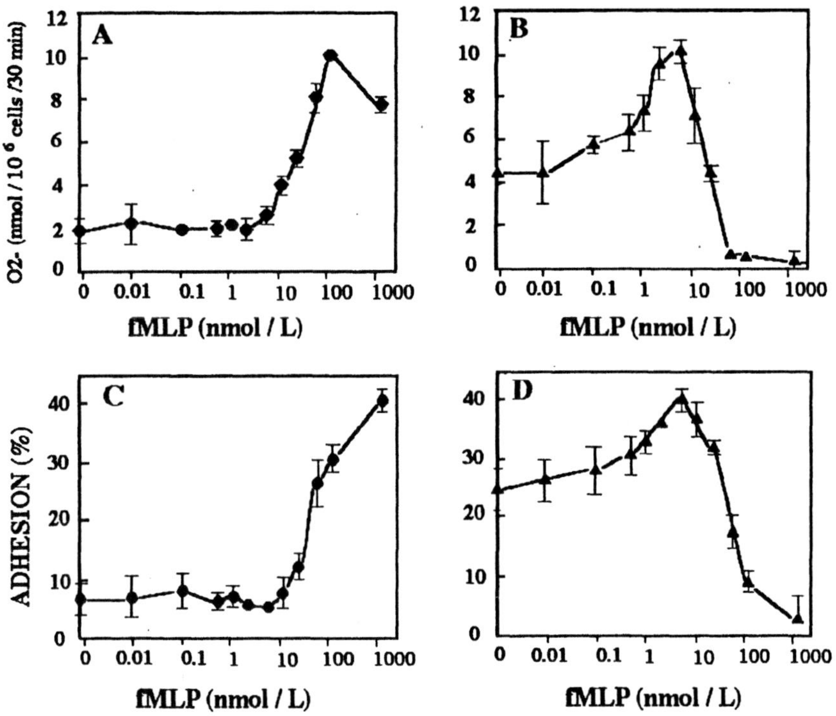

Figure 1. O2- production (A, B) and adhesion (C, D) of human neutrophils stimulated with different doses of fMLP. A, C: direct stimulation; B, D: effect of 100 nM fMLP on cells pretreated for 15 minutes with the indicated doses of fMLP. Blood neutrophils were obtained from healthy subjects by centrifugation of EDTA-anticoagulated blood over Percoll gradients (Metcalf *et al.*, 1986). O2- was measured by the reduction of ferricytochrome c in multiwell microplates and adherence to the serum-coated surface of microplates was quantitated by acid phosphatase assay (Bellavite *et al.*, 1992). Values are mean ± S.D. of triplicate determinations from a representative experiment of 12 performed.

TABLE 1. Effect of 15 minutes of pretreatment with different doses of fMLP on the O2- production induced by fMLP and PMA

Preincubation	O2- production (nmoles/106 cells/30 min)	
	Second stimulus fMLP (100 nM)	Second stimulus PMA (10 ng/ml)
Buffer (control)	4.1 ± 0.5	13.6 ± 0.3
5 nM fMLP	10.1 ± 0.9	18.4 ± 1.3
100 nM fMLP	0.4 ± 0.4	35.5 ± 2.2

Treatment of neutrophils with 1 nM fMLP for 15 minutes raised the membrane fMLP binding capacity by 60%, while treatment with 100 nM fMLP reduced the binding by over 90%, thus providing a clear molecular explanation for the desensitization. Low
doses of fMLP (0.5 nM and up) were able to cause an increase of intracellular free Ca2+ and of intracellular cAMP (Bellavite *et al.*, 1993a,b; Biasi *et al.*, 1993b). FMLP dose-response for intracellular free Ca2+ was markedly shifted to the left with respect to the dose response for O2- production: ED50 (dose causing 50% stimulation) were 1 nM for intracellular free Ca2+ and approximately 50 nM for O2- production respectively, indicating that the former is not a sufficient signal for the activation of the oxidative metabolism.

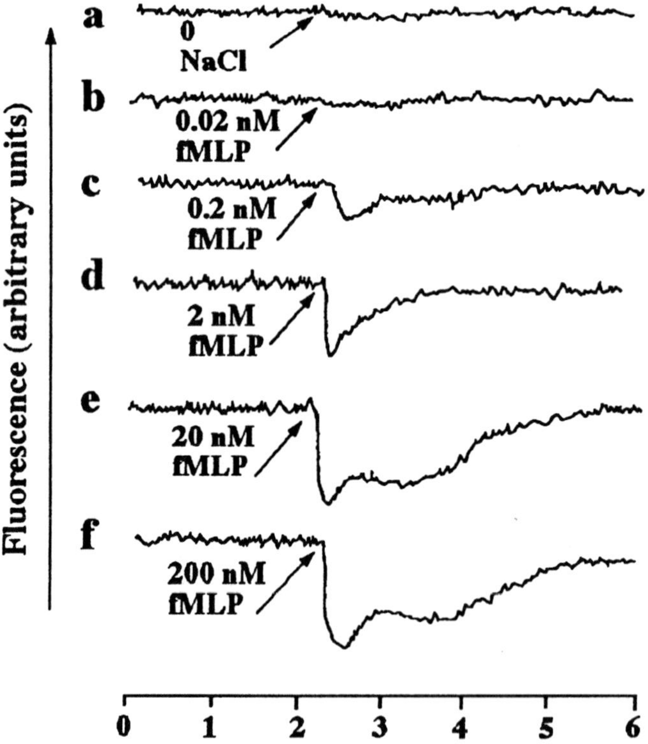

Figure 2. Recording traces of right-angle light scatter response (actin polymerization) of neutrophils to fMLP, assessed fluorimetrically according to Omann *et al.* (1989).

These data may indicate that the priming effect is the result of an increase in the number of receptors or of other mechanisms of post-receptor regulation, particularly linked to the increase of intracellular Ca2+. It should be noted, however, that many models of priming exist and that an increase of Ca2+ does not appear to be an absolute requirement (Biasi *et al.*, 1993b). By no means extraneous to priming are other biological events occurring at post-receptor level, such as the phosphorylation-dephosphorylation of specific proteins, the state of assembly of cytoskeletal proteins, and the constitution of membrane lipids. The latter point appears to be particularly relevant for the *in vivo* regulation of neutrophil responsiveness, as documented by our recent findings that the O2- release in response to fMLP is positively correlated with the cellular content of arachidonic acid and inversely correlated with linoleic acid and palmitic acid content (Bellavite *et al.*, 1995). These results suggest that the fatty acid composition of blood neutrophils may be a critical factor determining the capability of releasing free radicals in response to formylpeptides and are of relevance in view of a possible manipulation of the lipid composition of cell membranes by diet changes.

Another cell response that appeared to be sensitive to very low doses of fMLP was the actin polymerization, that is associated with cytoskeletal and morphological changes of the cells (figure 2). Doses of fMLP as low as 0.2 nM triggered a rapid actin polymerization response, that was already maximal at 20 nM. Of particular interest are also the time-dependent oscillations of scattering, induced by medium-high doses of fMLP, a finding in agreement with previous ones (Omann *et al.*, 1989).

4. Inverse Effects of Different fMLP Doses on Neutrophil Adhesion

We then studied the metabolic and adhesion responses to fMLP of neutrophils which were treated with bacterial endotoxin (lipopolysaccharide, LPS) or which were harvested from in a *in vivo* experimental inflammation (24-h exudation through a skin-window procedure).

Both LPS treatment and inflammation enhance the neutrophil's oxidative burst in response to fMLP, i.e. they induce cell priming (table 2). Moreover, by investigating the dose-response relationships of the adhesion response in these primed cells, we have observed an unexpected phenomenon (table 2 and figure 3): a) priming augments cell adhesion to serum-coated culture wells in the absence of further stimulation; b) addition of low, substimulatory doses of fMLP (0.5 nM to 5 nM) inhibits and reverses the spontaneous adhesion, c) high fMLP doses (> 100 nM) increase the adhesion and are additive to the spontaneous adhesion induced by priming. In conclusion, the chemotactic agent fMLP, which is considered to be an activator of neutrophil adhesion, paradoxically inhibits the same cell response at low doses when used in primed cells.

5. Discussion

We have investigated several different neutrophil functional activities triggered by contact with chemotactic peptides. The evidence here reported demonstrates that "early" responses of neutrophils to low doses of fMLP involve both structural (actin polymerization) and biochemical (Ca2+ and cAMP increase) changes, which are associated with inhibition of adhesion, while adhesion and superoxide release should be

TABLE 2. A comparison of the effects of low doses and high doses of fMLP on O2- production and adhesion of normal and primed neutrophils

Source of neutrophils	O2- production (nmoles/106 cells/30 min)		
	No fMLP	Low-dose fMLP[a]	High-dose fMLP[b]
Blood	0.6 ± 0.6	1.6 ± 1.0	11.5 ± 3.0
LPS-primed	3.6 ± 0.3	2.8 ± 0.3	18.9 ± 1.7
Exudate-primed	0.8 ± 1.0	5.4 ± 2.0	23.9 ± 5.3
	Adhesion (% of total)		
	No fMLP	Low-dose fMLP[a]	High-dose fMLP[b]
Blood	5.6 ± 3.4	3.7 ± 1.6	27.5 ± 8.1
LPS-primed	20.9 ± 10.5	7.4 ± 6.1	40.3 ± 16.9
Exudate-primed	17.1 ± 10.3	8.4 ± 4.9	39.2 ± 14.7

Exudate neutrophils were isolated according to the skin-window method described by Senn, with modifications (Biasi *et al.*, 1993a). LPS-primed neutrophils were obtained by incubating blood neutrophils for 1h with 1μg/ml LPS. a,b: Low doses 1 nM to 5 nM, high doses 100 nM to 500 nM, according to different experiments. The table summarizes data from experiments reported in Bellavite *et al.* (1993b) and Bellavite *et al.* (1994). The values significantly different ($p < 0.01$, paired Student t test) from the respective values obtained in the absence of fMLP (no fMLP) are shown in bold.

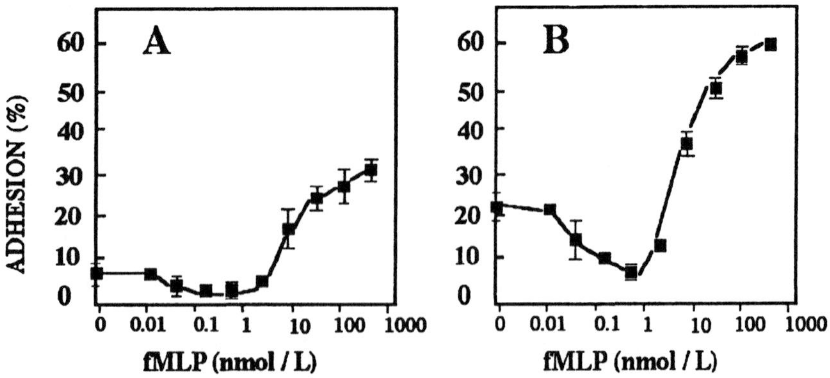

Figure 3. Effect of different doses of fMLP on the adhesion of human neutrophils. A: blood cells, B: cells isolated from a skin-window inflammatory exudate of the

considered as "late" or "extreme" responses to chemotactic factors, which require a 100-fold higher dose of this stimulant.

These results indicate that actin polymerization, a typically non-linear cellular phenomenon, reflects one of the earliest cellular events, occurring at very low agonist doses. Our data are in agreement with those reported by others, showing that half-optimal O2- production requires the occupancy of about 30% of the fMLP receptors, while intracellular Ca2+ increase and actin polymerization require occupancy of less than 3% and 0.1% receptors respectively (Sklar *et al.*, 1985).

A summary of the various responses which have been investigated and of their dependence on the fMLP dose is reported in table 3.

TABLE 3. Effects of different doses of fMLP on various neutrophil functions

Cell function	Low doses (0.1 to 5 nM)	High doses (10 to 1000 nM)
Actin polymerization	Activation	Activation and oscillations
Intracellular free Ca2+	Increase	Increase
Intracellular cAMP	Increase	Increase
fMLP receptors	Increase	Decrease
Adhesion (normal blood cells)	No effect	Activation
Adhesion (primed cells)	Inhibition	Activation
O2- production	No effect	Activation
Lysozyme release	No effect	Activation

These findings provide a reproducible *in vitro* model of the complex biological events occurring in leukocytes when they are treated with different doses of a bacterial product. In a wider perspective, these findings provide an example of how biologically active compounds may cause inverse effects on a system endowed with regulatory (feed-back) controls, when either the *doses* of the compound, or the *sensitivity* and the *responsiveness* of the system are varied by changes of environmental conditions. Neutrophil adhesion is the result of a fine regulation that enables the cell to move onto a surface by continuous adhesion/detachment events. Our current hypothesis is that low doses of fMLP cause inhibition of cell adherence because they increase cellular cAMP. In fact, adhesion is inhibited by adenosine, a physiologically relevant agent that increases intracellular cAMP (Bellavite *et al.*, 1992). Moreover, the adhesion of primed cells was inhibited also by addition of the membrane permeating cAMP analogue dibutirryl cAMP (1 mM) and of theophyllin (2 mM), which blocks the cAMP phosphodiesterase (Chirumbolo, unpublished observation).

The increase of adhesion and the oxidative burst are induced by much high doses of fMLP: it is conceivable that in order to trigger these responses, other intracellular messengers, besides cAMP and calcium, are required. Candidate molecules mediating the high-dose-dependent activation are diacylglycerols generated by phospholipid breakdown, that are activators of protein kinase C and whose production requires high fMLP doses (10 to 1000 nM) (Dougherty *et al.*, 1989).

Our data have shown the four different states of activation in neutrophils: a) resting state (dormant cells); b) homologous and heterologous priming, with increased O2- and decreased adhesion responses; c) full activation, where the cells are treated with high doses of agonists; d) specific desensitization to a second homologous stimulus. These types of tests are not simple laboratory artefacts, but allow us to reproduce a situation which occurs *in vivo*, i.e. where the cells in patients presenting bacterial infections or systemic inflammations are modified with respect to the cells in healthy people (Bloomfield and Young, 1988; Trautinger *et al.*, 1991; Smith, 1994). The fact that the disease conditions cause changes in receptor and transduction system sensitivity in various cells of the body is well known in many fields of medicine.

The concept is emerging that the leukocytes are involved in the cybernetic information networks of inflammation in a highly sophisticated and complex way: early activation of specific transduction systems, shape change, cell detachment from the stores, priming and chemotaxis require the occupancy of a minimum number of receptors, while the killing armamentarium (release of granule constituents, oxygen radicals) and the potentially harmful cell adhesion and spreading are triggered only when a massive engagement of membrane receptors is achieved. Finally, the effects of endotoxin and of the cellular lipid composition point to the existence of multiple control mechanisms and subtle synergisms between endogenous and exogenous compounds. All these factors are in connection with the individual's general state of health (infections, neuro-immuno-endocrine equilibrium, nutritional status).

Acknowledgments.

This work was supported by grants from Guna s.r.l. (Milano) and from Banca Popolare (Verona).

6. References

Bellavite, P., Chirumbolo, S., Mansoldo, C., Gandini, G., and Dri, P. (1992) A simultaneous assay for oxidative metabolism and adhesion of human neutrophils. Evidence for correlations and dissociations of the two responses, *J. Leukocyte Biol.* **51**, 329-335.

Bellavite, P., Chirumbolo, S., Lippi, G., Guzzo, P., and Santonastaso, C. (1993a) Homologous priming in chemotactic peptide stimulated neutrophils, *Cell Biochem. Funct.* **11**, 93-100.

Bellavite, P., Chirumbolo, S., Lippi, G., Andrioli, G., Bonazzi, L., and Ferro, I. (1993b) Dual effects of formylpeptides on the adhesion of endotoxin-primed human neutrophils, *Cell. Biochem. Funct.* **11**, 231-239.

Bellavite, P., Carletto, A., Biasi, D., Caramaschi, P., Poli, F., Suttora, F., and Bambara, L.M. (1994) Studies of skin-window exudate human neutrophils. Complex patterns of adherence to serum-coated surfaces in dependence on FMLP doses, *Inflammation* **18**, 575-587.

Bellavite, P., Guarini, P., Biasi, D., Carletto, A., Trevisan, M.T., Caramaschi, P., Bambara, L.M., and Corrocher, R. (1995) Correlations between the intensity of fMLP-dependent respiratory burst and cellular fatty acid composition in human neutrophils, *Brit. J. Haematol.* **89**, 271-276.

Biasi, D., Bambara, L.M., Carletto, A., Caraffi, M., Serra, M.C., Chirumbolo, S., and Bellavite, P. (1993a) Factor-specific changes of the oxidative metabolism of exudate human neutrophils, *Inflammation* **17**, 13-23.

Biasi, D., Bambara, L.M., Carletto, A., Caramaschi, P., Serra, M.C., Santonastaso, C., Guzzo, P., and Bellavite, P. (1993b) Molecular mechanisms of priming: exudate human neutrophils are primed to fMLP but have normal intracellular calcium and cAMP responses, *Bull. Mol. Biol. Med.* **18**, 225-236

Bloomfield, F.J. and Young, M.M. (1988) Enhanced chemiluminescence production by phagocytosing neutrophils in psoriasis, *Inflammation* **12**, 153-159.

Dougherty, R.W., Dubay, G.R., and Niedel, J.E. (1989) Dynamics of the diacylglycerol responses of stimulated phagocytes, *J. Biol. Chem.* **264**, 11263-11269.

Metcalf, J.A., Gallin, J.I., Nauseef, W.M., and Root, R.K. (1986) *Laboratory Manual of Neutrophil Function*, Raven Press Publisher, New York.

Omann, G.M., Porasik, M.M., and Sklar, L.A. (1989) Oscillating actin polymerization/depolymerization responses in human poly-morphonuclear leukocytes, *J. Biol. Chem.* **264**, 16355-16358.

Sklar, L.A., Hyslop, P.A., Oades, Z.G., Omann, G.M., Jesaitis, A.J., Painter, R.G. and Cochrane, C.G. (1985) Signal transduction and ligand-receptor dynamics in the human neutrophil. Transient responses and occupancy-response relations at the formyl peptide receptor, *J. Biol. Chem.* **260**, 11461-11467.

Smith, A. (1994) Neutrophils, host defense, and inflammation: a double edged sword, *J. Leukocyte Biol.* **56**, 672-686

Trautinger, F., Hammerie, A.F., Poschi, G., and Micksche, M. (1991) Respiratory burst capability of polymorphonuclear neutrophils and TNF-(serum levels in relationship to the development of septic syndrome in critically ill patients, *J. Leukocyte Biol.* **49**, 449-453.

Weiss, S.J. (1989) Tissue destruction by neutrophils, *N. Engl. J. Med.* **320**, 365-376.

THE ROLE OF THE BURSA OF FABRICIUS AND HIGHLY DILUTE BURSIN IN IMMUNONEUROENDOCRINE INTERACTIONS IN THE CHICKEN *

Biological Effects of Highly Dilute Bursin

B.J. YOUBICIER-SIMO [1], F. BOUDARD [1], M. GUELLATI [2] M. MEKAOUCHE [2] , J.D. BAYLÉ [2] , AND M. BASTIDE[1]
[1]*Laboratoire d'Immunologie & Parasitologie*
Faculté de Pharmacie, Université de Montpellier-1
15 Av. Ch. Flahault, 34060, Montpellier cedex 1, France
[2] *Laboratoire de Physiologie Générale*
Faculté des Sciences, Université de Montpellier-2
Pl. E. Bataillon, 34090 Montpellier cedex 5 (France)

1. Introduction

A huge set of data, pertaining to Mammals namely, supports the existence of two-way communication between the immune and neuroendocrine systems. For example, glucocorticoids released by the adrenals are immunodepressive; in turn, immune products act at different levels of the hypothalamo-pituitary-adrenal axis (HPA) (for review see Lilly and Gann, 1992). Also, the chief pineal hormone melatonin (MLT) is immunostimulatory (Maestroni *et al.*, 1989) whereas interferon-γ stimulates the release of MLT by cultured pineal glands (Withyachumnarkul *et al.*, 1990). The concept that products of the immune system act on the neuroendocrine system was suggested by Besedovsky and Sorkin (1977), based on their observation that in rats, serum corticosterone (CORT) increases during the course of immune response. However, in Mammals, most studies working out immunoneuroendocrine interferences deal with the T immune component mainly, therefore neglecting the involvement of the B lymphoid compartment, because the latter is rather anatomically diffuse. Therefore, it appears experimentally difficult to identify specific B immune signal (s) endowed with neuroendocrine competence.

Aves possess a species specific and anatomically distinct organ termed the bursa of Fabricius (Toivanen *et al.*, 1981). The latter is a diverticulum of the hindgut and it is also the primary site for B-cell differentiation (Toivanen *et al.*, 1981). The bursa anlage arises by the 5 th day of embryonic development as a swelling of the cloacal plate and it is invaded by B stem cells between the 8 th and 14 th day of embryonic life (Le Douarin *et al.*, 1975; Houssaint *et al.*, 1976; Lassila *et al.*, 1978). In the bursa follicles, the B precursor cells differentiate into mature lymphocytes which seed to the periphery by the end of the embryonic phase (Le Douarin *et al.*, 1975; Weill and Reynaud, 1987). Early removal of the bursa anlage leads to permanent B-deficient

---

* Communications presented at the 7 th GIRI Meeting, November 1993, Montpellier, France, and at the 8th GIRI Meeting, December 1994, Jerusalem, Israel.

M. Bastide (ed.), Signals and Images, 121–147.

chickens (Fitzsimmons *et al.*, 1973; Jankovic *et al.*, 1977; Granfors *et al.*, 1982; Jalkanen *et al.*, 1983). This surgical procedure provides a biological preparation suitable for the study of possible involvement of the B immune component in immunoneuroendocrine crosstalks. The function of the bursa of Fabricius is mediated by cellular and soluble factors that have been identified as integral parts of the bursa microenvironment (Heller and Friedman, 1979; Glick, 1984ab; Kuznik *et al.*, 1988). One of these factors is a low-molecular weight inducing agent termed bursopoietin that has been extracted from the chicken bursa (Brand *et al.*, 1976). This agent has been isolated, purified and termed bursin (Lys-His-Gly-NH_2) by the US team of Goldstein (Audhya *et al.*, 1986). Bursin has been immunohistochemically identified in the bursal epithelium (Viamontes *et al.*, 1989), as well as in avian and bovine bone marrow and in the intrahepatic bile ducts of bovidæ (Audhya *et al.*, 1990). A tetradecapeptide precursor named probursin and containing the sequence of bursin, tuftsin and the active site of somatostatin has also been discovered (Audhya *et al.*, 1991). Bursin induces B cells from their avian and mouse precursors *in vitro* (Goldstein *et al.*, 1977; Audhya *et al.*, 1986; Lassila *et al.*, 1989). However, it is still unclear whether the neuroendocrine function of the bursa of Fabricius also operates through bursin-dependent mechanisms. The present study was an attempt to disclose the possible involvement of the B immune component in immunoneuroendocrine interactions, using permanent B-deficient chickens (bursectomized chickens). This aim was achieved by testing:

1. the ability of bursectomized chickens to respond hormonally and immunologically to various environmental agents (ether vapour, antigen, photoperiod). The following parameters were assessed:

- the pituitary (corticotrophin: ACTH) and adrenal (corticosterone: CORT) responses to ether vapour exposure,
- the specific antibody (IgM, IgG), as well as the pituitary-adrenal (ACTH, CORT) and pineal (melatonin: MLT) responses to immunization against porcine thyroglobulin (Tg),
- the nature of the endocrine responses displayed by porcine Tg-sensitized chickens,
- the circadian rhythms of both pituitary-adrenal (ACTH, CORT) and pineal (MLT, N-acetyltransferase: NAT, hydroxyindole-o-methyltransferase: HIOMT) activities.

2. the effectiveness of the bursa-derived signal (bursin) in reversing the effects of bursectomy, by means of *in ovo* administration of different amounts of bursin (100 μg, 100 pg (5CH), 100 fg (7CH), 5 x 10^{-27} g (a pool of 15CH to 20 CH dilutions) to bursectomized embryos.

2. Materials and Methods

2.1. INCUBATION OF EGGS AND CHICKS MANAGEMENT

Eggs from New Hampshire strain were purchased from Couvoir des Cévennes (Ledenon 30, France) and incubated in a Maino Ladi France incubator (38 ± 1° C, 45-50% humidity, permanent darkness). An incorporated electric fan allowed homogeneity of the temperature and air inside the incubator. The eggs were permanently turned on automatically rotating drawers with 45° maximal inclination, so as to avoid sweating of the eggs opened during surgical bursectomy.

Newly-hatched chicks were housed under controlled photoperiod [12 h light (0700-1900), 12 h dark] and the environmental temperature was progressively reduced 1° C per day from 38° C to 22° C. Feed and water were available ad-libitum.

2. 2. SURGICAL METHODS

2. 2. 1. *Embryonic bursectomy*

The removal of the bursa anlage was performed at 80 h of incubation (4 th day of embryonic life) according to Fitzsimmons *et al.* (1973). The surgical operation was carried out under aseptic conditions. Briefly, the embryo was localized into the egg by candling and the shell was opened above it using fine forceps. Then a small slit was made in the chorion and amnios at the level of the caudal bud. A sterilized fine loop was passed over the caudal bud and tightened posterior to the leg buds, caudally to the cloacal membrane. The tail was cut off posterior to the nod using microsurgical scissors and the free fragment was discarded. The lips of the chorion and amnios were closed with forceps and the shell opening was overlaid with parafilm M and sealed with paraffin.

2. 2. 2. *Removal of the pineal glands and eyes*

The chickens were sacrificed by decapitation and the skull was opened dorsally between the cerebral hemispheres and the cerebellum. The meninges were slit above the pineal gland along the line of suture. The lips of the opening were drawn aside by means of slender forceps. The pineal body gripped by its stalk, withdrawn from the skull, immediately frozen in liquid nitrogen and stored at -80° C until assayed for the pineal enzymatic activities (NAT, HIOMT). The eyes from each bird were dissected out, wrapped individually in aluminium foil, immediately frozen in liquid nitrogen and stored at -20° C until time of assay. Then the retinae were isolated by the procedure described by Hamm and Menaker (1980).

2. 3. TRIPEPTIDES AND *IN OVO* TREATMENT

Bursin (Lys-His-Gly-NH_2) was synthesized *de novo* by fragment peptide coupling (Le Nguyen, 1987). The control tripeptide (Trp-Leu-Leu-NH_2) was a gift from Dr. Martinez, University of Montpellier, France). The injected solutions were prepared by sequential centesimal dilution steps including potent vertical stirring between two successive steps. On the 6 th and 9 th day of incubation, the embryos were treated *in ovo* by single administration of 100 µl of solution each day in the yolk sack , as follows: sham-bursectomized embryos (N) were given 100 µl of saline (N+S). The surgical treatment group (Bx) could receive either 100 µl of saline (Bx+S) or 100 µl of saline containing 5 CH of the control peptide (Bx+P_f) or 100 µl of saline containing different doses of bursin: 100 µg (Bx+B_{μ}) or 5 CH (Bx+B_p) or 7 CH (Bx+B_f) or a 15-20 CH pool (Bx+B^{-27}).

2. 4. EXPERIMENTAL GROUPS

The birds were distributed as follows:

N: Sham-bursectomized (the surgical handling was limited to opening and closing the chorion and amnios in 80 h old embryos),

N+S: sham-bursectomized birds injected in the yolk sack by single administration of 100 μl of saline on the 6 th and 9 th day of embryonic life,

Bx: chickens embryonically bursectomized at 80 h of incubation,

Bx+B: chickens embryonically bursectomized at 80h of incubation and supplemented on the 6th and 9th day of embryonic development by *in ovo* administration of 100 μl of saline containing different amounts of bursin: 100 μg (**Bx+**$\mathbf{B_{\mu}}$) or 5 CH (**Bx+**$\mathbf{B_p}$) or 7 CH (**Bx+**$\mathbf{B_f}$) or the 15-20 CH pool (**Bx+**$\mathbf{B^{-27}}$),

Bx+$\mathbf{P_f}$: chickens embryonically bursectomized at 80 h of incubation and receiving 100 μl of saline containing 5 CH of the control tripeptide (Trp-Leu-Leu-NH_2) on the 6 th and 9 th day of embryonic development.

2. 5. ANTIGENS

Porcine Tg or keyhole limpet hemocyanin (KLH) purchased from Sigma was injected subcutaneously. The dose of immunogen to be injected was determined in a preliminary experiment (data not shown). Chickens received three doses of 125 μg/100 g B.W. (750 μg/ml of saline) mixed (v/v) with complete (first immunization) or incomplete (second and third immunizations) Freund adjuvant.

2. 6. EXPERIMENTAL DESIGN

Four experimental protocols including a total of 10 experiments have been carried out during six years.

2. 6. 1. First protocol

Experiment 1. Five chicken samples were used: N (4), Bx (4), Bx+B_{μ} (4), Bx+B_p (4), Bx+B_f (3). In brakets is the number of animals. Eight week-old chickens were placed in an ether vapour saturated box for 1min between 0900 and 1000. Blood samples were withdrawn from the brachial vein 7 and 14 min after ether vapour application, and assayed for plasma ACTH and CORT respectively.

2. 6. 2. Second protocol

This protocol was designed to measure the hormonal and specific humoral immune responses to antigen challenge. Four expriments were performed (experiments 2, 3, 4, 5). In experiments 2, 3 and 4, young chickens were repeatedly immunized by injecting porcine Tg at 21, 30 and 39 days of age. Blood samples were withdrawn from the brachial vein, within 30 s, between 0900 and 1000, the day before immunization (d20) and at 9 day intervals after the first (d29), the second (d38) and the third (d47) antigen challenge. All the sera were blinded before hormones and antibody evaluation. The specificity of anti-Tg antibodies was also determined.

Experiment 2. The following groups of animals were used: N (8), Bx (7), Bx+B_{μ} (6), Bx+B_p (6) and Bx+B_f (7). The plasma levels of ACTH, CORT and MLT, as well as the serum titres of specific anti-Tg antibodies (IgM, IgG) were measured.

Experiment 3. It was carried out with 4 groups of chickens: N (8), Bx+S (6), Bx+B_f (9), Bx+B^{-27} (7). The plasma levels of CORT, as well as the serum titres of anti-Tg IgG antibodies were determined.

Experiment 4. Seven chicken samples were studied: N (7), N+S (6), Bx (11), Bx+S (7), Bx+P_f (4), Bx+B_f (7), Bx+B^{-27} (8). The following parameters were assessed: the plasma levels of ACTH, CORT and MLT, as well as the serum titres of specific anti-Tg IgG.

Experiment 5. This experiment was designed to determine if the hormonal responses of bursa-intact chickens sensitized to porcine Tg derived from metabolic reactions involving Tg or if these answers were linked to the immunization phenomenon. Tg is the endogenous prohormone of thyroid hormones which themselves modulate the pituitary-adrenal activity during the course of immune response (Besedovsky *et al.*, 1975; Trout *et al.*, 1988). Keyhole limpet hemocyanin (KLH) is a T-dependent antigen without any pharmacological activity. To attempt disclosing the nature of the hormonal response to porcine Tg, 17 bursa-intact chickens immunized against porcine Tg were compared with their KLH-sensitized counterparts (18 chickens) for CORT, MLT and IgG responses. The chickens were immunized at 21, 30 and 36 days of age by subcutaneous inoculation of either porcine Tg or KLH. Blood samples were collected between 0900 and 1000 at the age of 20 (d20), 29 (d29), 35 (d35) and 38 (d38). Then the plasma levels of CORT and MLT, as well as the serum titers of anti-Tg or anti-KLH antibodies (IgG) were measured.

2. 6. 3. Third protocol

Twelve week-old chickens raised since hatching under a $12L_{0700\text{-}1900}$-12D alternating light-dark regime underwent four experiments (experiments 6, 7, 8, 9).

Experiment 6. In this preliminary experiment, bursa-intact chickens (N, 13) were decapitated and their circulating, ocular and pineal MLT contents were measured between 1200 and 1300 and between 0000 and 0100 (under red light).

Experiment 7. Five samples were studied: N (9), Bx (3), $Bx+B_{\mu}$ (3), $Bx+B_p$ (3) and $Bx+B_f$ (9). The chickens were bled at two-point time, between 1200 and 1300 and between 2000 and 2100 (under red light) to assess plasma ACTH and CORT levels. MLT levels were measured in blood samples collected between 1200 and 1300 and between 0000 and 0100 (under red light).All the sera were blinded before hormones evaluation.

Experiment 8. Blood samples were withdrawn from the N (7), Bx (5), $Bx+B_f$ (4) and $Bx+B^{-27}$ (4) groups between 1200 and 1300 and between 0000 and 0100 (under red light) to assess the plasma levels of MLT. After blood sampling, the animals were immediately sacrificed by decapitation and their pineal tissues assayed for NAT and HIOMT activities.

Experiment 9. Blood samples were collected from the brachial vein at 4 h intervals around the clock (1000, 1400, 1800, 2200, 0200, 0600) to measure the plasma levels of ACTH, CORT and MLT. The chicken samples used were as follows: N (13), N+S (6), Bx (12), Bx+S (7), $Bx+P_f$ (4), $Bx+B_f$ (7), $Bx+B^{-27}$ (7). All the sera were blinded before hormones evaluation.

2. 6. 4. Fourth protocol

Experiment 10. This experiment was aimed at checking whether a succussed highly dilute solution of bursin was structurally different from the same dilution unsuccussed. To answer this question, we decided to measure the concentration of labelled bursin in serial dilutions of this compound. Since labelled bursin was not available on the market place, we used ^{3}H-Thymidine. ^{3}H-Thymidine (cpm) was measured in different concentrations of succussed thymidine solutions ranging from 10^{-7} to 10^{-41} M, and compared with equal dilutions of unsuccussed ^{3}H-thymidine and succussed solvent. Unsuccussed ^{3}H-thymidine solutions were prepared from an initial ^{3}H-thymidine solution (1 mCi/ml, specific activity 25 Ci/mmol., Amersham), by serial centesimal dilution in pure water. Succussed ^{3}H-thymidine solutions and the solvent (pure water) were prepared in the same way, but in addition, potent vertical shaking was performed between successive dilution steps. Finally, 200 µl from each solution were added to 3 ml of scintillation solution (Pico-Fluor 15) and counted in a ß-liquid scintillation analyzer (Packard).

2. 7. DETERMINATION OF SPECIFIC ANTI-Tg IgM AND IgG TITRES

The titres of specific antibodies against porcine Tg were determined in sera by indirect enzyme-linked immunosorbent assay (ELISA) technique. Microtitre plates (Immunoplates II, NUNC, Roskilde, Denmark) were coated with a solution of porcine Tg at 10 µg/ml in phosphate buffered saline (PBS) at pH 7.2, and the plates were incubated overnight at 4° C. The plates were washed three times with PBS containing 0.1% Tween 20 (V/V); then 100 µl of chicken plasma diluted by half from 1/20 were added to the plates and incubated at 37° C for 1 h. The plates were washed three times. For the determination of anti-Tg IgM, 100 µl of horseradish peroxidase-labelled anti-chicken IgM conjugate (Bethyl Laboratories, Montgomery, AL) was added and the plates were incubated at 37° C for 1 h. After the last washing, the enzyme substrate, o-phenylenediamine (Sigma) was added and the plates were icubated for 15 min at room temperature. The absorbance was measured at 450 nm using a Multiskan photometer (Flow Laboratories, Marseille, France). For the determination of anti-Tg IgG, 100 µl of alkaline phosphatase-labelled anti-chicken IgG conjugate (Sigma) were added and the plates were incubated for 1 h at 37° C. After the last washing, 4-nitrophenylphosphate (Sigma) was added and the plates were incubated for 30 min at 37° C. The absorbance was measured at 405 nm. The titres of anti-Tg IgG or IgM were defined as the reciprocal of the plasma dilution giving a absorbance equal to 1.0 by indirect ELISA. Results were expressed in titre log.

The same technique was used to measure anti-KLH antibodies. Briefly, microtitre plates were coated with 100 µl/well of a KLH solution (10 µg/ml PBS) and incubated overnight at 4°C. Then, the plates were washed three times with PBS containing 0.1% Tween 20 (V/V). Afterwards, 100 µl of chicken plasma dilute at 1/40 were added to each well and the mixture held under 37° C for 1 h. The plates were rinced three times. Then 100 µl of peroxydase-conjugated affinipure rabbit anti-chicken IgG Fc fragment specific (Jackson ImmunoResearch Laboratories) dilute at 1/1000 in PBS-0.1% Tween 20 were added to the wells. The plates were incubated again for 1 h at 37° C. After three washings, 100 µl of the enzyme substrate termed o-phenylenediamine (Sigma) were added to the wells and the plates were allowed to incubate 5 min at room temperature. The enzymatic reaction was stopped by adding 50 µl/well of H_2SO_4 and the absorbance was measured at 490 nm. The titres of anti-Tg IgM or IgG antibodies were defined as for anti-Tg antibodies and the results were expressed in the same way.

The specificity of anti-Tg antibodies was checked by either an inhibition technique or a cross reaction test, using self-chicken (ovalbumin: OVA, ß-actin: ACT, myosin: MYO) or foreign (BSA, porcine insulin: INS, porcine Tg: Tg) antigenic proteins (Sigma). For the inhibition technique, the sera collected from 38 day-old chickens were dilute according to their titres and they were incubated with 2 µg/ml of the different proteins overnight at 4° C. Then an ELISA test using either porcine Tg or KLH as the antigen was performed on each treated serum in comparison with the corresponding untreated serum. The cross reaction consisted in testing the binding capacity of the above-listed antigens to sera collected at d20, d29, d38 and d47, by means of ELISA test.

2. 8. DETERMINATION OF HORMONES

The plasma concentrations of CORT (ng/ml) were measured by competitive protein-binding assay. This technique was previously made suitable for chicken plasma samples (Buckland *et al.*, 1974). Intra and inter-assay coefficients of variation were 3.1 and 5.6% respectively. The sensitivity of the assay was 0.05 ng/ml. ACTH was assessed using a RIA kit (ACTH K-PR) purchased from CEA (Paris, France) (Ramade *et al.*, 1985). Intra- and interassay coefficients of variation were 7.5 and 9.0 % respectively ; the sensitivity of the assay was 10 pg/ml. Plasma MLT concentrations (pg/ml) were determined by RIA according to the method developed by Rollag and Niswender (1976) and adapted to chicken by Cogburn *et al.* (1987). Intra and inter coefficients of variation were 14 and 12% respectively. The sensitivity of the assay was 3 pg/ml.

2. 9. ENZYMATIC ACTIVITIES

NAT and HIOMT activities were measured according to the techniques of Voisin and Colin (1986) and Voisin *et al.* (1988), respectively. Individual pineal glands were homogenized by sonication in 60 µl sodium phosphate buffer (0.05 M, pH 7.9). The enzyme substrates [20 µl tryptamine (20 mM), 20 µl of ^{3}H-acetylcoenzyme A and H-acetylcoenzyme A mixture (specific activity: 4 Ci/mole)], 20 µl of sodium phosphate buffer (0.05 M, pH 7.9) and 40 µl of pineal homogenate, for a total volume of 100 µl were introduced in Eppendorf tubes. The blank Eppendorf were filled with 20 µl of tryptamine (20 mM), 20 µl of a mixture of ^{3}H-acetylcoenzyme A and H-acetylcoenzyme A (specific activity: 4 Ci/mole) and 60 µl of sodium phosphate buffer (0.05 M, pH 7.9), total volume 100 µl. The Eppendorf contents were mixed and incubated for 30 min at 37° C. Thereafter, 1 ml chloroform was added into the tubes at 4° C to stop the reaction. Then the tubes were shaken, centrifuged (1000 g, 1 min) and the upper phase discarded. The chloroform phase containing the reaction product(s) was washed with 200 µl sodium phosphate buffer , centrifuged (10,000 g, 1 min) and the upper phase discarded. The chloroform phase (500 µl) was evaporated (2 h, 50° C), 7 ml of scintillant added to the residual powder and the radioactivity measured in a ß scintillation counter.

For HIOMT activity, 100 µl containing the substrates [25 µl N-acetylserotonine (25 mM), 25 µl of a mixture of ^{3}H-S-adenosyl-methionine and H-S-adenosyl-methionine (specific activity: 25 Ci/mole)], 40 µl borate buffer (0.3 M, pH 10) and 10 µl of pineal homogenate. In blank tubes, 25 µl of N-acetylserotonin (25 mM), 25 µl of a mixture of ^{3}H-S-adenosyl-methionine and H-S-adenosyl-methionine (specific activity: 25 Ci/mole) and 50 µl borate buffer (0.3 M, pH 10) were mixed. The tubes were treated similarly to those used to determine NAT activity, except that the reaction was stopped using 1 ml of chloroform and 20 µl of borate buffer.

2. 10. STATISTICAL ANALYSIS

The data represent mean ± S.E.M and were processed by two-way ANOVA, followed by one-way ANOVA and Mann-Whitney tests.

2. 11. HISTOLOGICAL CONTROLS

The birds were sacrificed at the end of the experiments. After autopsy, the absence of bursal remnants was ascertained by thorough ocular inspection and Cleveland-Wolff stained serial sections of the whole cloacal region were examined to verify completeness of bursectomy. Only the animals which met this creteria were used.

3. Results

3. 1. PITUITARY-ADRENAL RESPONSES TO ETHER STRESS

In the first protocol (Figure 1), ACTH and CORT levels were at resting values prior to ether stress application, irrespective of experimental groups. Bursa-intact chickens (N) exhibited strong hormonal responses after ether vapour exposure. No ACTH response was noticed in bursectomized birds (Bx) who rather raised a significant CORT response. The effectiveness of bursin treatment in reversing the effects of bursectomy appeared inversely proportional to the amount of the tripeptide administered: 100 µg induced a moderate increase in CORT but not ACTH levels and only after stress application, while either 5 CH or 7 CH restored quite normal pituitary-adrenal responses.

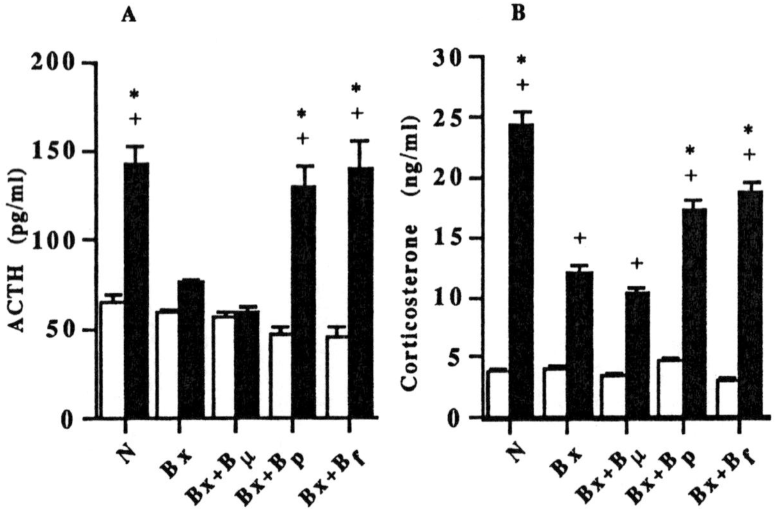

Figure 1. Plasma ACTH (A) and Corticosterone (B) levels before (white columns) and after (dark columns) ether stress application to sham-operated (N), bursectomized (Bx) and bursectomized chickens supplemented with decreasing amounts of bursin: 100 µg (Bx+Bµ) or 100 pg (Bx+Bp) or 100 fg (Bx+Bf) . +P<0.01 vs Rest (white columns); *P<0.01 vs Bx

3. 2. HORMONAL AND SPECIFIC HUMORAL IMMUNE RESPONSES TO IMMUNIZATION

In the second protocol, 20 to 47 day-old chickens underwent an immunization program to check the hormonal (ACTH, CORT, MLT) and specific antibody (IgM, IgG) responses. The results are outlined in Figures 2, 3 4. Regardless of experimental groups, hormonal levels (Figures 2A, B; 3A; 4A, B, C) were at resting values prior to immunization (d20) and hardly shifted 9 days after the first antigen challenge (d29). However, in sham-operated birds (N or N+S), hormonal responses reached a zenith at d38 and finally dropped to basal values after at d47. Assessed in parallel to endocrine

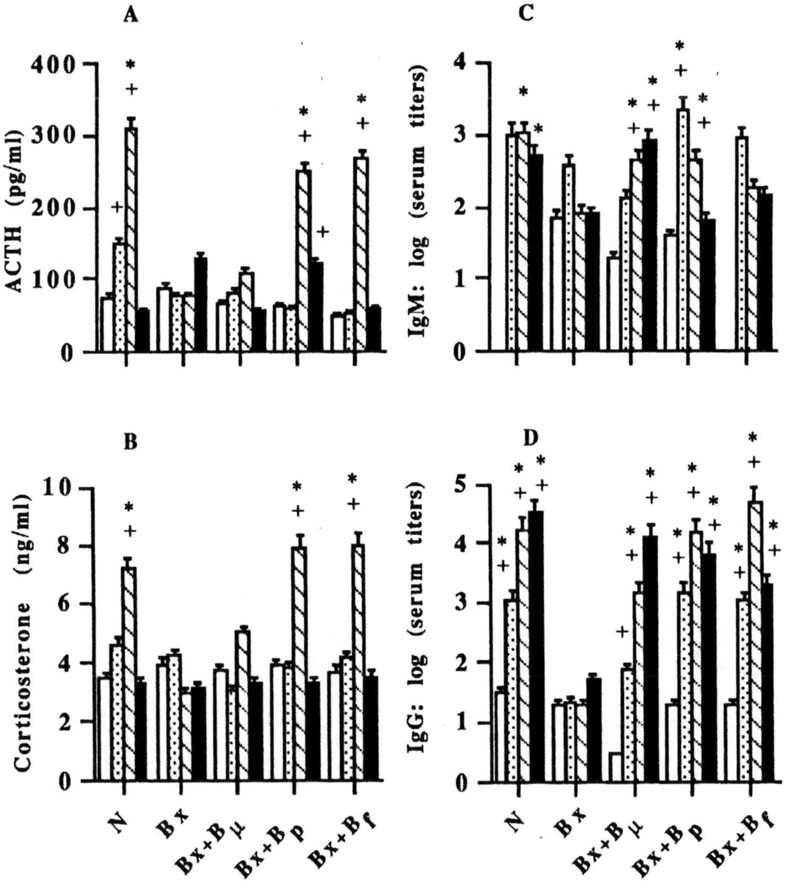

Figure 2. Hormonal (A, B) and specific antibody responses (C, D) after immunization with porcine Tg. Sham-operated birds (N) or bursectomized chickens (Bx) treated in ovo with different amounts of bursin (Bx+Bµ; Bx+Bp; Bx+Bf) were sequentially immunized at 21, 30 and 39 days of age. Then plasma levels of ACTH (A) and corticosterone (B), as well as the serum titers of anti-Tg IgM (C) and anti-Tg IgG (D) were measured the day before (d20: white columns) and at the age of 29 (d29: dotted columns), 38 (d38: hatched columns) and 47 (d47: dark columns) days. +P<0.01 vs d20; *P<0.01 vs Bx.

parameters, the specific humoral immune response (Figures 2C, D; 3B, C; 4 D) presented a different feature: especially in sham-operated samples, not only antibody production arose earlier, but it varied increasingly from the first to the third immunization.

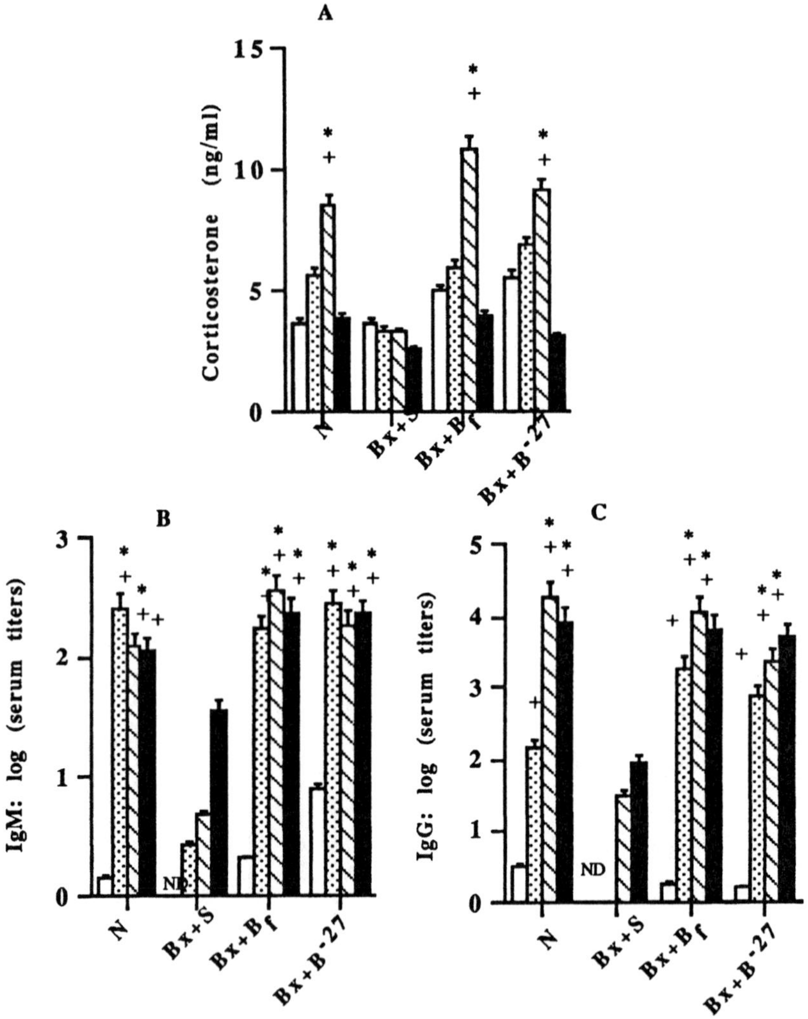

Figure 3. Corticosterone (A) and specific anti-Tg IgM (B) and IgG (C) responses to immunization against porcine Tg as measured in sham-operated (N) and bursectomized chickens supplemented with either the saline (Bx+S) or 100 fg (Bx+Bf) or a high dilution (5 x 10-27 g) of bursin. The birds were sequentially immunized against porcine Tg at the age of 21, 30 and 39 days and Corticosterone and antibody levels were evaluated the day before the first immunization (d20: white colums) and at the age of 29 (d29: dotted columns), 38 (d38: hatched columns) and 47 (d47: dark columns). +p<0.01 vs d20; *p<0.01 vs Bx. ND: Non-determined antibody titers (d20 for IgM, d20 and d29 for IgG).

On the other hand, hormonal levels remained steadily low in bursectomized chickens (Figures 2A, B; 3A; 4A, B, C) who also failed to mount specific antibody production in spite of repeated immunization (Fig. 2C, D; Figures 3B, C; 4D) and whether they had received the saline or the control tripeptide (Figure 4D).

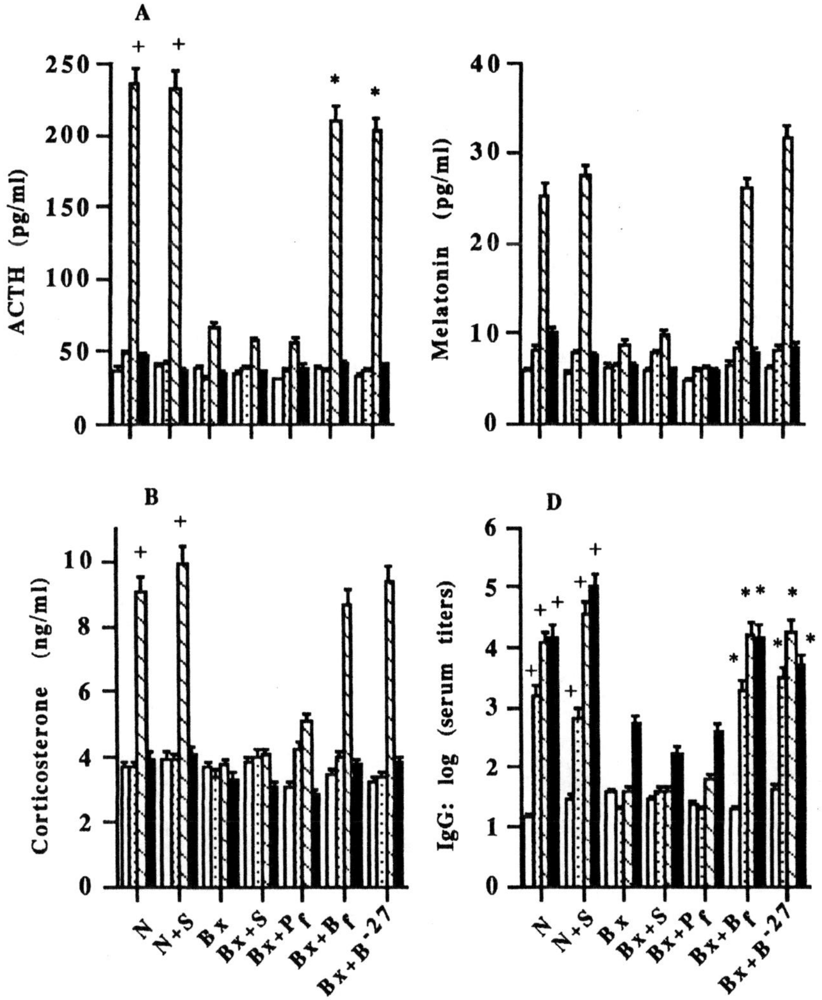

Figure 4 a:. ACTH (A), corticosterone (B), melatonin (C) and anti-Tg IgG (D) responses to immunization against porcine Tg. Seven experimental groups were used (N, N+S, Bx, Bx+S, $Bx+P_f$, $Bx+B_f$, $Bx+B^{-27}$. The animals were immunized at the age of 21, 30 and 38 days. Blood samples were collected the day before the first immunization (d20: white columns) and at the age of 29 (dotted columns), 38 (hatched columns) and 47 (dark columns) to check the plasma levels of hormonal (ACTH,corticosterone, melatonin) and specific antibody (IgG) responses.
$^{+}P<0.01$ vs Bx; $^{*}P<0.01$ vs Bx+S.

The administration of 100 μg of bursin could no longer reverse of the alterations caused by bursectomy. Conversely, either discrete amounts (5 CH, 7CH) or a

high dilution (15-20 CH pool) of bursin induced the recovery of normal endocrine and specific humoral immune performances in bursectomized recipients.
The specificity of anti-Tg antibodies was ascertained by an inhibition technique (Figure 4E) or a cross reaction test (Figure 4F) using either chicken self-proteins (OVA, ACT, MYO) or xeno-antigens (BSA, INS, Tg). Clearly, at the exclusion of the other proteins used, porcine Tg specifically reacted with the tested sera.

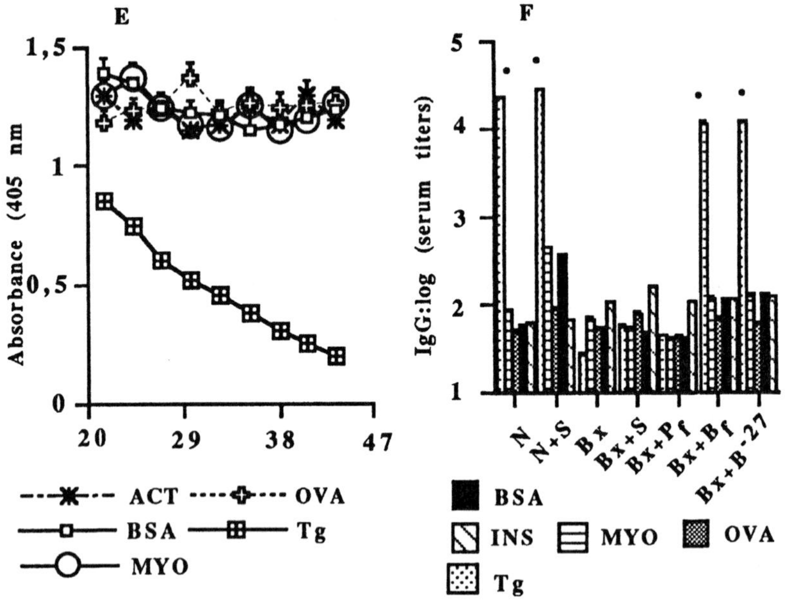

Figure 4 b. The specificity of anti-Tg IgG was assessed either by an inhibition technique (E) or a cross reaction test (F) using self-chicken (ovalbumin: OVA, myosin; MYO, ß-actin: ACT) or foreign (BSA, porcine insulin: INS, porcine Tg: Tg) proteins. In the inhibition technique, the sera collected from 38 day-old chikens were tested. The cross reaction test included sera harvested at the age of 20, 29, 38 and 47 days. ·P<0.01 vs other proteins.

In the study designed to elucidate the nature of endocrine responses triggered by porcine Tg (experiment 5), bursa-intact chickens exhibited the same profile of CORT response (Figure 5A), irrespective of the sensitizing agent (porcine Tg or KLH): CORT levels increased significantly at d35 and crested at d38. Even though antibody responses (IgG) started earlier (d29) than endocrine responses (d35), KLH- and porcine Tg-challenged chickens displayed the same feature of humoral immunity: IgG levels rose from prime to third immunization (Figure 5B). It is worth noting that in each case (KLH or porcine Tg), the humoral immune response (IgG) was contemporary of the endocrine response (CORT), both cresting simultaneously.

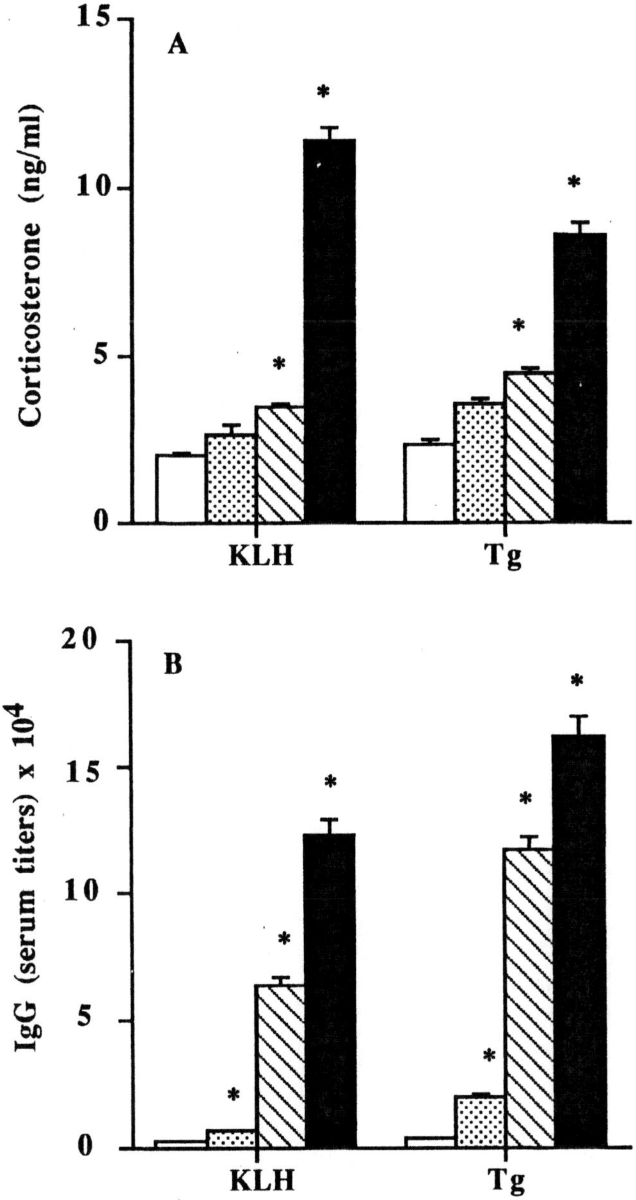

Figure 5. CORT (A) and IgG (B) responses of bursa-intact chickens sensitized to either keyhole limpet hemocyanin (KLH) or porcine Tg. Both parameters were measured the day before immunization (d20: white columns) and at the age of 29 (dotted columns), 36 (hatched columns) and 38 (dark columns) days.*$P<0.01$ vs d20. The samples comprised 18 and 17 chickens for KLH and porcine Tg respectively.

3. 3. CIRCADIAN RHYTHMS OF THE PITUITARY-ADRENAL AND PINEAL ACTIVITIES

In the third protocol (Table 1; Figures 6; 7), birds experienced a 12L-12D light-dark regime.Sham-operated birds (N) exhibited diurnal rhythms of plasma ACTH (Figure 6A), CORT (Figure 6B) and MLT (Figures 6C; 6D), with low light-time levels and significantly higher night-time values, in phase with light-dark cycle. Hence, broadpeaks were recorded during the dark phase (ACTH: 75.5 ± 3.42 pg/ml and CORT: 8.9 ± 0.25 ng/ml at 2100; MLT: 154.84 ± 8.21 pg/ml at 0000), whereas trough levels occurred during photophase (ACTH: 46 ± 3.49 pg/ml; CORT: 3.46 ± 0.28 ng/ml; MLT: 14.5 ± 3.06 pg/ml at 1200). Pineal and ocular MLT contents (Table 1) were also higher at night (13960 ± 2514 pg/pineal and 612 ± 102 pg/eye at 0000) than during daytime (1662 ± 349 pg/pineal and 292.5 ± 74.8 pg/eye at 1200).

TABLE 1. Day/night differences in melatonin levels

Organ	Time	
	1200	0000
Pineal gland	1,662 ± 349	13,960± 2514*
Retina	292 ± 75	612 ± 10*
Plasma	14.54 ± 3.06	154.84 ± 8.21*

Melatonin concentration measured at 1200 and 0000 in the pineal gland (pg/pineal), retina (pg/retina) and plasma (pg/ml) of 13 intact chickens P < 0.01 vs 1200.

This pattern of variation was in keeping with the rhythm of pineal NAT (Figure 6E) activity (1.3 ± 0.14 nmol/pineal/h at 1200 versus 24 ± 2.2 nmol/pineal/h at 0000). On the other hand, HIOMT (Figure 6F) did not vary significantly between the photo- and scotophase (13.62 ± 1.43 nmol/pineal/h at 1200; 18.05 ± 1.64 nmol/pineal/h) and irrespective of experimental groups. After bursectomy (Bx), the daytime levels of plasma ACTH and CORT. (Figure 6A, B) were no longer significantly different from nighttime ones (ACTH: 38 ± 0.46 to 47 ± 2.82 pg/ml; CORT: 4.24 ± 0.32 to 4.44 ± 0.29 ng/ml). Persistent diurnal rhythms were noticed in plasma MLT (5.7 ± 0.38 pg/ml at 1200 versus 65.77 ± 2.78 pg/ml at 0000) as well as NAT activity (0.73 ± 0.07 nmol./pineal/h at 1200 versus 5.13 ± 0.28 nmol./pineal/h at 0000), even though reduced by 50% in magnitude. In our attempt to reverse the effects of bursectomy, different amounts of bursin were administered in the yolk-sack of 6 and 9 days old bursectomized embryos.

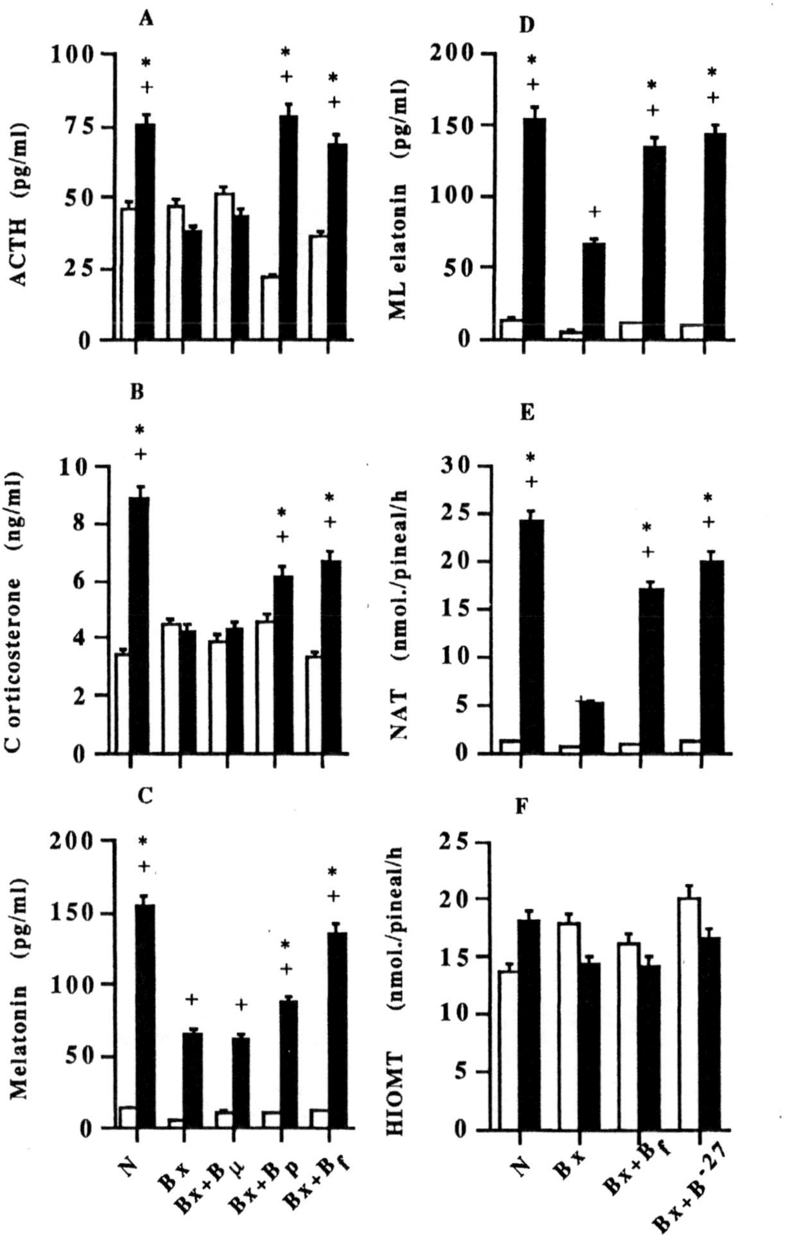

Figure 6. Pituitary (A), adrenal (B) and pineal (C, D, E, F) daily activities measured at daytime (white columns) and at night (dark colmuns). Six groups of chickens (N, Bx, Bx+Bμ, Bx+BP, Bx+Bf, Bx+B-27) were submitted to a 12L-12D light-dark cycle. Blood samples were collected during daytime (1200-1300) and at night (0000-0100) to measure plasma ACTH (A), corticosterone (B) and melatonin (C, D).The pineal glands were assayed for enzymatic activities of NAT (E) and HIOMT (F). + $P<0.01$ vs 12.00; * $P<0.01$ vs Bx.

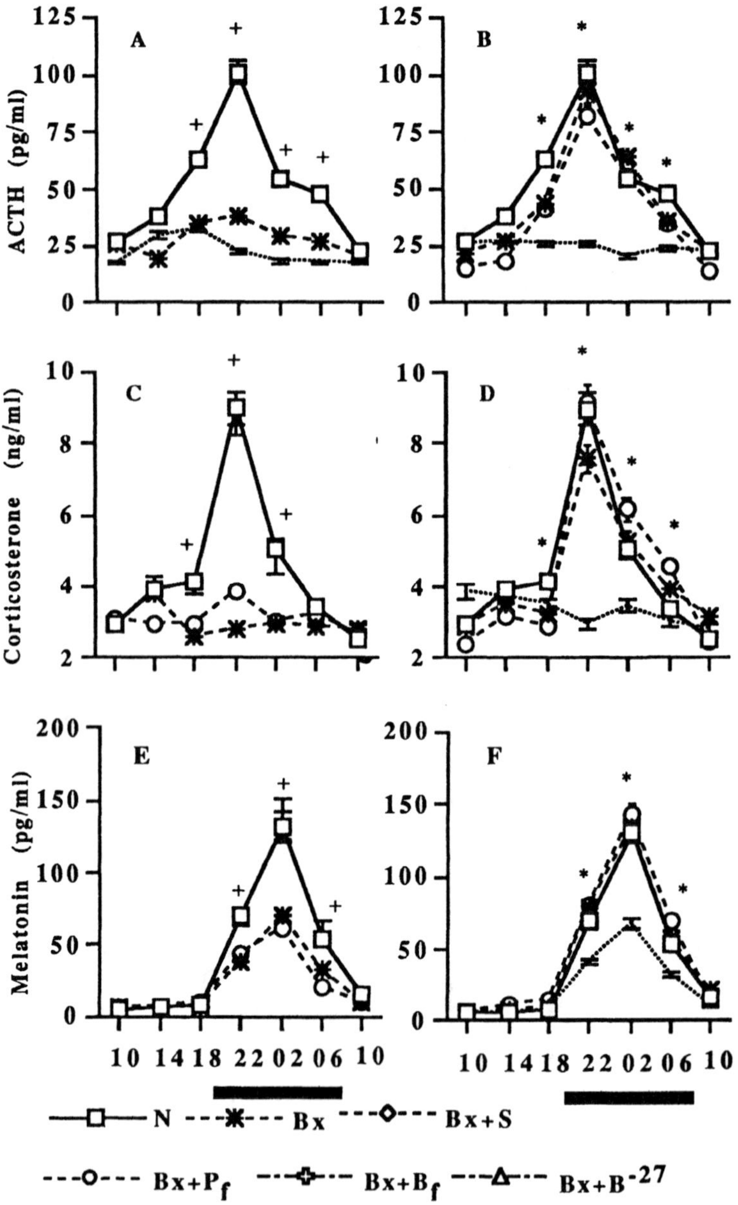

Figure 7. Circadian rhythms of the pituitary (A, B), adrenal (C, D) and pineal (E, F) hormones. The 7 groups of animals listed above were submitted to a 12L-12D alternating light-dark cycle, with lights off from 0700 to 1900 (black bar below the figures) .The plasma samples were assayed for ACTH (A, B), corticosterone (C, D) and melatonin (E, F). +P<0.01vs Bx; *P<0.01 vs Bx+S.

As results, the highest dose of bursin (100 µg) could not induce the recovery of the daily rhythms of plasma ACTH and CORT, whose levels remained steadily low and close to values observed in Bx animals (Fig. 6A, B). Likewise, 100 µg of bursin could no longer restore the normal amplitude of plasma MLT (Figure 6C). Conversely, lower amounts of bursin (5 CH, 7 CH, 15-20 CH pool) elicited significant daily hormonal and enzymatic rhythms, in an inverse dose-dependent manner (Figure 6A, B, C, D).
The hormonal parameters were also assessed every 4 h during a single light-dark cycle (Figure 7). Bursa-intact birds displayed pronounced circadian rhythms of plasma ACTH (Fig. 7A, B), CORT (Figure 7C, D) and MLT (Figure 7E, F), phase-locked with photoperiod and reaching a sharp peak at midscotophase. Bursectomy completely abolished circadian rhythms of the pituitary-adrenal hormones (Figure 7A, C), whereas MLT rhythm persisted, though markedly reduced in height (Figure 7E). Once more, neither the control tripeptide (Figure 7A, C, E) nor the saline (Figure 7B, D, F) could correct the effects of bursectomy. Administration of 100 µg of bursin has been reported also harmless in doing so (Youbicier-Simo, 1990). On the other hand, lower amounts of bursin (5 CH, 7 CH, 15-20 CH pool) allowed restoration of quite normal hormonal rhythms, in an inverse dose-dependent manner (Figure 7B, D, F).

3. 4. SERIAL DILUTIONS OF ^{3}H-THYMIDINE

In the fourth protocol, we demonstrated that in sequentially dilute ^{3}H-Thymidine solutions, the radioactivity decayed linearly, without showing any difference between successed and unsuccussed solutions (table 2).

TABLE 2: Quantification of ^{3}H-Thymidine

Concentrations (Mole/l)	cpm		
	Unsuccussed	Successed	Successed solvent
10^{-7}	1554340 ± 13317	1542857 ±23444	00
10^{-9}	16101± 221	15901 ± 380	00
10^{-11}	52 ± 11	153 ± 7	00
10^{-13}	16 ± 0.6	28 ± 3	00
10^{-15}	00	00	00
10-17 to 10-41	00	00	00

^{3}H-Thymidine (cpm) was quantified in different concentrations of succussed thymidine solutions ranging from 10^{-7} to 10^{-41} M and compared to unsuccussed thymidine and solvent. The data represent means ± SEM of three independent experiments. As result, a plummet in ^{3}H-Thymidine concentration was recorded and no residual tritiated compound could be detected above 10^{-15} M.

4. Discussion

4.1. BURSECTOMY AND EMBRYONIC MORTALITY

Our experimentation lasted 6 years during which more than 10.000 chicken embryos have been studied. In the sham-bursectomized groups, fetal loss (5-10%) was essentially restricted to the end of the incubation period. A possible explanation to this time-limited mortality is that the end of incubation corresponds to the preparation of hatching and represents a critical period during which the conditions of an extremely severe stress are gathered, weakening the fetus (Bauman and Bauman, 1977). Besides, two main critical periods characterized by a high death rate were identified during the development of the bursectomized chicken embryo: 20 to 30% of the operated embryos died between the day of bursectomy (4 th day of incubation) and the first day of *in ovo* treatment (6 th day of incubation), whereas 40 to 65% mortality occurred around hatching. A third critical phase arised during the couple of weeks following hatching, due to either cloacal malformations or immunodepression (Belo *et al.*, 1985). Finally, we recorded 8-10% survival two weeks after hatching. Performing surgical bursectomy at 60 h of embryonic life, Fitzsimmons *et al.* (1973) reported 6% survival after hatching; Belo *et al.* (1985) and Corbel *et al.* (1987) reported 5 and 7 % survival respectively, operating on 5 day-old chicken embryos. Grafting of 9.5 day-old chicken bursae (period of colonization of the bursa by B stem cells) does not improve the survival of bursectomized recipients (Ramade *et al.*, 1985; Abdul-Karim *et al.*, 1987), whose mortality rather got worse after the twofold administration (6 th and 9 th days of incubation) of bursin in the yolk sack. Finally, embryo handling appears as the prominent factor of mortality.

4.2. BURSECTOMY AND HUMORAL IMMUNITY

In order to evaluate the physiological consequences of both embryonic bursectomy and *in ovo* treatment with bursin, we tested the ability of bursectomized and bursin-treated bursectomized chickens to respond immunologically and hormonally to some environmental stimuli such as ether vapour, antigen challenge and photoperiod.
Control birds exhibited strong and increasing immunoglobulin production (IgG) during the course of immune response. These antibodies were specific to porcine Tg, since they reacted with neither self-chicken proteins (OVA, ACT, MYO) nor foreign antigens (INS, BSA). Unlike their bursa-intact counterparts, bursa-lacking chickens failed to raise specific anti-Tg antibodies, in spite of repeated immunization. These data are consistent with earlier reports indicating that bursa-lacking animals are able to produce immunoglobulins of the IgM, IgG and IgA classes, but fail to mount specific antibody responses against various antigens, despite iterative immunization (Jankovic *et al.*, 1977; Granfors *et al.*, 1982; Jalkanen *et al.*, 1983). This lack of specificity is correlated with decreased number of Ig bearing cells: IgG positive cells in the spleen, thymus and blood (Jalkanen *et al.*, 1984). The finland team of Toivanen also demonstrated that the number of B-cell clones is strikingly reduced in bursectomized chickens (Mansikka *et al.*, 1990). This oligoclonality of the B-cell compartment is paralleled by a 10 fold decrease in serum IgG titres, IgM and IgA levels remaining unchanged (Jalkanen *et al.*, 1983). It is worth noting that our results match these data, at least as far as the serum titres of IgG are concerned. By contrast, in bursectomized chickens, the IgM titres

assessed by us appeared rather lessened. Nevertheless, our analysis must be shaded, because the level of Ig depends on age: in fact, if the levels of IgM, IgG and IgA are markedly low in the 10 day-old bursectomized chickens, the latter recover normal Ig titres by the age of 10 weeks (Eerola *et al.*, 1983). Now, our measurements were performed half-way (3 to 7 weeks), between 10 days and 10 weeks. Evidence has been presented that at protein level, the isoelectric spectrum of immunoglobulins derived from bursectomized chickens displays normal gamma chains, but altered light chains, due to predominance of basic versus deficit in acid amino acids, which results in reduced antigen-antibody binding affinity (Granfors *et al.*, 1982, Jalkanen *et al.*, 1984). In addition, a low Ig gene conversion rate (V-J and V-D-J gene rearrangements) leading to a poorly diversified immunoglobulin repertoire has been reported in bursectomized chickens (Mansikka *et al.*, 1990). Collectively, these data suggest that the bursal microenvironment plays a crucial role in the elaboration and diversification of the specific antibody repertoire. These bursa-dependent events might occur very early during embryonic life, probably at the beginning of the colonization of bursa follicles by B precursors cells (Weill and Reynaud; 1986, Reynaud *et al.*, 1992). The bursa anlage develops on the 5 th day of embryonic life and is invaded by a single wave of B stem cells between the 8 th and the 14 th day of incubation (Houssaint *et al.*, 1976; Lassila *et al.*, 1978; Le Douarin *et al.*, 1985). We performed surgical bursectomy at 80 h of embryonic development, so as to obtain complete and permanent B-deficient animals.

4.3. BURSECTOMY AND IMMUNO-NEUROENDOCRINE RESPONSIVENESS

If the bursa of Fabricius is undoubtlessly a key organ of the B immune component in Birds, it has been for a long time suspected to also serve various neuroendocrine functions: reduced CORT levels and enlarged adrenals, have been observed in newly-hatched chicks bursectomized at 68 h of embryonic life (Pedernera *et al.*, 1980); Bursectomy performed in 2 week-old chicks increases the adrenal ascorbic acid response to ACTH (Perek and Eckstein, 1959, Perek and Eilat, 1966), whereas the hyperglycemic response following intramuscular injection of ACTH is lessened (Freeman, 1971); CORT levels increase after bursectomy (Mashaly, 1984). Analogous results have been previously obtained in our Laboratory. We reported that early surgical bursectomy (4 th day of embryonic life) results in various alterations in pituitary and adrenocortical functioning: ether stress-induced ACTH and CORT stimulations can be detected more precociously in bursectomized than in bursa-intact embryos; the non-stress responsive period of newly-hatched bursa-intact birds does not arise in bursa-lacking chicks (Guelatti, 1990; Guelatti *et al.*, 1991); young bursa-lacking chickens exhibit a smaller CORT response to stress than controls (AbdulKarim *et al.*, 1987); the delayed pituitary-adrenal responses of adenohypophysectomized chickens to Brucella abortus does not occur when embryonic bursectomy preceedes hypophysectomy (Baylé *et al.*, 1991). In line with our previous results, we now present evidence of inability for bursa-deprived young chickens to cope with ether stress.

Thyroglobulin (Tg) which was used by us to sensitize the chickens is also the endogenous precursor of thyroid hormones (Teppermann and Teppermann, 1987), and the latter are correlated to CORT during the course of immune response (Besedovsky *et al.*, 1975; Trout *et al.*, 1988). Unlike Tg which is a bioactive compound, KLH is a T-dependent antigen devoid of any pharmacological activity. We found that bursa-intact chickens sensitized to porcine Tg displayed the same profiles of pituitary-adrenal and humoral immune responses as their KLH-treated counterparts. The same feature of MLT response was recorded when immunizing bursa-intact chickens against porcine Tg or KLH (data not shown). Accordingly, it can be reasonably inferred that the changes of ACTH, MLT and CORT levels subsequent to porcine Tg treatment were not modulated by thyroid hormones derived from the sensitizing agent (porcine Tg), but were strictly linked to the immunization phenomenon. The fact that porcine Tg failed to induce noticeable endocrine answers in bursa-lacking chickens supports this point of view.

Therefore, the physiological meaning of the concomitance of hormonal and antibody responses to porcine Tg should be discussed in the context of two-way communication between the immune and neuroendocrine systems. The hormonal response to antigen challenge is always contemporaneous with humoral immunity: in rats or mice primed intraperitoneally with three antigens, concomitant peaks of CORT and antibody responses occurred 5 to 6 days after immunization (Besedovsky *et al.*, 1975); in the same way, both responses crested simultaneously 3 h after intravenous administration of *Brucella abortus* to immature chickens (Trout *et al.*, 1988). We recorded analogous pituitary-adrenal and IgG responses when immunizing chickens against porcine Tg: ACTH and CORT levels, as well as IgG titers crested 18 days after prime immunization (d38). It is worth noting that in the aforementioned studies (Besedovsky *et al.*, 1975; Trout *et al.*, 1988), the delay of onset between hormonal and antibody responses was short (several minutes to three days). This was not the case in our experiments: the antibody response arose earlier (d29) than the hormonal response which occurred only 9 days later (d38). This discrepancy probably comes from methodological differences, leading to specific response kinetics. In the experiments of Besedovsky *et al.* (1975) or Trout *et al.* (1988), the antigens were injected free of adjuvant and directly into the general circulation (intraperitoneally or intravenously); hence they were rapidly conveyed into the secondary lymphoid organs to induce antibody production and cytokine release. In our study, not only porcine Tg was mixed with either complete (prime immunization) or incomplete (second and third immunizations) Freund adjuvant, but it was applied subcutaneously. As a result, the antigen molecules diffused slowly and had to pass through the lymphatic circulation before reaching the site of immune response. Some cytokines such as IL-1 or IL-2 have potential to increase blood glucocorticoid level via stimulation of the HPA axis (Glick, 1984b; Besedovsky *et al.*, 1981; Lilly and Gann, 1992; Hu *et al.*, 1993). Considering above arguments and despite induction of humoral immunity at d29, it is likely that the level of cytokines reached at d29 was not high enough to feedback on the HPA axis and elicit noticeable ACTH and CORT responses. Accordingly, we assume that after prime immunization, these responses had been initiated later than d29, at a moment when the effective concentration of cytokines was attained. A few days after subcutaneous administration of porcine Tg, we observed inflammatory nodules at the site of immunization. Evidence has been presented suggesting that cytokines released by inflammatory cells play a major role in HPA axis immunoregulation (Lilly and Gann, 1992; Besedovsky and Del Rey, 1987). Therefore, an alternative explanation to the lack of endocrine responses at d29 is that these responses were missed, because they had been induced earlier than d29 by inflammatory cytokines.

Another distinctive feature of our results comes from the fact that endocrine and antibody responses were associated to repeated rather than unique immunization as reported by other authors (Besedovsky *et al.*, 1975; Trout *et al.*, 1988). Especially, CORT crested at d38, whether this sampling point was assessed after the second (experiments 2, 3, 4, 5) or the third (experiment 6) antigen challenge. Moreover, we noticed either a decline (experiment 2, 3, 4, 5) or a peak (experiment 6) of CORT following the third immunization. Also, the CORT level measured at d36 was half-way between the values at d29 and d38. For these reasons, the hormonal profiles might depend, not on iterative immunization, but on the time elapsed after prime immunization. This explanation seems more in agreement with the assumption of retarded arousal of ACTH and CORT responses (after d29). The concomitance of the peaks of CORT and IgG (d38) can be intrepreted as part of an immuno-neuroendocrine regulatory loop: the activation of B-cells, in conjunction with T-cell activation, induces the release of cytokines which increase glucocorticoid level via the HPA axis (Besedovsky *et al.*, 1981; Lilly and Gann, 1992; Hu *et al.*, 1993); this is a possible explanation to the peaks of ACTH and CORT observed at d38. This circuit is enhanced by MLT (Maestroni, 1993) which have been demonstrated by us to crest contemporaneously with pituitary-adrenal hormones. In turn, increased glucocorticoid levels inhibit MLT rise (Besedovsky *et al.*, 1975; Poon *et al.*, 1994), and also impede cell-mediated immunity, leading to both reduction of cytokine levels and decline of CORT level (Lilly and Gann, 1992; Hu *et al.*, 1993; Besedovsky *et al.*, 1985) as observed by us at d47. Finally, iterative immunization seemingly played the role of amplifying endocrine and immune responses.

However, the HPA axis and the pineal gland not only mediate adaptation to nociceptive situations, but they are also sensitive and responsive to light stimuli which acts as zeigteber on biological rhythms in most Vertebrates (Binkley, 1993). Therefore, we wondered if the bursa of Fabricius possibly influences the circadian rhythms of both HPA and pineal gland in the chicken.

4.4. BURSECTOMY AND CIRCADIAN RHYTHMS OF HPA AXIS AND PINEAL GLAND

The circadian activity of the pituitary-adrenal axis is governed by the rhythmic activity of CRF-containing neurons of the hypothalamus (Peczely and Antoni, 1984; Jozsa *et al.*, 1984). These CRF-neurons project in the median eminence, in the vicinity of pituitary portal capillaries where CRF is rhythmically released (Peczely and Antoni, 1984; Jozsa *et al.*, 1984) and triggers the circadian secretion of ACTH and downstream CORT. In the present study, bursa-intact chickens exhibited broad circadian rhythms of ACTH and CORT under a 12L-12D light-dark cycle. Some reports indicate that lesioning the hypothalamic paraventricular nucleus in the rat (Makara *et al.*, 1981) or the suprachiasmatic nucleus in the pigeon (Bouillé and Baylé, 1973) abolishes the circadian variations of plasma ACTH and CORT. Now the withdrawal of the bursa anlage from 4 day-old chicken embryos was also accompanied by the suppression of the daily rhythms of plasma ACTH and CORT in young chickens, despite exposure to a 12L-12D light-dark regime.

The daily synthesis and release of MLT is catalyzed *in vivo* by two key enzymes, NAT and HIOMT (Binkley *et al.*, 1973). In 8 week-old chickens raised under 12L-12D light-dark cycle, pineal NAT and MLT night-time levels are respectively 27 and 10 fold higher than light-time values, whereas HIOMT activity varies by 20 % only (Binkley *et al.*, 1973). Likewise, ocular and blood MLT levels display prominent daily rhythms with a midscotophase crest (Cogburn *et al.*, 1987). Our control chickens exhibited similar patterns of enzymatic and hormonal rhythms: night/daytime ratios were 20, 8, 2 and 10 for pineal NAT activity, pineal MLT content, and ocular and circulating MLT, respectively. Little change occurred in HIOMT activity, probably due to a slow turnover of this enzyme (Voisin *et al.*, 1988). Very striking is the observation that in bursectomized chickens, the magnitude of NAT and MLT rhythms was markedly reduced, in spite of persistent cyclicity. Phototransduction is anatomically supported by a complex network comprising several elements: photoreceptors (retinal, hypothalamic, pineal) perceive light information and convey it either directly or via the suprachiasmatic nucleus (SCN) to the pineal gland where the message is converted into MLT (Binkley, 1993). The visual suprachiasmatic nucleus (equivalent of mammalian SCN) is connected to the pineal gland through sympathetic catecholaminergic fibers and governs the circadian release of noradrenaline in the vicinity of pinealocytes (Cassone *et al.*, 1990). Noradrenaline turn-over is higher during the photo- compared to scotophase (Cassone *et al.*, 1990). The activation of α-2 adrenergic receptors on pinealocytes upon daytime increase in noradrenaline level inhibits the synthesis of NAT molecules and, therefore, MLT production, via an intracellular increase of cAMP (Voisin and Colin, 1986; Bylund *et al.*, 1988). Since the amplitude of pineal NAT activity and MLT production is controlled by the rhythmic delivery of noradrenaline, itself driven by the circadian function of the vSCN (Cassone *et al.*, 1990), the vSCN-SCG-pineal complex appears as the most probable locus of interferences between bursal signals (or the absence of bursal signals) and pineal rhythms. This hypothesis is strengthened by the fact that the excision of the bursa anlage (4 th day of embryonic life) preceeded the beginning of pineal sympathetic innervation which occurs by day 20 of embryonic life (Wight, 1971), as well as the onset of pineal NAT and MLT rhythms which are observed in 18 day-old embryos (Binkley and Geller, 1975). Similarly, the chronology of the onset of bursal and pituitary-adrenal functions might help understanding how the precocious influence of the bursa of Fabricius on pituitary-adrenal functioning occurs: the bursa anlage arises by the third day in the embryo (Toivanen *et al.*, 1981); it was excised (4 th day of embryonic life) prior to the onset of the HPA axis functioning which in the developing chicken embryo occurs during the second half of embryonic phase (Scanes *et al.*, 1987). Therefore, it is likely that as a result of precocious bursectomy, the normal development of pituitary-adrenal activities was hindered, due to the lack of adequate inductive bursa cue (s).

However, the exact locus of the interferences between bursa signals and both the HPA and the pineal gland have yet to be identified. The studied systems (bursa of Fabricius, HPA and pineal gland) arise during the embryonic phase which in chickens is also the critical period of sexual differentiation of the brain (Wilson and Glick, 1970). Accordingly, the mechanisms of brain development are likely to be involved in the functional maturation of bursa-pituitary-adrenal-pineal interrelationships. In fact, during the critical period, sex steroids can imprint parts of the brain that are associated with the cyclic control of hormonal release in such a way as to prevent their rhythmic function (Pilgrim *et al.*, 1994). Since embryonic bursectomy leads to significantly increased plasma testosterone levels (Pedernera *et al.*, 1980; Ramade *et al.*, 1986), such a rise might affect the cyclic activity of hypothalamic pacemakers. Among the latter is the SCN neurally coupled to CRF-containing nuclei which govern the circadian release of ACTH and CORT (Peczely and Antoni, 1984; Jozsa *et al.*, 1984) and to the pineal gland which secretes MLT in a cyclic manner (Binkley, 1993). As working hypothesis, it can be assumed that similar mechanisms occur in bursectomized embryos in the absence of specific bursal cue (s); in intact embryos, such signal (s) must normally prevent the unsettling of the hypothalamic and pineal pacemakers.

These dyschrogenic effects emphasize the crucial role held by signals originating in the bursa microenvironment in the ontogeny and functioning of immune, HPA and pineal activities. Further support to this assertion comes from the fact that bursectomized chickens embryonically grafted with 9 day-old embryonic bursae recover normal immune responsiveness and normal levels of plasma CORT (Abdul-Karim *et al.*, 1987) and testosterone (Pedernera *et al.*, 1980). In this regards, we demonsrated that the bursa-derived signal termed bursin is an important mediator of bursa-dependent functions.

4.4. BURSIN AS MEDIATOR OF BURSA-DEPENDENT FUNCTIONS: EFFECTIVENESS OF HIGHLY DILUTE BURSIN

Differents concentrations of bursin were tested, aimed at reversing the effects of bursectomy. *In ovo* administration of very low doses (5 CH, 7 CH) of bursin to bursectomized chickens allowed recovery of the ability to cope with ether stress, antigen challenge and restored normal pituitary-adrenal and pineal rhythmicity. Chorio-allantoic grafting of bursal rudiment from 9.5-day-old donor embryos has been reported to alleviate gonadal (Pedernera *et al.*, 1980) and adrenal (Abdulkarim *et al.*, 1987) changes caused by bursectomy. Also bursa extracts have ben reported to restore specific humoral immunity in bursectomized chickens (Baba and Okuno, 1976). Kuznik *et al.* (1988) demonstrated that peptides from bursal origin correct both immunity and homeostasis in bursectomized chickens. Increasingly, the data strengthen the assumption of specific inductive properties for minute doses of bursin on immune and neuroendocrine functions during embryonic life: native as well as synthetic bursin (Lys-His-Gly-NH_2) stimulates an intracellular increase of cAMP and cGMP in the human Daudi B cell line, at threshold concentrations of 0.1 to 1 μg/ml (Audhya *et al.*, 1986). On the other hand, either the inverse sequence of bursin (Gly-His-Lys-NH_2) or other variants of this tripeptide (Gly-His-Lys; Lys-His-Gly) appear poorly inductive, and only when administered in massive amounts (Audhya *et al.*, 1986; Lassila *et al.*, 1989), which makes this action physiologically irrelevant. Using either the saline or a control tripeptide with unknown biological activity (Trp-Leu-Leu-NH_2), we could not overcome the effects of bursectomy; *in ovo* administration of 100 μg of bursin was no longer effective. Hence, bursin worked in an inverse dose-dependent manner, the lower doses being the most effective. Underlying mechanisms remain unraveled. Bursin triggers intracellular increase of cAMP and cGMP (Goldstein *et al.*, 1977; Audhya *et al.*, 1986). Like homologous tripeptides (Gly-His-Lys, His-Gly-Lys) known to enhance cell growth, bursin contains histidine which binds to and promote copper uptake in a variety of cells (Pickart and Thaler, 1973; Pickart, 1981). Bursin is likely a growth factor (Lassila *et al.*, 1989).

Much more striking is that highly dilute bursin (15-20 CH pool) improved the performances of bursectomized chickens, to the same extent as did 7 CH of bursin. Highly dilute bursin seemed to transmit specific "information" to bursectomized recipients. Bursin's specific inductive properties were ascertained by replicating the data pertaining to the immunization protocol: the previous results (Youbicier-Simo *et al.*, 1993) were exactly reproduced (Youbicier-Simo *et al.*, 1996b). Furthermore, highly dilute bursin wholly restored normal circadian rhythms of the pituitary-adrenal axis (Youbicier-Simo, 1994; Youbicier-Simo *et al.*, 1996b) and pineal gland (Youbicier-Simo, 1994; Youbicier-Simo *et al.*, 1996a). Another striking observation is that the pharmacological effect induced by the saline in saline-supplied chickens (N+S) was suppressed after embryonic bursectomy: in figure 4 B, the bursa-intact chickens supplimented with the saline (N+S) exhibited a slightly higher CORT response than their untreated counterparts; on the other hand, this effect did not occur when bursa-lacking chickens (Bx) received the saline (Bx+S). How do highly dilute solutions work remains enigmatic, since they are theoretically devoid of active material compounds. It has been shown by nuclear magnetic resonance that the structure of high dilutions is different from that of solvents (Demangeat *et al.*, 1992). Several theories have been put forward, but a deciding factor of the efficiency of high dilutions seems to lie in the way such media are prepared, using sequential centesimal dilution steps with potent vertical stirring between successive steps. We demonstrated that in sequentially diluted ^{3}H-thymidine solutions, radioactivity decayed linearly, without showing any difference between successed and unsuccussed solutions (table 2). This testifies to a homogeneous molecular dispersion as result succussion, therefore precluding the assumption of heterogeneous molecular distribution sustained by the theory of clusters. We suggest that a non-molecular specific information signal corresponding to the original molecule (bursin) is conveyed to the recipient organism (Bastide and Lagache, 1992). Indeed, the activity of high dilutions is sensitive to electromagnetic fields (Hadji et al., 1992) and had been transferred to unshaken water through a coil (Endler *et al.*, 1995) and may be probably electromagnetically transmitted. Currently, an increasing number of experiments aimed at substantiating the biological activities of high dilutions are under way, and some reports are compelling (Bastide *et al.*, 1985; Bastide *et al.*, 1987; Daurat *et al.*, 1988; Jacobs *et al.*, 1994; Reilly *et al.*, 1994; Bastide and Boudard, 1995).

Finally, using a relevant experimental model, we have provided evidence that a compound found in the bursal microenvironment can functionally mimic *in vivo* some biological functions of the bursa Fabricii. We assert that in chickens, the bursa-derived signal (bursin) which is probably an ontogenic organizer, mediates early bursa-dependent functions. The originality and novelty of the present work lie in the underscoring of the inverse dose-dependent effectiveness of high dilutions, in contrast with the classical dose-dependent effect which underlies the molecularist theory. We introduce a new conceptual trend of the information processing termed the "paradigm of signifiers"; the latter is rather based on the encoding of the "specific information" contained in bioactive compounds by highly dilute media.

Aknowledgement.
The present work was supported by grants from Dolisos Laboratories (Paris).

5. References

Abdul-Karim, A.K., Ramade, F., and Baylé, J.D. (1987) Development of basal and stress-induced adrenocortical function in early bursectomized embryos and chickens, *Life Sci. Adv.* **6**, 121-128.
Audhya, T., King, R., and Goldstein, G. (1991) Bovine probursin tetradecapeptide contains amino acid sequence from somatostatin, tuftsin and bursin, *Life Science* **48**, 773-780.
Audhya, T., Kroon, D., Heavner, G., Viamontes, G., and Goldstein, G. (1986) Tripeptide structure of bursin, a selective B-cell differentiating hormone of the bursa of Fabricius, *Science* **231**, 997-998.
Audhya, T., Viamontes, G., Babu, U., and Goldstein, G. (1990) Bursin localization in mammalian bone marrow and epithelial cells of intrahepatic bile ducts, *Scand. J. Immunol.* **31**, 199-204.
Baba, T. and Okuno, Y. (1976) Effect of bursa of Fabricius extracts on antibody production in bursectomized or bursal autografted chickens, *Immunology* **31**, 533-539.
Bastide, M. and Boudard, F. (1995) An alternative concept of immunomodulation, in M. Guenounou (ed.), *Forum on immunomodulators*, John Libbey Publisher, Paris, pp. 303-316.
Bastide, M. and Lagache, A. (1992) *The paradigm of signifiers*, Atelier Alpha Bleue Publisher, Paris.
Bastide, M., Daurat, V., Doucet-Jabœuf, M., Pelegrin, A., and Dorfman, P. (1987) Immunomodulatory activity of very low doses of thymulin in mice, *Int. J. Immunotherapy* **3**, 191-200.
Bastide, M., Doucet-Jaboeuf, M., and Daurat, V. (1985) Activity and chronopharmacology of very low doses of physiological immune inducers, *Immunol. Today* **6**, 234-235.
Bauman, F.H. and Bauman, R. (1977) Respiratory function of embryonic chicken hemoglobulin, **31**, 331-343.
Baylé, J.D., Guellati, M., Ibos, F., and Roux, J. (1991) *Brucella abortus* stimulates extra-pituitary adrenocorticotropic-adrenocortical function through B lymphoid system, *J. Prog. Neuro-endocrino-immunol.* **4**, 99-105.

Belo, M., Martin, C., Corbel, C., and Le Douarin, N.M. (1985) A novel method to bursectomize avian embryos and obtain quail-chick bursal chimeras. I. Immuynocytochemical analysis of such chimeras by using species-specific monoclonal antibodies, *J. Immunol.* ;**135**, 3785-3794.

Besedovsky, H.O. and Del Rey, A.E. (1987) Neuroendocrine and metabolic responses induced by interleukin-1, *J. Neurosci. Res.* **18**, 172.

Besedovsky, H.O., Del Rey, A.E., and Sorkin, E. (1981) Lymphokine-containing supernatants from Con A-stimulated cells increase corticosterone blood levels, *J. Immunol.* **126**, 385-387.

Besedovsky, H.O., Del Rey, A.E., and Sorkin, E. (1985) Immune-neuroendocrine interactions, *J. Immunol.* **135**, 750s.

Besedovsky, O.H. and Sorkin,E. (1977) Network of immune-neuroendocrine interactions, *Clin. Exp. I mmunol.* **27**, 1-12.

Besedovsky, O.H., Sorkin, E., Keller, M., and Müller, J. (1975) Changes in blood hormone levels during the immune response, *Proc. Soc. Exp. Biol. Med.* **150**, 466-470.

Binkley, S. (1993) Structures and molecules involved in generation and regulation of biological rhythms in Vertebrates and Invertebrates, *Experientia*, **49**, 648-653.

Binkley, S. and Geller, E.B. (1975) Pineal enzymes in chickens: Development of daily rhythmicity, *Gen. Comp. Endoc.* **27**, 424-429.

Binkley, S., Macbride, S.E., KLEIN, D.C., and RALPH, C.L. (1973) Pineal enzymes: Regulation of avian melatonin synthesis, *Science* **181**, 273-275.

Bouillé, C. and Baylé, J.D. (1973) Experimental studies on the adrenocorticotropic area in the pigeon hypothalamus, *Neuroendocrinology* **11**, 73-91.

Brand, A., Gilmour, D.G., and Goldstein, G. (1976) Lymphocyte-differentiating hormone of the bursa of Fabricius, *Science* **193**, 319-321.

Buckland R.B., Blagrave, K., and Lague, P.C. (1974) Competitive protein-binding assay for corticosteroids in the peripheral plasma of the immature chicken, *Poultry Science* **53**, 241-245.

Bylund, D.B., Rudeen, P.K., Petterborg, L.J., and Ray-Prenger, C. (1988) Identification of a-2 adrenergic receptors in chicken pineal gland using [^{3}H]Rauwolscine, *J. Neurochem.* **51**, 81-86.

Cassone, V.M., Forsyth, A.M., and Woodle, G.L. (1990) Hypothalamic regulation of circadian noradrenergic input to the chick pineal gland, *J. Comp. Physiol.* **167**, 187-192.

Cogburn, L.A., Wilson-Placentra, S., and Letcher, L.R. (1987) Influence of pinealectomy on plasma and extraretinal melatonin rhythms in young chickens *(Gallus domesticus)*, *Gen. Comp. Endocrinol.* **68**, 343-356.

Corbel, C., Belo, M., Martin, C., and Le Douarin, N.M. (1987) A novel method to bursectomize avian embryo and obtain quail-chick chimæras. II. Immune response of bursectomized chicks and chimæras and post-natal rejection of the grafted quail bursa, *J. Immunol.* **138**, 2813-2821.

Daurat, V., Dorfman, P., and Bastide, M. (1988) Immunomodulatory activity of low doses of interferon α, β in mice, *Biomed. and Pharmacother.* **42**, 197-206.

Demangeat, J.L. Demangeat, C., Gries, P., Poitevin, P., and Constantinesco, A. (1992) Modification du temps de relaxation RMN à 4MHz des protons du solvant dans les très hautes dilutions salines de silice/lactose, *J. Med. Nucl. Biophys.* **16**, 135-145.

Eerola, E., Lassila, O., Gilmour, D.G., and Toivanen, A. (1983) Characteristics of chicken intraembryonic cells that express B-L (Ia-like) antigen under the influence of cultured bursal epithelium, *Scand. J. Immunol.* **18**, 175-183.

Endler, P.C., Pongratz, C.W., Smith, C.W., and Schulte, J. (1995) Non-molecular information transfer from thyroxine to frogs with regard to homeopathic toxicology, *Vet. Hum. Tox.* **37**, 259-260.

Fitzsimmons, R.C., Garrod, E., and Garnett, I. (1973) Immunological response following early embryonic surgical bursectomy, *Cellullar Immunology* **9**, 377-383.

Freeman, B.M. (1971) Stress and the domestic fowl, *J. World Poultry Science* **27**, 263-275.

Glick, B. (1984a) Interrelation of the avian immune and neuroendocrine systems, *The J.Exper.Zool.* **232**, 671-682.

Glick, B. (1984b) Ontogeny and microenvironment of the avian thymus and bursa of Fabricius with emphasis on the contribution of specialized cells to the avian immune response, *Adv. Vt. Sci. Comp. Med.* **30**, 67-97.

Goldstein, G., Sheid, M., Boyse, E.A., Brand, A., and Gilmour, D.G. (1977) Thymopoietin and bursopoietin: induction signals regulating early lymphocyte differentiation, in *Origins of Lymphocyte Differentiation* **41**, Cold Spring Harbor Symposium, pp. 5-8.

Granfors, K., Martin, C., Lassila, O., Suvitaival, R., and Toivanen, P. (1982) Immune capacity of the chicken bursectomized at 60 h of incubation. Production of the immunoglobulin and specific antibody, *Clin. Immunol. Immunopathol.* **23**, 459-463.

Guellati, M. (1990) Les relations immuno-corticotropes: Effets de la bursectomie embryonnaire précoce et du traitement substitutif par la bursin,*Thèse de diplôme de Doctorat*, Université de Montpellier II. pp. 75.

Guellati, M., Ramade, F., Le Nguyen, D., Ibos, F. and Baylé, J.D. (1991) Effects of early surgical bursectomy and opotherapic substitution on the functional development of the adrenocorticotropic axis, *J. Dev. Physiol.* **15**, 357-363.

Hadji, L., Arnoux, B., and Benveniste, J. (1992) Effect of dilute histamine on coronary flow of guinea-pig isolated heart. Inhibition by a magnetic field, *FASEB J.* **5**, 1583-1586.

Hamm, H.E. and Manaker, M. (1980) Retinal rhythms in chicks: circadian variation in melatonin and N-acetyltransferase activity, *Proc. Natl. Acad. USA* **77**, 4498-5002.

Heller, D.E. and Friedman , A.R. (1979) The effects of crude bursa of Fabricius extracts on the humoral immune response and its recovery in bursectomized chickens, *Dev. Comp. Immunol.* 3, 667-673.
Houssaint, E., Belo, M., and Le Douarin, N.M. (1976) Investigations on cell lineage and tissue interactions in the developing bursa of Fabricius through interspecific chimeras, *Dev. Biol.* 53, 250-264.
Hu, Y., Dietrich, H., Herold, M., Heinrich, P.C., and Wick, G. (1993) Disturbed immuno-endocrine communication via the hypothalamo-pituitary-adrenal axis in autoimmune disease, *Int. Arch. Allerg. Immunol.* 102, 232-.
Jacobs, L., Jimenez, M., Glowd, S.S., Gale, G. L., and Crothers, D. (1994) A randomized clinical trial in Nicaragua: Treatment of acute childhood diarrhea with homoepathy, *Pediatrics* 5, 719-726.
Jalkanen, S., Granfors, K., Jalkanen, M., and Toivanen, P. (1983) Immune capacity of chichen bursectomized at 60 hours of incubation. Failure to produce immune, natural and autoantibodies in spite of immunoglobulin production, *Cell Immunol.* 80, 363-373.
Jalkanen, S., Korpela, R., Granfors, K., and Toivanen, P. (1984) Immune capacity of the chicken bursectomized at 60h of incubation: cytoplasmic immunoglobulins and histological findings, *Clin. Immunol. Immunopathol.* 30, 41- 50.
Jankovic, B. D., Isakovic, K., Markovic, B.M., and Rajcevic, M. (1977) II. Immunological capacity of the chicken embryo. Humoral immune responses in embryos and young chickens bursectomized and sham-bursectomized at 52-64 hours of incubation, *Immunol.* 32, 689-697.
Jozsa, R., Vigh, S., Shally, A.V., and Mess, B. (1984) Lcalization of corticotropin-releasing factor-containing neurons in the brain of the domestic fowl, *Cell Tissue Res.* 236, 245-.
Kuznik, B.I., Stepanov, A.V., Tsybikov, N.N., and Khavinsov, V.G. (1988) Correction of immunity and homeostasis by peptides from the bursa of Fabricius and bone marrow of embryonically bursectomized chicks, *Farmacologiia i Toksykologiia* 5, 53-55.
Lassila, O., Eskola, J., Toivanen, P. Martin, C., and Dieterlen -Lievre, F. (1978) The origin of lymphoid stem cells studied in chick yolk sack embryo chimaeras, *Nature* 272, 353-354.
Lassila, O., Lambris, J.D., and Gisler, R.H. (1989) A role for Lys-His-Gly-NH_2 in avian and murine B cell development, *Cellular Immunology* 122, 319-328.
Le Douarin, N.M. , Houssaint, E., Jotereau, F., and Belo, M. (1975) Origin of hematopoietic stem cells in the embryonic bursa of Fabricius and bone marrow studied through interspecific chimeras, *Proc. Natl. Acad. Sci. USA* 72, 2701-2705.
Le Nguyen, D. (1987) Renin substract: II. Rapid solid phase synthesis of the ratine sequence tetradecapeptide using BOC reagent, *J. Chem. Soc. Perkins Trans.* I, 1915-1919.
Lilly, M.P. and Gann, D.S. (1992) The hypothalamic-Pituitary-Adrenal-Immune Axis, *Arch. Surg.* 127, 1463-1474.
Maestroni, G.J.M. (1993) The immunoendocrine role of melatonin, *J. Pineal Res.* 14, 1-10.
Maestroni, G.J.M., Conti, A., and Pierpaoli, W. (1989) Melatonin, stress and the immune system, *Pineal Res. Rev.* 7, 203-223.
Makara, G.B., Stark, E., Karteszi, M., Fellinger, E., Rappy, G., and Szabo, D. (1980) Effect of electrical stimulation of the neurohypophysis on ACTH release in rats with hypothalamic lesions, *Neuroendocrinology* 31, 237-
Mansikka, A., Jalkanen, S., Sandberg, M., Granfors, K., Lassila, O., and Toivanen, P. (1990) Bursectomy of chicken embryos at 60 hours of incubation leads to an oligoclonal B cell compartment and restricted Ig diversity, *J. Immunolol.* 145, 3601-3609.
Mashaly, M.M. (1984) Bursectomy and its influence on circulating corticosterone, triiodothyronine, and thyroxine in immature male chickens, *Poultry Science* 63, 798-800.
Peczely, P. and Antony, F.A. (1984) Comparative localization of neurons containing ovine Corticotrophin Releasing Factor (CRF)-like and neurophysin-like immunoreactivity in the diencephalon of the pigeon (*Columbia livia domestica*), *J. Comp. Neurol.* 228, 69-80.
Pedernera, E.A., Romano, M., Besedovsky, O.H., and Maria Del Carmen Aguilar (1980) The bursa of Fabricius is required for normal endocrine development in chicken, *Gen. Comp. Endocrinol.* 42, 413-419.
Pereck, M. and Eckstein, B. (1959) The adrenal ascorbic acid content of molting hens and the effect of ACTH on adrenal ascorbic acid content of laying hens, *Poultry Science* 38, 996-999.
Pereck, M. and Eilat, A. (1966) The bursa of Fabricius and acorbic acid depletion following ACTH injections in chicks, *J. Endocrinology* 20, 215-255.
Pickart, L. (1981) The use of GlyHisLys in culture systems, *In Vitro* 17, 459-466.
Pickart, L. and Thaler, M.M. (1973) Tripeptide in human serum which prolonges survival of normal liver cells and stimulated growth in neoplastic liver, *Nature New Biol.* 243, 85-87.
Pilgrim, C. and J.B. Hutchison. 1994. Developmental regulation of sex differences in the brain: can the role of gonadal steroids be redefined?, *Neurosci.*.60, 843.
Poon, A.M., Liu, Z.M., Tang, F., and Pang, S. F. (1994) Cortisol decreases 2[125 I] iodomelatonin binding sites in the duck thymus, *Eur. J. Endocrinol.* 130, *320-*
Ramade, F., Abdulkarim, A.K., and Baylé, J.D. (1985) Influence de la bursectomie précoce sur l'activité basale et la réponse au stres de la glande surrénale de l'embryon et du jeune poulet, 110è congrès national des Sociétés Savantes, Montpellier, Sciences Fasc. IV 67.
Ramade, F., Abdulkarim, A.K., Mas, N., and Baylé, J.D. (1986) Evolution de la testostéronémie chez l'embryon et le jeune poulet après bursectomie précoce, *C. R. Soc. Biol.* 180, 64-68.
Ramade, F., Malaval, F., and Baylé, J.D. (1985) Radioimmunoassay of plasma adrenocorticotropic hormone levels in the pigeon, *I. R. C. S. Med. Sci.* 123, 517-518.

Reilly, D., Taylor, M.A., Beattie, N.G.M., Campbell, J.H., McSharry, C., Aitchison, T.C., Carter, R., and Stevenson, R.D. (1994) Is evidence for homeopathy reproducible?, *The Lancet* **344**, 1601-1606.

Reynaud, C.A., Imhof, B.A., Anquez, V., and Weill, J.C. (1992) Emergence of committed B lymphoid progenitors in the developing chicken embryo, *EMBO J.* **11**, 4349-4358.

Rollag, M.D. and Niswender, G.D. (1976) Radioimmunoassay of serum concentrations of melatonin in sheep exposed to different lighting regimes, *Endocrinology* **98**, 482-489.

Scanes, C.G., Hart, L.E., Decuypere, E., and Kuhn, E.R. (1987) Endocrinology of the avian embryo: an overview, J. *Exp. Zool. Suppl.* **1**, 253-264.

Tepperman, J. and Tepperman H.M., (1987) The Male, in S.S Scott (ed.), *Metabolic and Endocrine Physiology.*, Year Book Medical Publishers, Inc. Chicago-London, pp. 103.

Toivanen, A., Toivanen, P., Eskola, J. and Lassila, O. (1981) Ontogeny of the chicken lymphoid system, in M. E. Rose, L. N. Payne, and B. M. Freeman (eds.), *Avian Immunology*, British Poultry Science Ltd, Edinburgh, pp. 45-62.

Trout, J.M., Mashaly, M.M., and Siegel, H.S. (1988) Changes in the profiles of circulating white blood cells, corticosterone, T_3 and T_4 during the initiation of humoral immunity in immature male chickens, *Dev. Comp. Immunol.* **12**, 331.

Viamontes, G.I., Audhya, T.K., Babu, U., and Goldstein, G. (1989) Immunohistochemical localization of bursin in epithelial cells of the avian bursa of Fabricius, *J. Histochem. Cytochem.* **6**, 793-799.

Voisin, P. and Colin, J.P. (1986) Regulation of chicken pineal arylalkylamine-N-Acetyltransferase by postsynaptic α2-adrenergic receptors, *Life Sci.* **39**, 2025-2032.

Voisin, P., Guerlotte, J., and Colin, J.P. (1988) An antiserum against chicken hydroxyindole-o-methyltransferase reacts with the enzyme from pineal gland and retina and labours pineal modified photoreceptors, *Brain Res.* **464**, 53-61.

Weill, J.C. and Reynaud, C.A. (1987) The B cell compartment, *Science* **238**, 1094-1098.

Wight, P.A.L. (1971) Development of sympathetic innervation of the pineal of the domestic fowl, *Acta Morphol. Neerl. Scand.* **9**, 47-56.

Wilson, J.A. and Glick, B. (1970) Ontogeny of mating behavior in the chicken, *Amer. J. Physiol.* **218**, 951.

Withyachumnarnkul, B., Nonaka, B. O., Santana, O., Attia, A.M., and Reiter, J.R. (1990) Interferon-γ modulates melatonin production in rat pineal glands in organ culture, *J. Interferon Res.* **10**, 403-411.

Youbicier-Simo, B.J. (1990) Approche chronobiologique des interrelations entre systèmes immunitaire et neuro-endocrine, *Diplôme d'Études Approfondies, Université de Montpellier II*, pp. 15.

Youbicier-Simo, B.J. (1994) Etude cinétique et chronobiologique de l'influence de la bursectomie et de la bursine sur les activités corticotropes, épiphysaires et immunitaires chez le poulet, *Thèse de diplôme de Doctorat*, Université de Montpellier II. pp. 180.

Youbicier-Simo, B.J., Boudard, F., Mekaouche, M., and Bastide, M. (1996) A role for Bursa Fabricii and bursin in the pineal biosynthetic activity in the chicken, *J. Pineal Res., (in press).*

Youbicier-Simo, B.J., Boudard, F., Mekaouche, M., and Bastide, M. (1996) Specific abolition reversal of pituitary-adrenal activity and control of the humoral immunity in bursectomized chickens through highly dilute bursin, *Intern J Immunopathol.Pharmacol..*, **9**, (*in press.*)

Youbicier-Simo, B.J., Boudard, F., Mekaouche, M., Bastide, M., and Baylé, J.D. (1993) Effects of embryonic bursectomy and *in ovo* administration of highly diluted bursin on adrenocorticotropic and immune responses of chickens, *Int. J. Immunother.* **9**, 169-180.

NON-MOLECULAR INFORMATION TRANSFER FROM THYROXINE TO FROGS*

By Means of Homoeopathic Preparation and Electronic Processing

P.C. ENDLER[1], W. PONGRATZ[1], C.W. SMITH[2], J. SCHULTE[3], F. SENEKOWITSCH[4] AND M. CITRO[5]
[1]*Ludwig Boltzman Institut für Homöopathie, Durerg.4, 8010 Graz, Austria*
[2]*Department of Electric and Electronic Engineering, University of Salford, 827221 Salford, UK.*
[3]*National Superconducting Cyclotron Laboratory, Michigan State University, East Lansing, USA*
[4]*Institute of Bioinformatics, Graz, Austria*
[5]*Research Institute Alberto Sorti (IDRAS), Turin, Italy*

1. Introduction

Research on the scientific basis of homoeopathy can generally proceed on two lines: *(a)* with regard to the principles underlying inversion effects of specially prepared dilutions (van Wijk and Wiegant, 1994; Bastide, 1994; Endler *et al.*, 1994a) and *(b)*with regard to the biophysical properties of the information transfer process. This paper is dedicated to *(b)*.

Substances so much diluted that no original molecule is present still exert biological effects, i.e. molecular bio-information may be transduced via water (Bastide *et al.*, 1987; Hadji *et al.*, 1991; Benveniste *et al.*, 1992; Youbicier-Simo *et al.*, 1993; Smith, 1994; Endler *et al.*, 1994b; Pongratz *et al.*, 1995; for a survey, see Endler and Schulte (1994), and hence inhibit climbing activity of juvenile frogs (Endler *et al.*, 1994b). In the work presented here, we have studied the thyroxine-controlled metamorphosis of the amphibian *Rana temporaria* from the two- to the four-legged stage and further on to the juvenile stage.

Thyroxine, added to the water of a basin (final concentration 10^{-9} or 10^{-8} part by weight, 2 or 20 nM), induces or accelerates respectively, the development of tadpoles (Pitt-Rivers and Trotter, 1964; Endler and Schulte, 1994). It was postulated that perimolecular energetic fields are involved in the interaction of biologically active substances and the organism (Endler and Schulte, 1994).

---

* Communications presented at the 7 th GIRI Meeting, November 1993, Montpellier, France, and at the 8th GIRI Meeting, December 1994, Jerusalem, Israel.

M. Bastide (ed.), Signals and Images, 149–159.

The blind experiments on non-molecular information from thyroxine presented in this paper were carried out between 1990 and 1994. In experiments A, the influence of a high dilution of thyroxine was tested. In experiments B, it was attempted to electronically digitize information from this high dilution. In experiments C, it was attempted to transfer information from molecular thyroxine by means of a high gain amplifier. In experiments D, the high dilution of thyroxine was sealed in glass vials hung into the basin water.

2. Methods

In previous experiments, it was observed that a homoeopathically prepared dilution such as 10^{-30}M of thyroxine can exert an inhibiting effect on tadpole metamorphosis (Endler and Schulte, 1994). However, the relation between the seasonal time the experiment is performed at and the frequency of application of the dilutions was found to be crucial. In spring, more frequent stimulation is needed than in autumn (Table 1).

TABLE 1. Relation of seasonal time and frequency of application with regard to the effect of homoeopathically prepared thyroxine (10^{-30}) on amphibian metamorphosis. strong, weak, none = strong, weak, no inhibiting effect

Application at intervals of	spring	summer	autumn
0 h /8 h	strong	weak	none /opposite
48 h	weak	weak	strong

This relation between seasonal time and frequency of application (as was derived from protocol A) was also taken into consideration in the further experiments B, C and D.

2.1. PROTOCOL A: USE OF AN AGITATED DILUTION

2.1.1. *Animals and Staging*

The first experiments were performed in summer with Rana temporaria from an Austrian pool about 1500 m above sea level. For the main study, Rana temporaria were taken with the transition from the two-legged stage onwards in autumn. We chose only those tadpoles where the hindlegs slightly started to be straddled (stage 31, Gosner's staging table - Gosner, 1960 -). The two-legged tadpoles were observed at regular intervals for 1-2 weeks until (a) the forelegs, preformed under the skin, broke through. After a developmental period of 1 - 2 weeks, the forelegs broke through within a few minutes. Thus, this parameter seems adequate to define a final stage. Furthermore (b), the animals were observed until they reached the juvenile stage (stage of tail reduction).

2.1.2. *Further conditions*
The transitions were observed in disposable basins (8l). The room temperature was 19°C. Indirect natural light was used. The tadpoles were fed with a standardized aquarium product ad libidum. Two sets of basins were always used for the treatment with the two test solutions (see below). The number of animals per basin was alike (10 - 20) in each experiment. The basins were rotated in the course of the experiment.

2.1.3. *Preparation of testing solutions and exposition to probes*
The tadpoles were observed under the influence of tetra-iodo-thyronine sodium pentahydrate (Sigma) specially prepared in an aqueous solution 1 : 10^{30} part by weight. The stock solution had a concentration of 1 : 10^{4} part tetra-iodo-thyronine sodium pentahydrate of weight; it was diluted in steps of 1 : 10. The diluted solution (10^{-5}, 10^{-6} etc.) was agitated according to standardized instructions (Homöopathisches Arzneibuch, 1991): at every step, a sterile bottle was filled with the dilution, and was pushed down at short regular intervals (e.g. against a rubber impediment) to create mechanical shocks. Disposable pipettes were used. Water was similarly "diluted" (water 10^{-30}) as control. Check for hormone-active contamination was described previously (Endler and Schulte, 1994). Always 3 drops of the test solutions were added to the basin water at intervals of 48 h.

2.1.4. *Evaluation of the data*
The cumulative frequencies (fa, fb) of (a) four-legged tadpoles and (b) juveniles were compared by a chi-square test to those of two- or three-legged animals at the respective time intervals. In order to give a further survey on the results, from the graphs the delay was extrapolated with which the test animals reached the fa- and fb- level of the control animals. Further statistical evaluation is discussed elsewhere (Endler and Schulte, 1994; Endler and Pongratz, 1994).

2.1.5. *Laboratories/researchers*
For the first study (summer 1990), experiments were performed in three independent laboratories (P.C.E., W.P., R.V.W. see List of Authors and Acknowledgements). For the main study (autumn 1993), experiments were performed by W.P.

2.2. PROTOCOL B: COMPACT DISC RECORD /REPLAY

2.2.1. *Animals, staging and further conditions* are described in protocol (A).

2.2.2. *Preparation of testing solutions and exposition to probes*
Vials of thyroxine (10^{-30} M, or water were placed on an input coil linked to a filter and to an amplifier with a gain of E6. Frequencies below 80 kHz were digitized at the Nyquist frequency, buffered in a RAM and multiplexed on to a CD. After noise reduction and filtering, the signal was attenuated by E6 to restore the original analog level.

Water vials were placed on an output coil for 4 min (device: Fa. Hoefler, Austria). At intervals of 48 h, always 3 drops of the test solutions were added to the basin water. For further details, see experiments (A).

2.2.3. *Laboratory /researcher*
Experiments were performed in one laboratory by W.P.

2.2.4.*Evaluation* was done as described in protocol (A).

2.3. PROTOCOL C: USE OF AN ELECTRONIC CIRCUITRY

2.3.1. *Animals*
Rana temporaria were taken from Austrian pools about 400 m above sea level (spring). Staging and further conditions are described in protocol (A).

2.3.2. *Preparation of testing solutions and exposition to probes*
Vials of thyroxine (1 mM) or water were placed on an input coil, one end of which was connected to a special amplifier (linear from DC to HF) by a single insulated wire (devices used: Fa. Brügemann, FRG - main study - ; Fa. Medtronik, FRG). Water vials were placed on a similar output coil for 4 min. (Brügemann - device) or 15 min. (Medtronik - device) respectively. At intervals of 8 h always 8 ml of basin water were replaced by the propes (test or control) produced at the output coil. For standardisation of this method, see (Citro, 1991; 1992: Citro *et al.*, 1994; Aissa *et al.*, 1993; Benveniste *et al*,. 1994).

2.3.3. *Laboratory /researcher*
The experiments were performed in different laboratories by two independent researchers (W.P., C.V.).

2.3.4. *Evaluation* was done as described in protocol (A).

2.4. PROTOCOL D: USE OF AN AGITATED DILUTION IN SEALED VIALS

2.4.1. *Animals*
The first experiments were performed both in summer with *Rana temporaria* and with *Bufo bufo* as well as in autumn with *Rana temporaria*. For the main experiments with Rana temporaria, animals were taken from the pool 400 m above sea level with the transition from the two-legged stage onwards in spring as described in (C). Staging and further conditions are described in protocol (A).

2.4.2. *Preparation of testing solutions*
It was done as described in protocol (A). Exposition to probes: Eight ml of the thyroxine or water dilutions were sealed in hardglass vials with an optical transmission spectrum > 350 nm. The coded vials were first placed in a common water bath to make sure that they were not individually contaminated outside and then hung into the corresponding basins. The method of application of test substances in sealed ampoules had been used in therapeutical practice and by different researchers before (Endler and Schulte, 1994)).

2.4.3. *Laboratory /researcher*
Experiments were performed in different laboratories by three independent researchers (P.C.E., W.P., K.W.). The evaluation was done as described in protocol (A).

3. Results

3.1. PROTOCOL A: USE OF AN AGITATED DILUTION

First experiments were performed in summer by P.C.E. (with a total of 545 animals), by W.P. (230 animals) and by R.v.W. (360 animals). At some, but not all measuring points the f(a,b)-values were significantly below those for reference. This is true for the data from the tree laboratories both when pooled ($P<0.01$) as well as when treated separately ($P<0.01$).

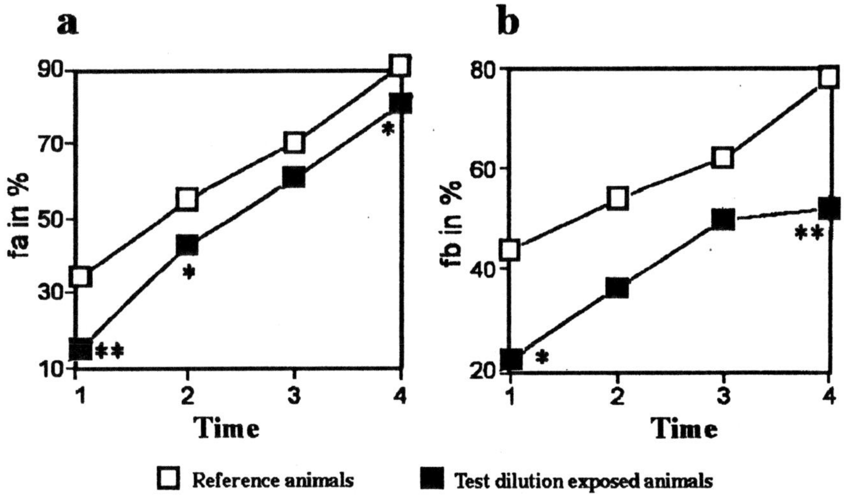

Figure 1. The influence of the high dilution of thyroxine, (a) on the transition from the two-legged to the four-legged stage and (b) on the transition to the juvenile stage of *Rana temporaria* tadpoles. "100%" refers to 140 (fa) or 50 animals (fb), respectively, in each group. Time categories 1 - 4, depending on the experiment: normally 2-3 days (see methods). 1 SD < 10%; * , $P < 0.01$. For absolute numbers of animals, see Table 2. For further explanation, see the text.

TABLE 2. Absolute numbers of animals in Figure 1

Time category	1	2	3	4
N_{fa} (control)	048	077	099	127
N_{fa} (test probe)	021	060	085	113
N_{fb} (control)	022	027	031	039
N_{fb} (test probe)	011	018	025	026

The main experiments (optimized protocol) were performed in autumn by W.P. (with a total of 280 animals). As is shown in Figure 1, at all measuring points in time, the f(a,b)-values for the animals exposed to the test dilution (black squares) were below those for reference (white squares) ($P<0.01$).

From these results it can be extrapolated that the tadpoles exposed to the test liquid reached the f(a,b)-levels of control animals after a delay of about 1 - 2 d. In other words, the chance of passing metamorphosis is generally smaller for the group treated with the high dilution of thyroxine, added to the basin water at intervals of 48 h than for the water control group.

3.2. PROTOCOL B: COMPACT DISK RECORD /REPLAY

Several experiments were performed in autumn by W.P. with a total of 468 animals. As is shown in Figure 2, at all measuring points in time, the f(a,b)-values for animals exposed to the test dilution (black circles) were below those for reference (white circles) ($P<0.01$).

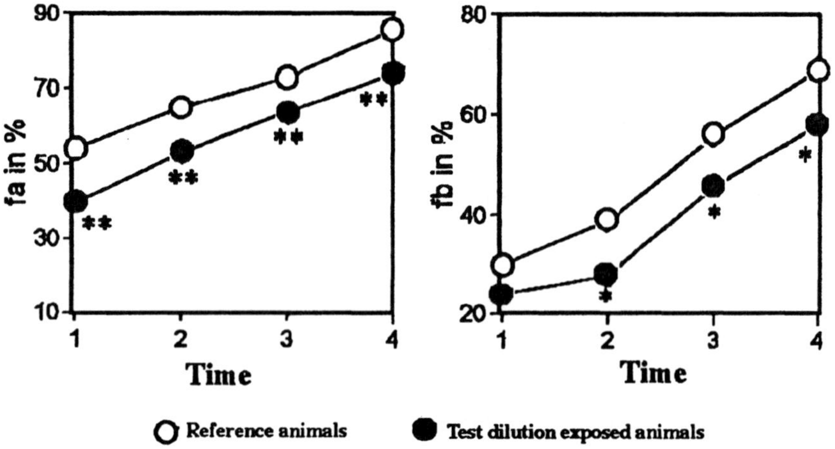

Figure 2. The influence of the liquid exposed to digitised and CD - stored thyroxine information. "100%" refers to 234 animals in each group. 1 - 4, time categories, depending on the experiment: normally 24 h. 1 SD > 10%; *, $P < 0.05$; **, $P < 0.01$. For absolute numbers of animals, see Table 3. For further details, see the text.

TABLE 3. Absolute numbers of animals in Figure 2

Time category	1	2	3	4
N_{fa} (control)	126	152	171	201
N_{fa} (test probe)	094	124	151	175
N_{fb} (control)	070	091	131	161
N_{fb} (test probe)	057	066	109	137

From these results it can be extrapolated that the tadpoles exposed to the test liquid reached the f(a,b)-level of control animals after a delay of about 1 d. In other words, the chance of passing metamorphosis is generally smaller for the group treated with the liquid exposed to the digitized and CD - stored thyroxine information.

3.3. PROTOCOL C: USE OF AN ELECTRONIC CIRCUITRY

Several experiments were performed in spring by W.P. with a total of 756 animals and one experiment by C.V. (180 animals). As is shown in Figure 3 (see next page), at all measuring points in time, the f(a,b)-values for animals exposed to the test dilution (black triangles) were below those for reference (white triangles) ($P<0.01$). This is true for the data from the two laboratories both when they were pooled as well as when they were treated separately.

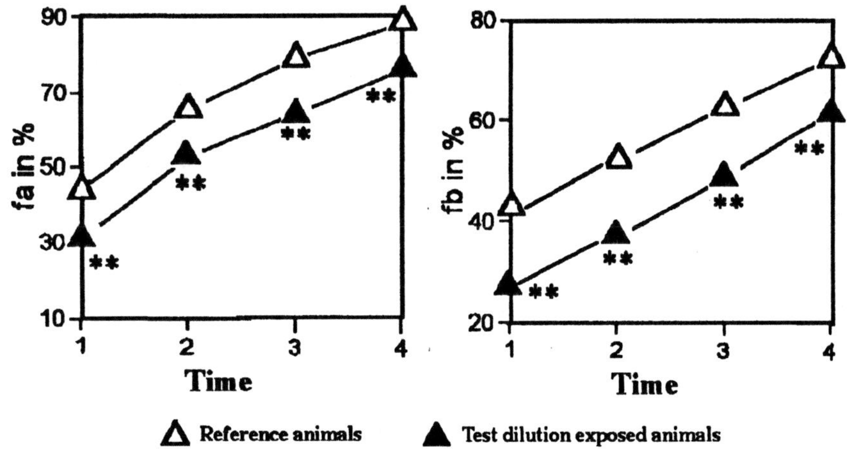

Figure 3. The influence of the liquid exposed to electronically transferred thyroxine information. "100%" refers to 468 animals in each group. Time categories 1 - 4, depending on the experiment: normally 24 h. 1 SD < 10%. For absolute number of animals, see Table 5. For further information, see text and the Figure 1

TABLE 4 Absolute numbers of animals in Figure 3

Time category	1	2	3	4
N_{fa} (control)	206	309	374	417
N_{fa} (test probe)	145	248	304	360
N_{fb} (control)	201	243	295	342
N_{fb} (test probe)	126	173	229	290

From these results it can be extrapolated that the tadpoles exposed to the test liquid reached the f(a,b)-levels of control animals after a delay of about 1 - 1,5 d. In other words, the chance of passing metamorphosis is generally smaller for the group treated with the liquid exposed to electronically transferred thyroxine information.

3.4. PROTOCOL D: USE OF AN AGITATED DILUTION IN SEALED VIALS

The first experiments were performed in summer with *Bufo bufo* (P.C.E., 347 animals and K.W., 104 animals) and with *Rana temporaria* (H.H., 217 animals). At some, but not all measuring points, the fa-values were significantly below those for reference. This is true for the data from the three researchers both when they were pooled as well as when they were treated separately ($P < 0.01$ or $P < 0.05$, respectively). Further experiments with *Rana temporaria* were performed in autumn (W.P. *et al.*, 1995). In general, the differences showed that the test dilution slowed down development to the four-legged stage. However, this could not be statistically proved (Endler and Schulte, 1994). The main experiments were performed in spring by W.P. (792 animals) and by K.W. (with a total of 432 animals). As is shown in Figure 4, at all measuring points in time, the f(a,b)-values for animals exposed to the test dilution (black squares) were below those for reference (white squares) ($P < 0.01$). This is true for the data from the two laboratories both when they were pooled as well as when they were treated separately.

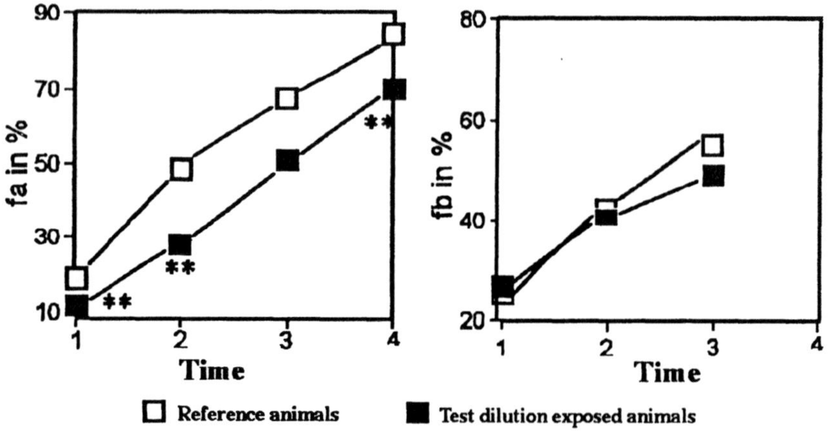

Figure 4. The influence of the high dilution of thyroxine, sealed in glass vials. "100%" refers to 612 (time categories 1 and 2, $f_{a,b}$) and 468 (time categories 3 and 4, f_a) or 216 (time categories 3 and 4, f_b) animals in each group. Time categories 1 - 4, depending on the experiment: normally 1 or 2 d. 1 SD < 10%; * , $P < 0.01$. For absolute numbers, see Table 5. For further details, see the text and the Figure 1

From these results it can be extrapolated that the tadpoles exposed to the test liquid sealed in glass vials reached the f(a,b)-levels of control animals after a delay of about 1 - 2 d. In other words, the chance of passing metamorphosis is generally smaller for the group treated with the high dilution of thyroxine, sealed in glass vials and hung into the aquaria than for the water control group.

TABLE 5. Absolute numbers of animals in Figure 4

Time category	1	2	3	4
N_{fa} (control)	113	295	315	390
N_{fa} (test probe)	072	168	236	324
N_{fb} (control)	150	258	118	
N_{fb} (test probe)	156	247	105	

4. Discussion

These results refer to frogs during their metamorphosis from the two-legged stage (a) to the four-legged stage and (b) to the juvenile stage, treated with the test water to which information from thyroxine was transferred by different means. This caused differences in the rate of metamorphosis as compared with reference (water to which information from water was analoguely transferred). In the blind experiments the test liquids significantly slowed down metamorphosis.

In experiments A, the influence of a high dilution (10^{-30} M) of thyroxine, prepared in a special process of stepwise dilution and agitation, or similarly diluted water was tested. These probes were directly added to the basin water. In experiments B, vials of the thyroxine dilution (10^{-30} M) or water were placed on an input coil linked to a filter and to an amplifier with a gain of 10^6. Frequencies up to 80 kHz were digitized at the Nyquist frequency, buffered in a RAM and multiplexed on to a CD. After noise reduction and filtering, the signal was attenuated by 10^6 to restore the original analog level. Water vials were placed for 4 min on an output coil. These probes were directly added to the basin water. In experiments C, vials of thyroxine (1 mM) or water were placed on an input coil, one end of which was connected to a special amplifier (linear from DC to HF) by a single insulated wire. Water vials were placed for 4 min on a similar output coil. These probes were directly added to the basin water. In experiments D, a thyroxine dilution (10^{-30} M, see protocol A) and similarly prepared water were sealed in hardglass vials hung in the basin water.

During the treatment, the cumulative frequencies of four-legged tadpoles (for fa, see Table 6) and of juvenile frogs dropped in all types of experiments.
From Table 6 it can be seen that the chance of entering the four-legged stage is generally smaller in the groups treated with the information from thyroxine than in the control groups. Information from molecular thyroxine can be transferred by a process of stepwise aqueous dilution and agitation or by means of an electronic circuitry, it can be stored on a compact disk and it can exert its effect through the wall of a sealed glass vial (for references, see Table 6). This leads to the assumption that this bio-information is electromagnetic in nature and can be processed by conventional electronic circuits and devices.

Recent physics research has revealed that water dipoles may develop phase coherent oscillations through radiation coupling (del Giudice *et al.*, 1988; del Giudice, 1994). It is proposed that these could be modulated as a time-ordered pattern of signals (Endler and Schulte, 1994) and induce the coherent wave propagation (electron propagation) in metals (Endler and Schulte, 1994). A further theory suggests that the phase coherent oscillations may originate information pattern formation through isotopicity effects in high dilutions (Endler and Schulte, 1994). The theoretical physical aspect of the experiments A - D referenced here has yet to be evaluated in order to test and develop a theory on bio-information storage and transfer in 'homoeopathically' prepared dilutions.

TABLE 6. Cumulative frequency of four-legged tadpoles

Protocol / Reference	drop of f_a	N of animals (test probe + control)	P (chi^2 test)	researchers
A Endler *et al.*; 1991	*10%*	*683 + 682*	*0.001*	*PCE, WP, RvW*
	19%	140 + 140	0.001	WP
B Endler *et al*,. 1994d, Senekowitsch *et al*,. 1995	14%	234 + 234	0.001	WP
C Citro *et al.*, 1995	13%	468 + 468	0.001	WP, CV
D Endler *et al.* ,1994c	*09%*	*303 + 313*	*0.01*	*PCE, KW, HH*
	21%	612 + 612	0.001	WP, K

Non-molecular information transfer from thyroxine to frogs. For initials, see List of Authors and Acknowledgements. *First experiments (the preliminary protocols);* Main experiments (optimized protocols). For further explanation, see the text. Main experiments of protocol A and experiments B were performed when natural metamorphosis proceeds relatively slowly (autumn), whereas experiments C and main experiments of D were performed when natural metamorphosis proceeds relatively fast (spring). A less differentiated table was presented in Endler *et al.*, 1995.

The possibility of storing (biologically /chemically active) information from molecules on data carrier such as a CD may open new ways both for therapeutical as well as for technical use. However, the authors wish to emphasize that they understand their work as a contribution towards a deeper understanding of the theoretical background of homoeopathy.

Acknowledgements

We are obliged to those who stimulated the experiments, especially to Thomas Kenner and Max Moser, Institute of Physiology, University of Graz and Max Haidvogl, Boltzmann Institute of Homeopathy, Graz. Furthermore, we would like to express thanks to those who performed independent experiments (see Table 6): Roeland Van Wijk, Department for Molecular Cell Biology, University of Utrecht, The Netherlands; Christina Vinattieri, Department of Holistic Medicine, University of Urbino & IDRAS, Turin, Italy; Karl Waltl, Boltzmann Institute of Homoeopathy, Graz; Helge Hilgers, Institute of Zoology, University of Vienna, Austria.

5. References

Aissa, J., Litime, M.H., Attis, E., and Benveniste, J. (1993) Molecular signalling at high dilution or by means of electronic circuitry, *J.Immunol.* **150**, A146.

Bastide, M. (1994) Immunological examples on ultra high dilution research, in P.C. Endler, J. Schulte (eds.): *Ultra High Dilution. Physiology and Physics*, Kluwer Academic Publishers, Dordrecht, pp 27-34.

Bastide, M., Daurat, V., Doucet-Jaboeuf, M., Pélegrin, A., and Dorfman, P. (1987) Immunomodulator activity of very low doses of thymulin in mice, *Int J Immunotherapy* **3**, 191-200.

Benveniste, J., Aissa, J., Litime, M.H., Tsaegaca, G.T., and Thomas, Y. (1994) Transfer of the molecular signal by electronic amplification, *FASEB J* **8**, A398.

Benveniste, J., Arnoux, B., and Hadji, L.(1992) Highly dilute antigen increases coronary flow of isolated heart from immunized guinea-pigs, *FASEB J* **6**, A1610.

Citro, M. (1991) Vom Pharmakon zur Frequenz; elektronischer Medikamententransfer, *Proc. II int. Symp. Biokybernetische Medizin*, Würzburg.

Citro, M.(1992) TFF dal farmaco alla frequenza. *Vivibios II* **3**, 25-30.

Citro, M., Endler, P.C., Pongratz, W., Vinattieri, C., Smith, and C.W., Schulte, J.(1995) Hormone effects by electronic transmission, *FASEB J* (Abstract 12161).

Citro, M., Smith, C.W., Scott-Morley, A., Pongratz, W., Endler, P.C.(1994) Transfer of information from molecules by means of electronic amplification, in P.C. Endler, J. Schulte (eds.): *Ultra High Dilution. Physiology and Physics*, Kluwer Academic Publishers, Dordrecht, pp 209-214.

Del Giudice, E. (1994) Is the 'memory of water' a physical impossibility?, in P.C. Endler, J. Schulte (eds.): *Ultra High Dilution. Physiology and Physics*, Kluwer Academic Publishers, Dordrecht, pp 117-120.

Del Giudice, E., Preparata, G., and Vitiello, G.(1988) Water as a free electric dipole laser, *Phys Rev Lett.* **61**, 1085-1088.

Endler, P.C., Citro, M., Pongratz, W., Smith, C.W., Vinattieri, C., and Senekowitsch, F.(1994d) Übertragung von Molekül-Information, *Proc. Int. Symp. Niederenergetische Bioinformation*, Bad Waltersdorf.

Endler, P.C., and Pongratz, W.(1994) *On effects of agitated highly diluted thyroxine (E-30)*, Comprehensive report, available at the Institute for Zoology, University of Graz, Universitätsplatz 2, A-8010 Graz.

Endler, P.C., Pongratz, W., Kastberger, G., Wiegant, F.A.C., and Schulte, J.(1994b) The effect of highly diluted agitated thyroxine on the climbing activity of frogs, *J Vet Hum Tox.* **36**, 56-59.

Endler, P.C., Pongratz, W., Smith, C.W., and Schulte, J.(1995) Non-molecular information transfer from thyroxine to frogs with regard to 'homoeopathic' toxicology, *J.Vet.Hum.Tox.*, **37**, 259-260.

Endler, P.C., Pongratz, W., Van Wijk, R., Kastberger, G., and Haidvogl, M. (1991) Effects of highly diluted sucussed thyroxine on metamorphosis of highland frogs, *Berlin J Res Hom* **1**, 151-160.

Endler, P.C., Pongratz, W., Van Wijk, R., Waltl, K., Hilgers, H., and Brandmaier, R.(1994c) Transmission of hormone information by non-molecular means, *FASEB J* **8**, A400.

Endler, P.C., Pongratz, W., Van Wijk, R., Wiegant, F.A.C., Waltl, K., Gehrer, M., and Hilgers, H.(1994a) A zoological example on ultra high dilution research, in P.C. Endler, J. Schulte (eds.): *Ultra High Dilution. Physiology and Physics*, Kluwer Academic Publishers, Dordrecht, pp 39-68.

Endler, P.C. and Schulte, J. (1994) *Ultra High Dilution. Physiology and Physics*, Kluwer Academic Publishers, Dordrecht.

Gosner, K.L.(1960) A simplified table for staging anuran embryos and larvae with notes on identification, *Herpetologica* **16**, 183-195.

Hadji, L., Arnoux, B., and Benveniste, J. (1991) Effect of dilute histamine on coronary flow of guinea-pig isolated heart, *FASEB J* **5**, A1583.

Homöopathisches Arzneibuch (1991) Stuttgart: Deutscher Apothekerverlag and Frankfurt Govi Verlag.

Pitt-Rivers, R. and Trotter, W.R.(1954) *The Thyroid Gland*, Butterworth Publisher, London.

Pongratz, W., Endler, P.C., Poitevin, B., and Kartnig, T.(1995) Effect of extremely diluted plant hormone on cell culture, *Proc. 1995 AAAS Ann. Meeting*, Atlanta.

Senekowitsch, F., Endler, P.C., Pongratz, W., and Smith, C.W.(1995) Hormone effects by CD record /replay, *FASEB J* Abstract 12025 .

Smith, C.W.(1994) Coherence in living biological systems. *Neural Network World* 4 **3**, 379-388.

Van Wijk, R. and Wiegant, F.A.C.(1994) *Cultured mammalian cells in homeopathy research: the similia principle in self-recovery*, University Utrecht Publisher, Utrecht.

Youbicier-Simo, B.J., Boudard, F., Meckaouche, M., Bastide, M., and Baylé JD (1993) The effects of embryonic bursectomy and in ovo administration of highly diluted bursin on adrenocorticotropic and immune response of chicken, *Int. J. Immunother.* **9**, 169-190.

PHARMACODYNAMIC EFFECTS OF VERY HIGH DILUTIONS OF BELLADONNA ON THE ISOLATED RAT DUODENUM*

A Pharmacological Study of High Dilution Effect

A. CRISTEA, S. NICULA AND V. DARIE
Laboratory of Pharmacology, Faculty of Pharmacy, Bucharest, Romania

1. Introduction

The modern data of molecular biology and pharmacology regarding the double - way (bidirectional) effects of some chemical mediators and drugs at various doses, the knowledge of special laws of information processing and storage, different from those of the other two forms of manifestation of matter, the substance and energy (Restian, 1981, 1989), as well as our concern for the cybernetic and informational pharmacology (Cristea, 1985, 1991), revealed us a theory which we called the "informational theory of doses" (Cristea, 1996).

In the light of this theory, we endeavoured to systematically develop an " informational pharmacology of dosages " (Cristea, 1994; 1995). For this purpose we initiated systematic researches on the pharmacodynamic effects of a drug, over a wide scale of dilutions.

Considering that the informational pharmacology of concentrations could be a scientific basis of linkage between the allopathic pharmacology and a future homeopathic pharmacology (Cristea, 1994), we concomittantly extended the study also to dilutions prepared according to the dynamization technique by succussion, used in the homeopathic pharmacy (Hahnemann 's technique).

For the first research, we adopted a classical pharmacodynamic method, the "in vitro" method which uses isolated organs (Turner, 1965), taking into account that in the allopathic experimental pharmacology, the isolated organ is acted up on, with high dilutions, situated at the inferior limit of the low homeopathic dilutions.

The first drug chosen was *Belladonna* (Cristea *et al.*, 1991), because it is used both in allopathy and in homeopathy. Thus in allopathy, atropina, the alkaloid extracted from *Atropa belladonna* is known to exert a spasmolytic action and is used in the treatment of colics on the basis of the therapeutic principle of contraries (*"Contraria contrariis curentur"*). As is known, in homeopathy, the *Belladonna* tincture at suballopathic dilutions exerts a spastic action and is also indicated at high dilutions in the treatment of colics, but on the basis of another terapeuthic principle, that of similitude (*"Similia similibus curentur"*).

---

* Communication presented at the 7th GIRI Meeting, November 1993, Montpellier, France.

M. Bastide (ed.), Signals and Images, 161–170.

2. Method

1.1. DRUGS AND DILUTIONS

The *Belladonna* (B) tincture (containing 0,03g% total alkaloids, expressed in atropine) was progressively diluted with water, at a geometrical ratio of 1:100. Thus, a scale of centesimal dilutions (C) between the first dilution 10^{-2} (1 C) and the 200th dilution 10^{-400} (200 C), with a content of 6.023 x 1017 and respectively 6.023 x 10^{-381} atropine molecules/100 ml, was obtained. The content of moles and atropine molecules/100 ml was calculated for all the 1 C - 200 C dilutions, according to Avogadro's number (6.023 x 10^{23} molecules/ mol) and taking into account the molecular weight of atropine (MW = 289.41g ~ 300g) (Table 1).

TABLE 1. Centesimal dilutions (CH) content in moles and molecules of atropine / 100 ml according to Avogadro's number and molecular weight

SUBSTANCE	CH DILUTIONS	MOL %	MOLECULES %
Belladonna	1	10^{-4}	6.023 x 1017
Belladonna	5	10^{-12}	6.023 x 109
Belladonna	6	10^{-14}	6.023 x 107
Belladonna	9	10^{-20}	6.023 x 101
Belladonna	10	10^{-22}	6.023 x 10^{-1}
Belladonna	20	10^{-42}	6.023 x 10^{-21}
Belladonna	25	10^{-52}	6.023 x 10^{-31}
Belladonna	30	10^{-62}	6.023 x 10^{-41}
Belladonna	40	10^{-82}	6.023 x 10^{-61}
Belladonna	45	10^{-92}	6.023 x 10^{-71}
Belladonna	50	10^{-102}	6.023 x 10^{-81}
Belladonna	55	10^{-112}	6.023 x 10^{-91}
Belladonna	60	10^{-122}	6.023 x 10^{-101}
Belladonna	70	10^{-142}	6.023 x 10^{-121}
Belladonna	75	10^{-152}	6.023 x 10^{-131}
Belladonna	80	10^{-162}	6.023 x 10^{-141}
Belladonna	90	10^{-182}	6.023 x 10^{-161}
Belladonna	95	10^{-192}	6.023 x 10^{-171}
Belladonna	100	10^{-202}	6.023 x 10^{-181}
Belladonna	110	10^{-222}	6.023 x 10^{-201}
Belladonna	120	10^{-242}	6.023 x 10^{-221}
Belladonna	125	10^{-252}	6.023 x 10^{-231}
Belladonna	130	10^{-262}	6.023 x 10^{-241}
Belladonna	140	10^{-282}	6.023 x 10^{-261}
Belladonna	150	10^{-302}	6.023 x 10^{-281}
Belladonna	160	10^{-322}	6.023 x 10^{-301}
Belladonna	170	10^{-342}	6.023 x 10^{-321}
Belladonna	200	10^{-402}	6.023 x 10^{-381}
Atropine	2	2.87 x 10^{-3}	6.023 x 10^{18}
Acetylcholine	2	5.57 x 10^{-3}	6.023 x 10^{18}

The dilutions were prepared in two ways: a) by simple homogeneization, without succussion (N dilutions) and b) by "dynamization" with manual succussion (knocking

slightly 100 times the vial against a firm substrate, rhythmically, at a frequency of 1 succussion/second for each dilution), according to Hahnemann's homeopathic technique (D dilutions).
Likewise, D dilutions 1 C - 5 C were prepared in water, starting from distilled water, as well as from alcohol 70°, the solvent with which the *Belladonna* tincture was prepared. These dilutions were used as control.
All the D and N solutions were prepared in the Homeopathic pharmacy in Bucharest, especially for this research. The reference substances, atropine sulphate (ATR) and acetylcholine chlorhydrate (ACH), were used in standard N aqueous solutions, 10^{-2}, with a content of 17.33×10^{18} and respectively 33.55×10^{18} molecules/100 ml.

1.2. ANIMALS

Male white Wistar rats, with an average weight of 150 g, fasting, with access only to water the day of the experience, were used. They were sacrificed by instantaneous decapitation.
The experiences were carried out in spring and autumn, between 10 h and 16 h.

1.3. TECHNIQUE

The "in vitro" classical technique on the isolated rat duodenum, mounted in the isolated organ bath (Turner, 1965), with special precautions for high dilutions, was used.
Duodenum fragments 2 cm long in situ were isolated and immediately mounted in the isolated organ bath containing Tyrode type physiologic salt solution, treated with glucose, regulated thermostatically at 37 °C and oxygenated by air bubbling at a constant output. The motor activity (motility) of the isolated organ, both the normal tone and isotonic contraction was recorded using a classical kimograph.

The traction (stress) constantly exerted upon the duodenum during the test was of 2 g and the inscription was permanently 10 times amplified.
The accommodation to the in vitro conditions occurred within 10 minutes. The rest interval between two tests was of 5 minutes. On a duodenum fragment was performed two tests. After each test followed the washing of the duodenum (by emptying and refilling the bath with Tyrode's solution), repeated four times. The volume of the solution making the object of the research, dropped with a pipette into the bath, was always of 0.2 ml / 20 ml of bath.

Separate baths were used for the tests with : *Belladonna*, 1 C - 9 C dilutions, 10 C - 200 C dilutions, N dilutions, D dilutions and atropine. During a working day, in the same bath, the low solutions were tested in an increasing order of concentrations and the high dilutions were tested at random.

1.4. TESTS

In each test, each dilution of *Belladonna* (B), as well as the solution of each reference substance, was tested on eight duodenum fragments (n = 8).

1.4.1. *Testing the effect on the normal tone of the rat duodenum*

The action of the test dilution of B, on the normal tone of the isolated rat duodenum, was recorded during 5 minutes. The values of the effects exerted by the B dilutions (tone decrease or increase) were expressed in percentage against the effects (relaxation or contraction) produced by the reference substances atropine sulphate (ATR) and acetylcholine chlorhydrate (ACH) respectively , at the standard dose of 10^{-6} g/ml of bath.

1.4.2. *Testing the effect on the ACH induced spasm*
A spasm of the isolated rat duodenum was induced by acethylcholine chlorhydrate (ACH) at the standard dose of 1 mg / ml of bath (10-6 g/ml of bath). Two minutes after the administration of ACH, during the maintenance "en plateau" of the maximum spasm, the test dilution of B was administered and its effect on the spasm was observed for 3 minutes.

The intensity of the effect exerted by B dilutions (decrease or increase of ACH spasm) was expressed in percentage against the contraction produced by ACH at the standard dose of 10^{-6} g/ml bath, on the same duodenum .

1.4.3. *Testing the identity principle ("Aequalia aequalibus curentur")*
The identity principle, on which isopathy is based, was tested.
The therapeutic effect of B dilutions, on the relaxation or contraction induced on normal tone by B itself at the dilutions which had an significantly effect on normal tone, namely 1 C - 5 C and 45 C respectively (according to the test described at point 1.4.1.), was studied. The number (%) of appearances of therapeutic effect (determined as the return to the value of normal tone ± 25 %) was established.

1.5. STATISTICAL ANALYSIS

The statistical significance of the results was calculated by using the Student's "t" and Wilcoxon tests for 1.4.1. and 1.4.2. tests and Chi square test for 1.4.3. test.
From the point of view of scientific requests of the method on the isolated rat duodenum and of statistical tests used, in correlation with the bioethical regulations of research on animals, we established the number of the determinations for each dilution of B, at eight duodenum fragments, from eight rats (n = 8).

3. Results

3.1. THE EFFECT OF BELLADONNA (B) DILUTIONS ON THE NORMAL TONE OF THE RAT DUODENUM

The effect of D water was a statistically insignificant rise in the tone. The D aqueous solutions of alcohol 70° had no effect.
The N aqueous solutions of B had the effect of lowering the normal tone of the isolated rat duodenum, only up to 6 C (unidirectional, monophasic effect) (Figure 1).
The D dilutions of B exerted an effect up to 45 C, i.e. relaxation between 1 C-20 C and contraction between 30 C-45 C (bidirectional, biphasic effect) (Figure 1).
At the same degree of dilution, the D dilutions had a more intense effect than the N solutions (Figure 1). The statistically significant values ($p<0.05$, at n = 8) are shown in Table 2.

BELLADONNA DILUTIONS ON ISOLATED DUODENUM

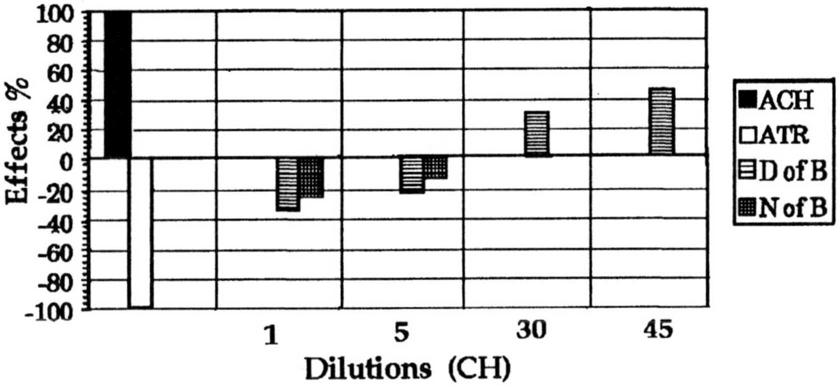

Figure 1. Statistically significant Belladonna (B) effects % ($p < 0.05$), on the normal tone of the isolated rat duodenum

TABLE 2. Statistically significant Belladonna (B) effects % ($p < 0.05$), on the normal tone of the isolated rat duodenum

Dilution	± Effects %	(Means ±SEM)
C	D	N
1	-34.88 ± 0.20	-25.62 ± 0.10
5	-22.48 ± 0.50	-14.22 ± 0.30
30	+28.77 ± 0.40	0
45	+44.38 ± 0.40	0

C = centesimal dilutions at the 1:100 ratio
D = dynamized dilutions, according to Hahnemann's technique
N = non-dynamized dilutions, prepared by sample homogenization
p = probability by Student's t test (n = 8)

3.2. THE EFFECT OF BELLADONNA (B) DILUTIONS ON THE SPASM INDUCED BY ACETYLCHOLINE (ACH)

The effect of D water consisted in a slight stimulation and was statistically non-significant. The D aqueous dilutions of alcohol 70° had no effect.

The N solutions of B had a spasmolytic effect only up to the 9 C dilution, with a content of 6.023×10^2 ATR molecules/100 ml (unidirectional, monophasic effect) (Figure 2).

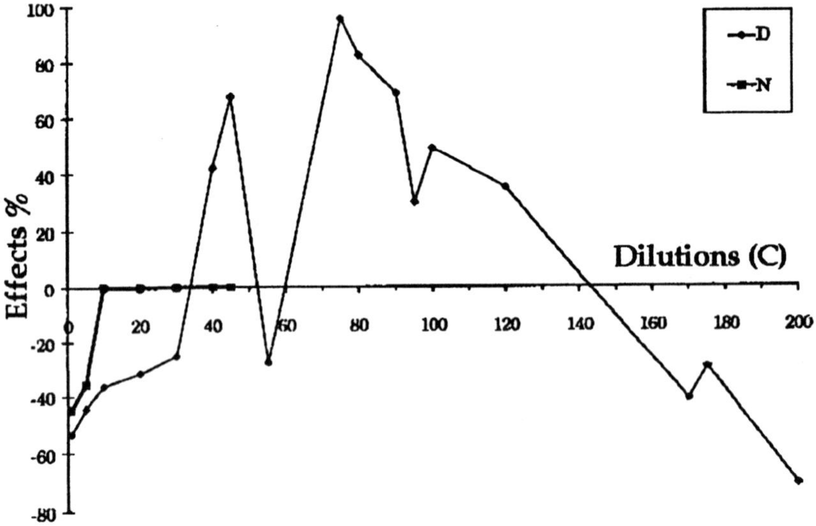

Figure 2. Statistically significant *Belladonna* (B) effects % (P < 0.05) on the spasm induced by acetylcholin (ACH)
D= dynalized solutions
N= non-dynamized solutions

TABLE 3. Statistically significant Belladonna (B) effects % (p < 0.05), on the spasm induced by acetylcholine (ACH)

DILUTION	± EFFECTS %	(MEANS ± SEM)
C	D	N
1	-53.30±1.10	-44.54±0.10
5	-44.30±1.20	-35.79±1.50
10	-36.55±1.30	0.00
20	-31.95±1.30	0.00
30	-25.40±0.50	0.00
40	+42.02±1.20	0.00
45	+67.60±0.10	0.00
55	-27.80±0.30	-
75	+95.45±1.20	-
80	+82.39±0.60	-
90	+68.74±1.30	-
95	+29.81±0.20	-
100	+48.92±1.10	-
120	+34.99±0.70	-
170	-41.15±1.10	-
175	-29.73±1.60	-
200	-71.12±2.00	-

C= centesimal dilutions at the 1:100 ratio
D= dynamized dilutions, according to Hahnemann's technique
N= non-dynamized dilutions, prepared by sample homogenization
p= probability by Student's t test (n = 8)

The D dilutions of B had an effect up to the 200 C, with a content of 6.023 x 10^{-381} ATR molecules/100 ml. On the scale of D dilutions 1 C - 200 C, the effect had a sinusoidal course, i.e. the decrease of the contraction at the 1 C - 25 C, 55 C - 60 C and 160 C - 200 C dilutions and the increase of the spasm at the 30 C - 50 C and 70 C - 150 C dilutions (bidirectional, multiphasic effect) (Figure 2).

The higher D dilutions were more efficacious. Thus, among the D dilutions which have antagonized the ACH spasm, the highest dilution studied, 200 C, had the most intense effect of 71.12 %. Among the dilutions which have amplified the ACH spasm, the 75 C had the highest effect of 95.45 % (Table 3).

Conversely, at the N dilutions, the efficacy decreased with the increase of the dilution, hence with the lowering of the concentration.

The statistically significant values ($p<0.05$, at $n = 8$) are listed in Table 3.

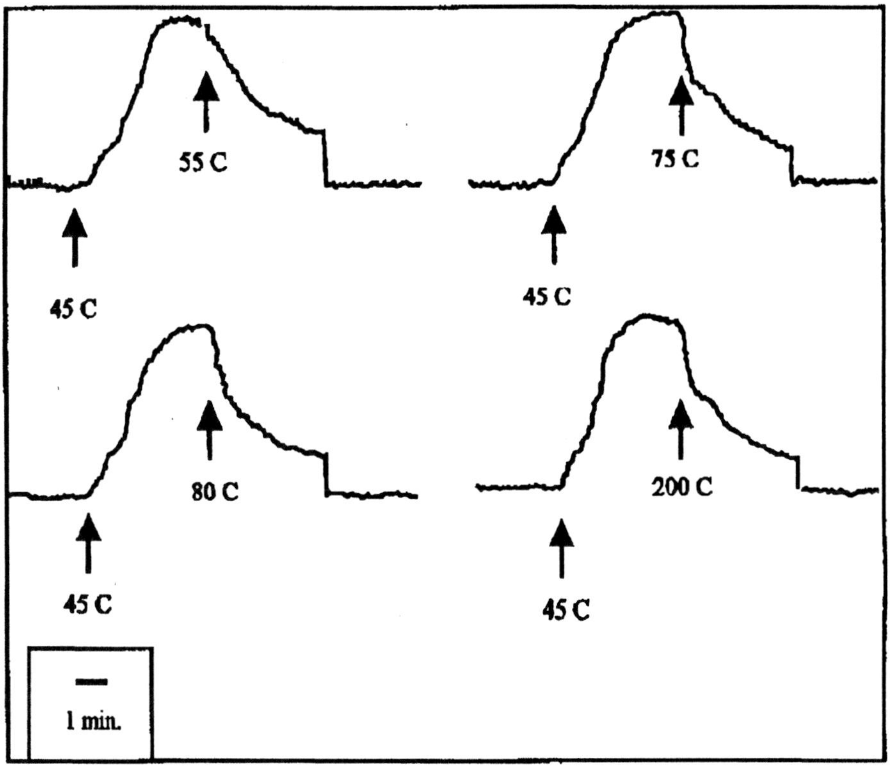

Figure 3. Demonstration of the identity principle, with dynamized (D) dilutions of Belladonna (B). The therapeutic effect of D dilutions of B, on the contraction induced by the D 45 CH dilution of B.
n = 8, p<0.05 (Chi square test)

3.3. THE EFFECT OF BELLADONNA (B) DILUTIONS ACCORDING TO THE IDENTITY PRINCIPLE ("AEQUALIA AEQUALIBUS CURENTUR") USED IN ISOPATHY

The N solutions of B had no curative effect, according to the identity principle, on the relaxation induced by the 1 C solution of B.
All the studied D dilutions of B had a therapeutic effect (in all the n = 8 cases), according to the identity principle, both on the relaxation induced by the N or D dilutions 1 C and on the contraction induced by the D dilution 45 C of B . The prerequisite was to use for the treatment, higher dilutions than the dilution used to induce the imbalance of tonicity. Thus, B in the D dilutions 5, 45, 75 and 200 C reduced the hypotonia induced by B in the 1 C dilution, both N and D. On the other hand, B in the D dilutions 55, 75 and 200 C diminished the contraction induced by B in the D dilution 45 C (Figure 3).

4. Discussion

The statistically significant results disclosed quantitative and especially qualitative differences in the pharmacodynamics of the non-dynamized (N) and dynamized (D) preparations according to Hahnemann's technique.
Quantitative differences :
-At the same degree of dilution, the D dilutions had a more intense biological effect than the N dilutions.
Qualitative differences :
-The N dilutions exerted a pharmacodynamic action only up to the limit of the molecular specificity, calculated on the basis of Avogadro's number at the 9 C dilution, with a content of 6.023×10^1 molecules/100 ml.
In contrast with the above, the D dilutions exerted a statistically significant biological effect also at high dilutions, over 10 C, which mathematically and according to Avogadro's number, do no more contain any molecule of the dissolved active substance.
-The N dilutions had unidirectional monophasic pharmacodynamic action on the 1 C-9 C scale.
The action of D dilutions, on the spasm, was bidirectional (double - way) multiphasic, with a sinusoidal course,on the 1 C - 200 C scale (Figure 2).
-Belladonna (B) in N dilutions exerted only a relaxant pharmacodynamic action on the duodenal smooth muscle, on the 1 C - 9 C scale, an effect known in the allopathic pharmacology.
In D dilutions, on the 1 C - 200 C scale, B exerted also a contracting pharmacodynamic action on the duodenal smooth muscle, action which in the allopathic pharmacology is unknown for B.
-In N solutions the effect decreases with the increase of the dilution degree and rises with the increase of the concentration and dose.
In D dilutions, the efficacy rises with the increase of the dilution degree.
-Only dynamized (D) dilutions (prepared according to Hahnemann's homeopathic technique) acted according to the identity principle ("*Aequalia aequalibus curentur*") used in isopathy.
The N solutions acted only according to the principle of contraries ("*Contraria contrariis curentur*") used in allopathy.

All these qualitative differences, detected between the N and D preparations, with respect to their biological effect, can be explained only by admitting that another form of information, qualitatively different from that existing in the N preparations, is also present in the D preparations (Cristea, 1991). The appearance of this new form of information in D dilutions can be due only to dynamization by rhythmical succussions according to Hahnemann's preparation technique.

Since in high D dilutions, exceeding 10 C, the molecule of dissolved active substance is no more present as an informational substrate, according to Avogadro's number, the information in D dilutions should be looked for in another informational substrate : a submolecular structure, a certain form of energy or, merely, a specific molecular organization of the solvent.

The clearly superior potency and efficacy of the D preparations demonstrate that the information conveyed by these preparations is much more specific to the living matter. Starting from the hypothesis that according to the laws of information processing and storage, the structural information carried by the active atropine molecule can be conveyed, under certain condition, by the solvent, we performed - concomitantly with the pharmacological study-, a research by physical methods, in order to detect in the solvent, the informational substrate of high B dilutions.

For this purpose, a calorimetric methodology was used, i.e. the high resolution mixing calorimetry (HRMC) and the differential scanning calorimetry (DSC), the sensitivity and reproductiveness of which in the study of aqueous solutions has been demonstrated on a wide series of cases (Dragan, 1987). The HRMC and DSC experiments disclosed a definite organisation of water in D dilutions, in comparison with N dilutions. Two distinct structures, with a sinusoidal multiphasic course, in accordance with the bidirectional multiphasic pharmacodynamic effects, were observed in D dilutions of B (Cristea *et al.*, 1987; Dragan, 1991).

-A systematic study on bidirectional rhythmic effect phenomenon was made (Husemann, 1992).

-The results, which we have obtained in calorimetric studies performed on aqueous *Belladonna* (B) solutions, highly diluted in a centesimal geometric progression on the 1 C - 200 C scale, are consistent also with those of other researchers which were carried out on high dilutions and lead to the same conclusions (Cabaner and Bastide, 1992; Endler and Schulte, 1994).

-The homeopathic drug, characterized by infinitesimal doses and by dynamization according to Hahnemann's preparation technique, has a marked informational character, acting in conformity with one of the laws of information action : "Minimum quantity, maximum significance, maximum effect" (Restian, 1971).

5. References

Cabaner, C. and Bastide, M. (1992) Isothermal microcalorimetric analysis of homeopathic dilutions, *6th GIRI Meeting*, Paris, France

Cristea, A. (1985) Farmacologia cibernetica (Cybernetical Pharmacology), *Nat.Conf. of Cybernetics*, Bucharest, Romania, 3.

Cristea, A., Dragan, G. and Darie, V. (1987) Potency of pharmacodynamic action and structure of highly diluted aqueous solutions of Belladonna, *Internat. Conf. on Water and Ions in Biological Systems*, Bucharest, Romania, 4.

Cristea, A., Nicula, S. and Darie, V. (1991) Pharmacodynamic effects of very high dilutions of Belladonna on the isolated rat duodenum, *5th GIRI Meeting*, Paris, France.

Cristea, A. (1991) Information - A hidden parameter of the homeopathic preparation, *Nat Conf. of Homeopathy*, Bucharest, Romania, 2.

Cristea, A. (1991) Pour une pharmacologie cybernétique et informationnelle, *Cybernetica* 2, 85-93.

Cristea, A. (1994) Homeopatia in lumina unei teorii informationale a dozelor (The homeopathy - in the light of the informational theory of doses), Abstr. Nat. Conf. of Homeopathy, Sibiu, Romania.
Cristea, A. (1994) Farmacologia informationala a dozelor, *10-th Nat. Congress of Pharmacy*, Cluj-Napoca, Romania.
Cristea, A. (1995) Conceptia informationala in farmacologie, *Farmacia*, 3-4, 1-14.
Cristea, A. (1996) For an informational theory of dosage. Illustration from cell and molecular biology and pharmacology, *Cybernetica* , (in press).
Dragan, G. (1987) Comparative study on molecular associations in solid and liquid phases of aqueous solutions. Presentation of high resolution mixing calorimetry (HRMC) and the differential scanning calorimetry (DSC), *Acta Polymerica* 4, 211-220.
Dragan, G. and Darie, V. (1991) Composite structure of homeopathic solutions of Belladonna as revealed by calorimetric measurements, *Abstr. 5-th GIRI Meeting*, Paris, France.
Endler, P.C. and Schulte, J. (Eds), (1994) Ultra high dilution. Physiology and physics, Kluwer Acad. Publ., Dordrecht.
Husemann, F. (1992) Rhythmusphanomene beim wirksamkeitsnachweiss potenzierter heilmittel - Nachgewiesen von Kolisko (1923) bis Cristea (1991), *Der Merkurstab* 2, 73-90.
Restian, A. (1971) Informatie minima, semnificatie maxima, *Rev. de Filosofie* 9, 1229-1235.
Restian, A. (1981) *Homo Ciberneticus*, St. Enc.Publisher, Bucharest, Romania.
Restian, A. (1989) *Integronica*, Ed. St., Bucharest, Romania.
Turner, R.A. (1965) *Screening methods in pharmacology*, Acad. Press, New York.

CHAMOMILLA HOMEOPATHIC DILUTION EFFECT ON CENTRAL NERVOUS SYSTEM *

An Experimental Pharmacological Study

A. CRISTEA, S. TEODORESCU-NEGRES AND V. DARIE
Lab. Pharmacology, Faculty of Pharmacy, Bucharest, Romania

1. Introduction

The concept concerning the informational causality of the drug effect (Cristea,1996) casts a new light on the importance of the biological variability and of the individual reactivity for the drug effect and pharmacotherapy. Therefore in the past years, we extended pharmacological researches even over the extremes, represented by healthy individuals, "sensitive" and "resistant" to a drug, selected from a Gauss - distributed normal population. Because for centuries, assessment of homeopathic drug effect was made on healthy humans "sensitive" to this remedy, that is why we made the studies on individual reactivity, using also a homeopathic remedy.

The *Chamomilla* remedy was especially selected by us for these studies. In allopathic therapy the *Matricaria Chamomilla* flowers are used for the following pharmacodynamical actions: analgesic, antispastic antiinflamatory, antihistaminic and antiseptical. In homeopathic therapy the homeopathic drug *Chamomilla* is used not only in unbearable colics and algia with different locations, accompanied by an excitation of the central nervous system (CNS) which results in agitation, irritability, fury, violence and aggression (including verbal), but also in convulsions consecutive to a state of furious agitation.

We studied the effects of homeopathic remedy *Chamomilla* on CNS (Cristea *et al.*, 1993), effects that are not used in allopathic medicine. The blind screening methodology, well known in allopathic pharmacology, was utilised for CNS action investigation. In our experiment, we have proceeded according to the homeopathic principle of individuals "sensitive" to certain remedy, principle which conditions the efficiency of homeopathic therapy method. In order to select the "sensitive" to *Chamomilla* remedy - type, we used a simple original procedure (Cristea, 1994; 1995).

2. Methods

2.1. TESTING SOLUTIONS

Low degree dilution, 5 C, and high degree dilution, 30 C, of homeopathic remedy *Chamomilla* (Cham), were prepared, starting from mother-tincture of Cham, by

* Communication presented at the 7th GIRI Meeting, November 1993, Montpellier, France.

M. Bastide (ed.), Signals and Images, 171–178.

successive water dilution at a 1% rate and dynamisation (D), according to Hahnemann homeopathic technique. It was processed comparatively to control groups (non-treated and treated whith the solvent represented by the corresponding dynamised (D) solution of 70 ° alcohol in distilled water).

The treatment with Cham remedy was made per os, four times at a one day-period with low degree dilutions and two times at three day-period with high degree dilutions. The tests were made three hours after the latest administration. The administration of reference drugs (amphetamine, diazepam, clonazepam, noraminophenasone) was made *per os*, in a single dose and the tests were made at a one hour-latence. Distilled water was used as solvent for the reference drugs. The volumes administered per os were 0.1ml/10 grams weight of mouse and 1 ml/100 g B.W.

2.2. ANIMALS

The chosen animal species were male albino mice and male wistar rats with an average weight of 20 and 200 g respectively.

2.3. EXPERIMENTS

The blind screening methodology, well known in allopathic pharmacology, was utilised for CNS actions investigation. Both biological actions in healthy animal (n = 20, for each group) and therapeutical effects in animals submitted to certain experimental morbide models (homologated in allopathic pharmacology) (n = 10, for each group) were studied for *Chamomilla*. The statistical significance of the results ($p < 0.05$) were highlighted in the following tests.

2.3.1. *In healthy animals, for the effect on the CNS tone:*

- actometric test using a photoelectric cells-actometer, for spontaneous motility, was expressed in number of movements / minute, for each individual,
- platform test using a circular platform, for evasion - investigation reflex, was expressed in number of emergences / minute, for each individual,
- wire test using a stretched wire, for redressement reflex,was expressed in number of animals redressed / group,
- rotational axe test using a classical rotational axe, for nimbleness, was expressed in number of nimble animals / group,
- group toxicity test, at crowd, without amphetamine, was expressed in number of dead animals / group, at two hours.

2.3.2. *Experimental morbide models*:

- the actometric test, platform test, wire test and rotational ax test were used for the tranquilizing effect,
- chemical nociceptive stimul test using acetic acid 0,6%, 0.1 ml / 10 g B.W., i.p., for analgesic effect,
- chemical - induced convulsions test using strychnine 0.8 mg / kg B.W., s.c., for anticonvulsivant effect,
- chemical - induced narcose test using hexobarbital 80 mg / kg B.W, i.p., for hypnotic effect.

International bioethic regulations of researches on animal were complied with.

2.4. PROCEDURES FOR SELECTION OF "SENSITIVE" TYPE

We found and developed a simple and safe methodology for selecting the healthy type "sensible" to the action of an allopathic or homeopatic drug (Cristea, 1994; 1995). This methodology consists of the following:

- establishing the physiological function appropriate for the pharmacological investigation of the drug being studied;
- determining the level of that physiological function in every individual from the group suitable for the experimental or clinical study (at least 20 individuals and optimally 100 individuals);
- tracing the Gauss bell-shaped curve, using the distribution frequency of the values of the investigated physiological function in the group subject to selection;
- dividing the individuals into normal ones, the function being at level f = average f±1 (about 67% of individuals) and extreme ones with f > average f+1 and f < average f-1 (33% of the individuals);
- selecting the healthy individuals "sensitive" to the drug under study, either those with the function f > average f+1 or those with f <average f-1 (16.5% of the individuals);
- classifying the "sensitive" healthy individuals by groups, for purpose of the pharmaceutical research.

We tested the described methodology for the selection of the "sensitive" healthy individuals, in both allopathic and homeopathic pharmacology.

We have chosen the actometric test, in order to select the "sensitive" type to Cham remedy, which test led us to divide the animal collectivity, according to Gauss bell shaped distribution curve, in animals with medium spontaneous motor activity ("normals") and in animals with hypo- or hypermotility ("sensibles") (Figure 1).

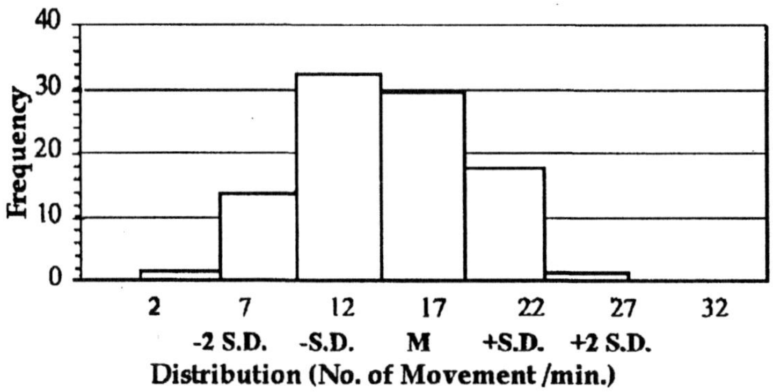

Figure 1. Gauss bell-shaped curve, in actometric test for the selection of the normal and the extreme types of CNS tone

Gauss' normal distribution	Theoretical	Experimental
Individuals between -1S.D. and +1S.D.	67%	65%
Individuals between -2S.D. and +2S.D.	95%	96%

Statistical analysis: N = 100; Mean = 16.684; Median = 17; Minimum = 2; S.D. = 5.089; SEM = 0.522; Sum = 1585; Variance = 25.899

2.5. STATISTICS

The results were statistically processed by the Student "t" and Wilcoxon tests for the values expressed in number of obervations (%) at each individual and by Chi square test for values expressed in number of indivuduals (%) being noticed in each group. The effects (%) were calculated compared to a control group treated with the solvent.
From the point of view of bioethics, the number of animals per group, was twenty for healthy animals (n = 20) and ten for animals submitted to experimental disease models (n = 10).

3. Results

The low degree dilution, 5 C, of Cham had an action in the stimulatory sense, on "sensitive" healthy mice, with spontaneous hypomotility, (n = 20), as it is shown in figure 2.

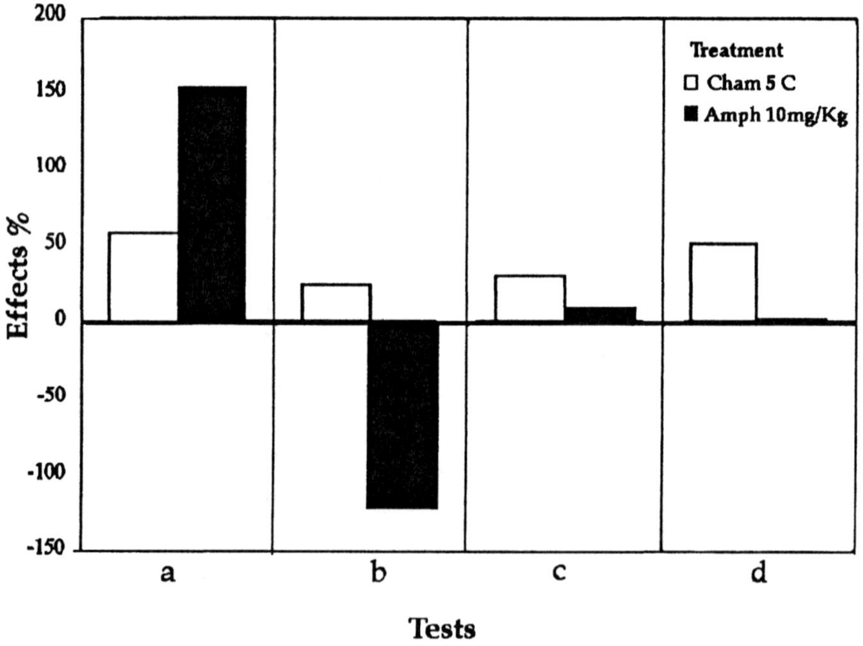

Figure 2. Stimulatory actions in healthy "sensitive" mice (with hypomotility); n = 20 (for each test)
a) Increase of motility (%); $p < 0.01$ (Student's t test)
b) Modification of evasion-investigation reflex (%); $p < 0.05$ (Student's t test)
c) Increase of redressement reflex (%); $p < 0.05$ (Chi square test)
d) Increase of nimbleness (%); $p < 0.05$ (Chi square test)

- increasing the motor activity by 58% ($p<0.01$) at 5 C (D) Cham; a dose of 10 mg/ kg B.W. per os of amphetamine (Amph) produced a greater increase in spontaneous motility (154.06% ; $p<0.001$);
-increasing the evasion-investigation reflex by 24.18% ($p<0.01$) at 5 C (D) Cham; to the contrary, a dose of 10 mg/kg weight per os of Amph produced an important decrease of evasion- investigation reflex (121.36%);
-increasing the redressement reflex by 30% at 5 C (D) *Chamomilla*; Amph increased redressement reflex of 10%;
-increasing the nimbleness by 50% at 5 C (D) *Chamomilla*; Amph had no effect.
The low degree dilution also induced in "sensitive" healthy mice with spontaneous hypomotility a state of excitement on CNS, reported in the group toxicity test (n = 10) without Amph, resulting in group lethality of 43.2% at 5 C (D) Cham, compared with 5 C (D) alcohol; Amph induced a 100% lethality, compared with water as solvent, at 1 1/2 hours.
The high degree dilution, 30 C, also induced the tranquilizing effect reported in the "sensitive" healthy mice with hipermotility (n =20), by way of (Figure 3)
-decreasing the motility by 31.68% at 30 C (D) Cham ($p<0.01$); a dose of 5 mg/ kg B.W. of diazepam (Diaz) per os produced a decrease in spontaneous motility of 29.41% ($p<0.01$);
-decreasing the evasion-investigation reflex by 56.03% at 30 C (D) Cham ($p<0.01$); the Diaz produced a decrease of evasion-investigation reflex (61.92%);
-decreasing the redressment reflex by 20% at 30 CHD Cham; Diaz produced a decrease of 20%;
-the nimbleness was not significantly modified at 30 C (D); Diaz decreased by 10% the nimbleness.

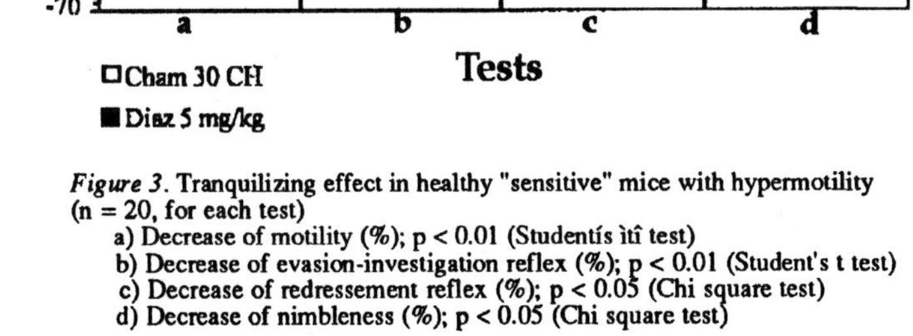

Figure 3. Tranquilizing effect in healthy "sensitive" mice with hypermotility (n = 20, for each test)
a) Decrease of motility (%); $p < 0.01$ (Studentís ìtî test)
b) Decrease of evasion-investigation reflex (%); $p < 0.01$ (Student's t test)
c) Decrease of redressement reflex (%); $p < 0.05$ (Chi square test)
d) Decrease of nimbleness (%); $p < 0.05$ (Chi square test)

The high degree dilution, 30 C, induced the following pharmacotherapeutical effects, in animals in which a disease syndrome has been experimentally induced:
-anticonvulsivant action in mice (n = 10) has been manifested by decreasing lethality by 30% at 30 C (D) *Chamomilla;* the clonazepam (Clonaz) 1 mg/kg B.W. decreased the lethality by 50%:

-analgesic action in mice (n = 10) was distinguished by decreasing the number of contorsions 33.8% at 30 C (D) *Chamomilla* (p<0.01); noraminophenasone (Noraminophen) 100 mg/kg B.W., per os, induced a decrease of 51.1% in the number of contorsions (p<0.001).

The effects on hexobarbital-induced narcose were:

-antagonizing, while using low degree dilution, 5 C, in "sensitive" healthy mice with hypomotility, (n = 10), as follows: 5 C (D) Cham decreased by 42.9% the percentage of narcosed mice, increased by 79.06% the latence of narcose (p<0.001) and decreased by 21.9% (p<0.05) the period of narcose;

-potencying, while using high degree dilution, 30 C, in "sensitive" healthy mice with hypermotility, (n = 10), as follows: 30 C (D) Cham decreased by 21.46% (p<0.01) the latence of narcose.

The Diaz decreased by 39.51% (p<0.01) the latence of narcose and increased by 175.76% (p<0.01) the period of narcose.

No neuroleptic action was noticed in the catalepsy test and postural test as well as in the palpebral ptosis test.

No miorelaxant action was recorded in the ipsilateral flexion reflex test and postural test. The cornean and pineal reflexes were also not been modified as well.

No opioid-like analgesic action was noticed in Hafner test. For this reason, the analgesic action, statistically significant performed in chemical nociceptive stimulation test, can only be of a minor analgesic type and not of morphine-like type or can result from a known peripheral antispastic action of Cham.

The results including their statistical significance are summarized in Table 1.

TABLE 1: Summary of the Results

Test	Type of motility	Treatment	n	Effect %±SEM	P	
Motility	Hypo	Cham 5 C	20	+58 ± 1.55	<0.01	Student's
		Amph	20	+154.06 ±1.18	<0.001	"
	Hyper	Cham 30 C	20	-31.68 ± 0.93	<0.01	"
		Diaz	20	-29.41 ± 1.38	<0.01	"
Evasion Investi-gation Reflex	Hypo	Cham 5 C	20	+ 24.18 ±1.48	< 0.05	Student's
		Amph	20	- 121.36 ± 2.43	< 0.01	"
	Hyper	Cham 30 C	20	- 56.03 ± 1.60	< 0.01	"
		Diaz	20	- 61.92 ± 1.72	< 0.01	"
Analgesic Reflex	Hypo	Cham 5 C	10	+33.8 ± 1.79	< 0.01	"
		Noraminophen	10	+ 51.1 ± 1.52	< 0.001	"
Narcose with hexobar-bital Latence/ Length	Hypo	Cham 5 C	10	+ 79.06±0.24	< 0.001	"
				-21.90± 1.54	< 0.05	"
	Hyper	Cham 30 C	10	-21.46± 0.06	< 0.01	"
				+23;68 ±4.55	<0.05	Wilcoxon
		Diaz	10	-39.51 ± 0.106	<0.01	Student's
				+175.76 ± 12.08	<0.01	Wilcoxon

4. Discussion

The results of these experimental pharmacological researches indicate, for the homeopathic remedy Cham, a stimulatory action on CNS, performed at low degree dilution, 5 C, in "sensitive" healthy animal organism with hypomotility; it also validates the stimulating effects on CNS described in the homeopathic pathogenesis of this remedy.

The comparison with amphetamine, considered as a reference substance, suggests that the stimulating effect on CNS by the homeopathic remedy Cham is not an amphetamine-like effect. The stimulatory action on CNS is usually related to the polinic derivates located in the aerial part of the plant and quoted for anticonvulsivant effect (Guermonprez *et al.*, 1989).

The results also show some inhibitory effects, namely inhibition of the CNS (tranquilizing, analgesic, anticonvulsivant), significantly performed at high degree dilutions, 30 C, either in "sensitive" healthy animal organism with hyperexcitability and hypermotility or in animal organism submitted to experimental morbid syndrome such as pains and convulsions.

Our experimental pharmacological researches complete the researches performed by Guillemain (1983). The cited author reported that, at 3 C and 4 C dilutions, the Cham remedy significantly diminishes the number of fights in rats after exposure to electric shocks. Moreover, using 3 C and 4 C Cham or 5 C *Coffea* treatment, the author recorded decreased coffein-induced motor hyperactivity in male rats.
These pharmacotherapeutical effects prove experimentally the validity of the homeopathic therapeutical principle of similitude ("*Similia similibus curentur*").

Concerning the value of this therapeutic principle, we discussed in another paper (Cristea, 1996), both informationally and cybernetically, the four physiological adjustement modalities, i.e. iso-, homeo-, allo- and enantio-regulation, in the biocybernetical mechanisms i.e. feed-back and feed-before (feed-forward).

It is in worth saying, that the effects observed in our experiments have had, without exception, a highly statistical significance ($p<0.01$) in "sensitive" extreme individuals and were statistically non significant ($p>0.05$) in "resistent" extreme individuals. Thus, we successfully demonstrated the quieting effect of the homeopathic remedy Cham, at high-degree dilution 30 C, in "sensitive" healthy mice, with spontaneous motor restfulness. We were also able to reproduce the experimental pathogenesis of the remedy Cham, in low-degree dilution of 5 C, refering to restfulness in "sensitive" healthy individuals, selected for their spontaneous motor hypoactivity. These results prove the objective value of experiments in healthy individual, though "sensitive" to certain remedy, promoted by Hahnemann.

In our opinion, the "sensitive" and "resistent" extreme individuals, within a population with normal Gauss-like biological variability, represent psycho-neuro-endocrine typologies which should be the object of future biological and pharmacological studies.

This assessment is based on our allopathic pharmacological studies on "sensitive" individuals (Cristea 1993, Cristea, 1994). Using our method in the allopathic pharmacology studies on the relationship between the "alarm" adrenergic system and the "modulator" endogenous opioid system, we were able to select the adrenergic type individuals (A) (described by Rosenman, 1970) and the opioid type individuals (O) (described by Cristea A., 1993), both experimentally and clinically. The experimental pharmacology research made on type A and type O individuals proved that type A individuals are "sensitive" to Clonidine (Clon) while type O individuals are "sensitive" to naloxone (Nalx). Also we emphasized on the dose-effect curves, the double-way effect of Clon and Nalx, on stress analgesia, with distinct inflexion points, for the normal and the two extremes individuals of type A and type O (Cristea *et al.*, 1995). Ignoring or analysing the biological variability represents a real problem evolving from european pharmacological debates (Black, 1995).

In conclusion, we consider that the results of these studies justify the opportunity of future researches concerning the intimate mechanism of the exciting action on CNS, of homeopathic drug Chamomilla, an action which explains the quieting effect, in accordance with the therapeutical principle of similitude.

5. References

Black, J. (1995) Biological variability: ignore, reduce or analyse?, *1st European Congress of Pharmacology*, Milan, Italy.

Cristea, A. (1993) Opioid and Adrenergic Types of behaviour to stress, VIth Congress of ECNP, Budapest, Hungary; *Europ. Neuropsychopharmacol.* 3, 446-447.

Cristea, A. (1994) Mechanisms of stress analgesia in opioid and adrenergic types of behaviour, VIIth Congress of ECNP, Jerusalem, Israel; *Europ. Neuropsycho-pharmacol* 3 (special issue).

Cristea, A.(1994) Highlighting the healthy individual type "sensitive" to a drug - Methodology, *8th GIRI Meeting*, Jerusalem, Israel.

Cristea, A.(1995) Metodologie pentru evidentierea tipului de individ sanatos, "sensibil" la un medicament (Highlighting the healthy individual type "sensitive" to a drug. Methodology), *Farmacia* 1-2, 17-19.

Cristea, A., Negres, S., Joean, D. (1995) The double-way effects of naloxon and clonidine, on experimental stress analgesia, *Rom. J. Physiology* 1-4, 137-143.

Cristea, A. (1996) Informational pharmacology, *X Intern. Congress of Cybernetics and Systems*, Bucharest, Romania.

Cristea, A. (1996) Iso-, Homeo-, Allo- and Enantio-regulation of the biocybernetic systems, *X Intern. Congress of Cybernetics and Systems*, Bucharest, Romania.

Guermonprez, M., Pinkas, M., and Torck, M.(1989) Matière Médicale Homéopathique, Boiron Publisher; Lyon, pp. 188-193.

Guillemain, J. (1983) La recherche pharmacologique appliquée à l'Homéopathie, sa grandeur, ses faiblesses, *Homeop. Fr.* 71, 191-200.

Rosenmann, R.H. (1980) Personality type A behaviour pattern and coronary heart disease, *R. Soc. Med.* 14, 65-70.

HIGH DILUTIONS OF ANTIGENS MODULATE THE IMMUNE RESPONSE TO KLH *

Modulation of the Specific Immune Response

Z. WEISMAN, M. OBERBAUM, R. TOPPER, N. HARPAZ, AND Z. BENTWICH
R. Ben Ari Institute of Clinical Immunology, Kaplan Hospital, Rehovot; The Hebrew University Hadassah Medical School, Jerusalem, Israel

1. Introduction

Homeopathy, despite being practiced by large number of people around the world (Einsenberg *et al.*, 1993; Kewal, 1995; Fisher and Ward, 1994) is not acceptable by the established medical and scientific community mainly because of the lack of acceptable scientific explanation, or theory, that would account for its claimed effects (Delbanco, 1994; Campion, 1993; Skrabanek, 1986). The possibility that extremely diluted solutions, that do not contain any trace of the solute, would still retain any biological effect of that solute is practically considered heretical, or at least unacceptable, according to the existing scientific paradigms. It is therefore of utmost importance to try and address this general issue by accepted conventional methods, and this has motivated the present study. In this study we have tested the possibility that highly diluted "homeopathic" antigen solutions retain the ability to generate, or influence, a specific immune response to such antigen in an animal model.

2. Material and Methods

2.1 MICE

C57Bl/6J, pathogen-free 6-8 wk old females, were obtained from The Jackson Laboratory (Bar Harbor, ME). For some experiments inbred mice were used (Ehud Hepher, Hulda, Israel). The animals were permitted access to food and water *ad libitum*, and 12 h light (0700-1900), 12 h dark / day .

2.2 ANTIGENS

Keyhole Limpet Hemocyanin (KLH) (Calbiochem Corporation, La Jolla, CA 92037, cat. no. 3748411) and Porcine Thyroglobulin (Tg) (Sigma, Israel, cat. no.T-1126) were dialyzed , and sterile stock solutions of 3mg/ml in saline were kept at 4°C (sterilized by using 0.45 filters).

* Communication presented at the 8th GIRI Meeting, December 1994, Jerusalem, Israel.

M. Bastide (ed.), Signals and Images, 179–189.

2.2.1 *Dilutions of the antigens*
KLH, Tg, and saline (as a control solution) were diluted in steps of 1:100. For the first dilution, designated 1 C, 5 μl of the stock solution was added to 5 ml sterile water (pyrogen free) in conical 10ml plastic sterile tubes (Sterilin). The tubes were shaken vigorously for 3 min by a Dolisos dynamisator, followed for 1 min. by a Boiron dynamisator. This procedure was repeated until one dilution before the last. The last dilution was prepared similarly using sterile saline instead of water.

2.3 EXPERIMENTAL DESIGN

2.3.1 *Preconditioning of the mice*
The animals were divided at random into groups of 8-10 animals per cage. The mice were injected intraperitoneally (i.p) with 0.3 ml of either antigen or saline, at different dilutions, 3 times per week during 8 weeks. In some experiments the animals were preconditioned orally by adding antigens or saline dilutions to the drinking water (1:100) and mixing vigorously for 2 min by dynamisator.

2.3.2 *Immunization*
At the end of the preconditioning period all the mice, including sex and age matched naive controls that were kept untreated during the preconditioning time, were immunized with 1 μg KLH per mouse in Complete or Incomplete Freund's adjuvant (CFA, ICFA) diluted v/v with Phosphate Buffer Saline (PBS). Immunization with CFA was given in the tail base while ICFA was injected in the foot pads.

2.3.3 *Bleedings*
The mice were bled, by orbital sinus puncture, at intervals during the preconditioning period, and after immunization. Serum was separated and kept at -20 °C until assayed for antibodies titer.

2.4 IMMUNOLOGICAL ASSAYS

KLH IgM and IgG antibodies titers were assayed by ELISA in serum of individual mice. Flat bottom microtiter plates (Nunc-Immuno Plate MaxiSorp) were coated overnight at 4 °C with 5 μg/100 μl/well of KLH, Tg or Rabbit Serum Albumin (RSA) (Sigma, Israel) dissolved in PBS. After blocking with 2.5% fetal calf serum in DMEM medium containing 0.02% Hepes and 0.05% sodium azide, serum dilutions (1:1000 or 1:10000 for IgM and IgG respectively) were added, and the plates were incubated for 2 h. The reaction was developed with alkaline phosphatase-conjugated goat anti mouse IgM or IgG (μ or γ-chain specific respectively)(Sigma, Israel), followed by alkaline phosphatase substrate (Sigma 9389). After 30 min incubation the optical density was recorded at 405 nm. The results are represented by the mean value of each experimental group, and are expressed in optical density units or as percentages of normal untreated mouse serum (NMS).

Lymphocyte proliferation was determined by plating 5×10^5 spleen cells in 0.2ml RPMI medium enriched with 1% fresh NMS, 2 mM glutamine, 1% Non Essential Amino Acids, 1 mM Sodium Pyruvate, 5×10^{-5}M 2-Mercaptoethanol, and antibiotics in flat bottom microtiter plates (Nunc). The cells were cultured for 5 days in medium alone or in medium containing one of the following: Concavalin A (con A)

0.8μg/ml, lipopolysacharide (LPS) 4 and 8 μg/ml, KLH/Tg 5, 25, and 50 μg/ml. The cultures were set in triplicates. ^{3}H Thymidine (0.5 Ci/well) was added to the cultures for the last 18 h, the cells were harvested, and the radioactivity was counted by a β-counter (cells with con A and LPS were labeled and harvested at day 3). Proliferation was expressed as stimulation index (S.I). S.I = The maximum counts per minute (cpm) in the presence of antigen divided by the cpm in medium alone.

2.5 STATISTICAL ANALYSIS

The Anova, Multiple measurements analysis of variance (Anova) and the Student's t-test were used. $P < 0.05$ was considered as significant.

3.Results and discussion

3.1 THE CONCENTRATION OF ANTIGENS

The concentration of antigens and the total number of molecules injected to each mice during the 8 weeks of preconditioning with the various dilutions are summarized in table 1. As can be seen in the table, no antigen is present any more in the 15 C dilution which is beyond the Avogadro number.

TABLE 1: The number of molecules of antigen injected to each mice at different dilutions

Dilution	Molarity	# of injections	total # of molecules/mouse
4 C	10^{-14}	24	48,106
6 C	10^{-18}	24	4800
7 C	10^{-20}	24	48
15 C	10^{-36}	24	0

3.2 THE EFFECTS OF PRECONDITIONING BY INJECTIONS

3.2.1 *Antibody generation during preconditioning*
Repeated i.p injections of high dilutions of KLH, Tg or saline induced generally a similar increase in serum IgM (Fig.1). The response was not antigen specific since similar titers were observed with assays for antibodies to KLH, RSA or Tg (Data not shown). There was no induction of IgG antibodies by the preconditioning treatment.

3.2.2 *Antibody response to immunization with KLH*
IgM response. In 2/6 experiments the IgM response to immunization with KLH was significantly different in mice preconditioned by injections of high KLH dilutions. In the first experiment (immunization with CFA), all the KLH pretreated mice had showed significantly lower IgM titers compared with the control naive mice, 14 days post immunization ($P<0.05$ student 's t-test) (data not shown). These results

could not be repeated in a following similar experiment in which all the mice, whether pretreated with KLH or not, responded similarly to immunization with KLH. Therefore, the immunization procedure was changed, and ICFA was used instead of CFA, hoping to slow down the immune response of the animals, so it will be able to distinguish between variations in the kinetics of the immune response. However, in only 1/4 such experiments did we find a significant difference in the kinetics of IgM response by KLH preconditioned mice ($P<0.05$ anova). As can be seen in Figure 2, mice that were pretreated with the 7 C and 15 C KLH dilutions show an earlier and higher IgM anti KLH response compared with saline injected mice or naive mice.

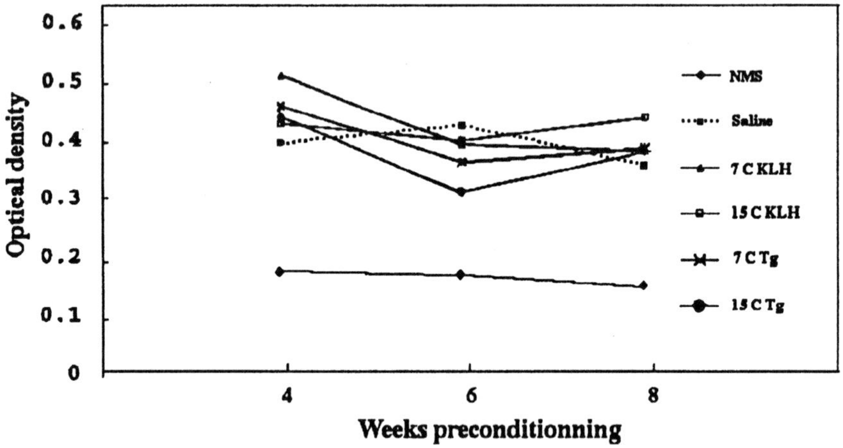

Figure 1. Anti KLH IgM response. Mice were treated by repeating i.p injections of saline, KLH and Tg. The results represent the mean antibody titers of each group.

In all the other experiments the IgM response was similar in all the groups. Since repeated i.p injections of mice evoked mostly a polyclonal IgM response (see figure 1), a possible modulation of a specific anti KLH response by the treatment could be obscured.

IgG response. In contrast to the IgM response, preconditioning with high diluted antigen solutions clearly affected the specific anti KLH secondary response of the animals in 4/6 experiments. A very dramatic effect was observed in the first experiment. As can be seen in Figure 3, at 14 days post immunization, most of the mice that were preconditioned by injections of high dilutions of KLH switched to IgG response, while the control naive mice still show low titers of IgG and high titers of IgM (data not shown).

In 3/4 additional experiments, in which ICFA was used for immunization, a significant increase in IgG response was observed in the 15C KLH preconditioned group, and in two experiments also in the 7C KLH preconditioned group (Figures 4 and 5).

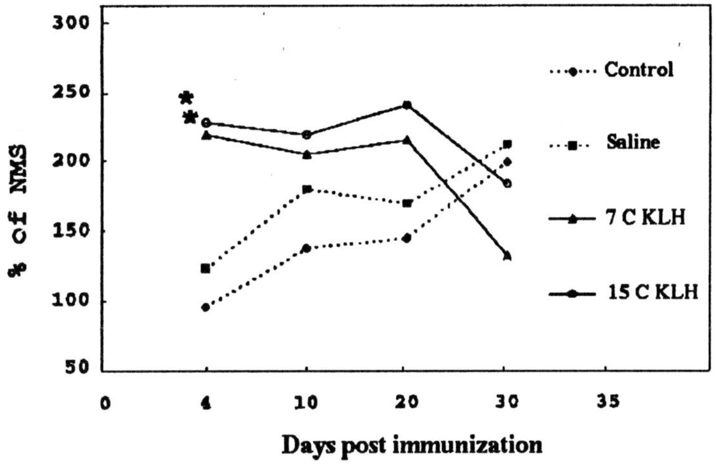

Figure 2. Kinetics of anti KLH IgM post immunization response. Mice were preconditioned for 2 mo. by i.p injections with 7 and 15 C KLH and saline dilutions. The treated and naive mice (control) were then immunized in the foot pads. The results are expressed as percentage of IgM titer of untreated mice.
* significant different kinetics from control and saline groups (Anova).

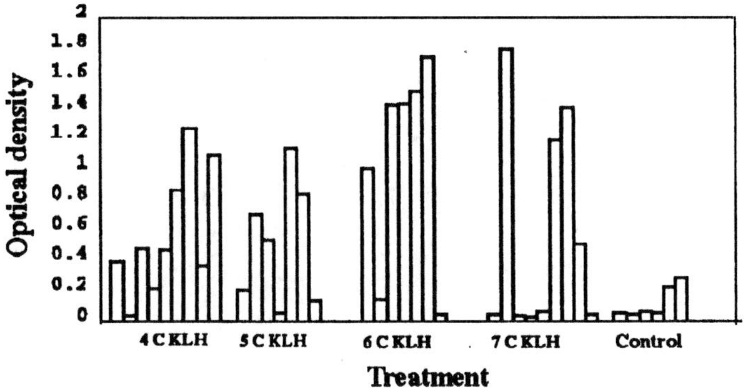

Figure 3. Anti KLH IgG response 14 days post immunization. Mice were injected i.p with several dilutions of KLH for 2 months. The mice were immunized with KLH/CFA in the tail. Each bar represent one animal.

To check for the specificity of the effect of KLH, the antigen Tg was included in some experiments. Different groups of mice were treated with 7 C and 15 C dilutions of KLH and Tg, as well as with the 7C dilution of Saline (Figures 4 and 5).

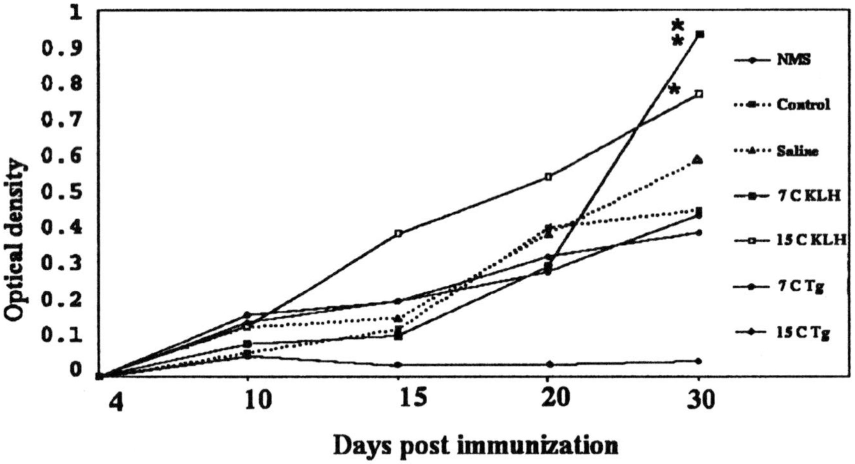

Figure 4. Kinetics of anti KLH IgG response. Mice were preconditioned with repeated i.p injections of saline, KLH and Tg dilutions for 2 months and subsequently immunized with KLH/ICFA.
* - significantly different from the control and saline preconditioned mice (Anova).
**- Student t-test.

In Figure 4, it can be seen that the kinetics of the IgG response was significantly different in the 15 C KLH pretreated mice ($P=0.055$, anova multiple measurements analysis of variance, and $p<0.01$ student's t-test at day 15). The effect of preconditioning by 7 C KLH dilution was evident only 30 days post immunization. Saline or Tg dilutions did not affect the subsequent immune response to KLH. The response was anti KLH specific since the ELISA assays on RSA or Tg coated plates were negative. However, as can be seen in Figure 5, preconditioning with 15 C Tg also increased significantly the IgG response similar to that of the KLH preconditioned mice. The differences in the kinetics of the response of all the preconditioned groups, with the exception of 7 C Tg , compared to the control groups, were highly significant ($P<0.02$ Anova).

Taken together, it can be summarized that preconditioning with high dilution of antigen accelerated significantly the secondary IgG response to the antigen. The central issue of the specificity of this response is not yet resolved. It is possible that KLH and Tg cross react with each other, both being very large molecules. Such experiments need to be repeated, using additional antigens both in the preconditioning treatment and in immunization.

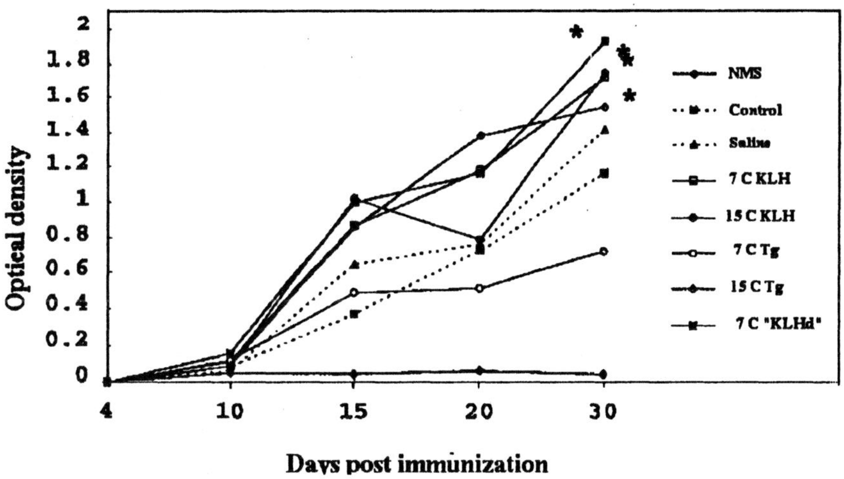

Figure 5. Kinetics of anti KLH IgG response. Mice were preconditioned with repeated i.p injections of saline, KLH and Tg dilutions for 2 mo.nths and subsequently immunized with KLH/ICFA. Group " KLH 7d " was orally preconditioned by adding 7 C KLH to the drinking water.
* - significantly different from the control and saline preconditioned mice (Anova).

3.3 THE EFFECT OF ORAL PRECONDITIONING

Oral antigen treatment was given through the drinking water (3 experiments).

3.3.1 *IgM response*

There was no IgM response in the animals during the pretreatment period. After immunization with KLH, there was no difference between the different groups, whether preconditioned or not.

3.3.2 *IgG response*

In 2/3 experiments the kinetics of the anti KLH IgG response was different in the preconditioned groups compared with the control naive mice or the saline preconditioned one. In Figure 5 it can be seen that mice receiving the 7 C KLH in the drinking water ("KLH 7d") showed a significantly faster increase in IgG titers compared with the control groups, similarly to the kinetics shown by the groups which were preconditioned by injections. In another experiment (Figure 6) oral treatment by 7 C KLH , as well as by 7 C Tg , increased significantly the IgG response to immunization with KLH at day 11 post immunization (the increase in IgG in the 15 C KLH was not significant).

Taken together, these results highly suggest that preconditioning of mice by high dilution of antigens, delivered orally, can transfer signals to the immune system and modulate its response to subsequent immunization.

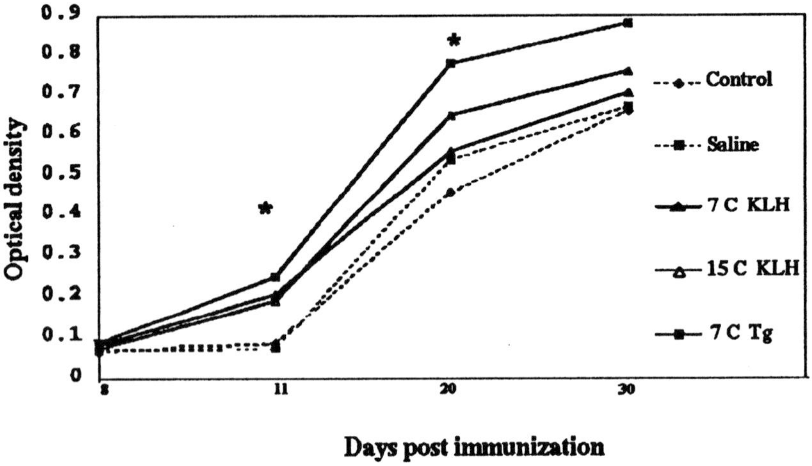

Figure 6. Kinetics of anti KLH IgG response. Mice were preconditioned orally with KLH and Tg dilutions in the drinking water for 2 months and subsequently immunized with KLH/ICFA.
* significantly different from the control naive and saline preconditioned mice ($p<0.001$, student's t- test).

3.4 THE EFFECT OF PRECONDITIONING ON THE PROLIFERATIVE RESPONSE

A different approach to study the effects of preconditioning by high dilutions of antigens is by testing the proliferative responses of lymphoid cells, in vitro, to the immunizing antigen. The results of two such experiments are described.

Spleen cells, obtained from animals preconditioned with high dilutions of antigens and from naive animals, 10 days post immunization with KLH/ICFA were cultured in the presence of KLH and Tg for 5 days. (con A and LPS were added to the cultures as positive controls for testing the activity of the cells).The proliferative response was measured by the uptake of labeled thymidine, and was expressed as stimulation index; a stimulation index higher than 2 was regarded as a positive response. In Figure 7 it can be seen that spleen cells from naive or saline preconditioned mice (4 mice) did not respond to KLH or Tg added to the cultures. However, 5/6 KLH (7 C and 15 C) preconditioned mice responded to KLH and 1 responded also to Tg. It can be also seen that 1/3 mice preconditioned with Tg (7 C) responded to both KLH and Tg.

In the second experiment the animals were preconditioned by oral treatment. The results are summarized in Figure 8. In this experiment, at 10 days post immunization with KLH, 15/16 animals responded already to KLH stimulation in vitro (S.I > 2). However, the highest proliferative responses to both antigens were observed in preconditioned mice.

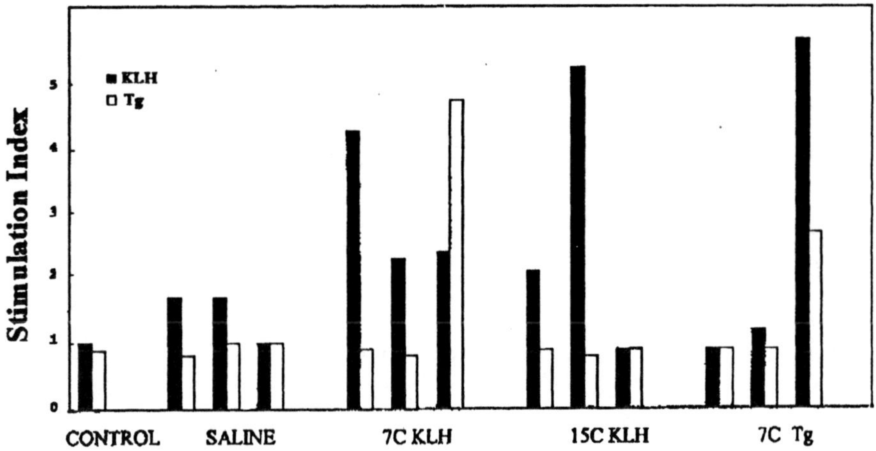

Figure 7. Proliferative response of spleen cells to KLH and Tg. Mice were preconditioned for 2 mo by repeated i.p injections of high dilutions of KLH, Tg and saline, and then immunized with KLH/ICFA. Ten days post immunization the spleen cells were cultured in the presence of the antigens KLH and Tg. The results are expressed as stimulation index, each bar represent the response of one mouse.

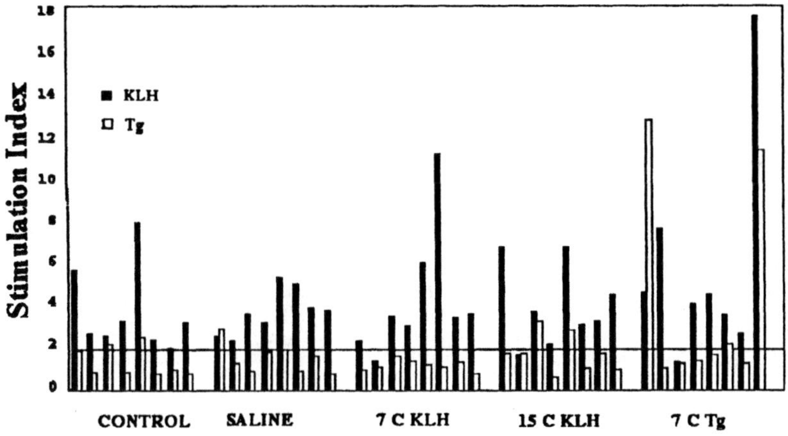

Figure 8. Proliferative responses of spleen cells to KLH and Tg. Mice were oral preconditioned for 2 mo by adding high dilutions of KLH (7 and 15C), Tg (7C) and saline (7C) to the drinking water, and then immunized with KLH/ICFA. Ten days post immunization the spleen cells were cultured in the presence of the antigens KLH and Tg. The results are expressed as stimulation index, each bar represent the response of one mouse.

In both experiments, preconditioning of the mice with high dilution of antigen probably modulated the response of the mice to subsequent immunization. This was reflected by an increased response of the spleen cells to the secondary challenge with the antigen in vitro. The significance of the response to Tg by spleen cells from KLH immunized mice is not clear. It may suggest cross reaction between the 2 antigens, but it can also be the effect of preconditioning with high dilution of the Tg. More experiments are needed to define the best conditions for this type of experiments, and the kinetics of the response.

The cumulative results of the experiments described above strongly support the possibility that antigen administered in high diluted form, whether orally or parenterally, does have a modulating and specific effect on the immune response to that antigen. This effect was observed for both humoral (IgG) and cellular responses, and was observed in animals preconditioned even with antigen diluted beyond Avogadro's number. This in itself is of great importance in lending support to the basic concept , that highly diluted substances retain their biological activity even in this form. Similar results in other experimental systems have been published in the last decade (Bastide *et al.*, 1985; Bastide *et al.*, 1987; Daurat *et al.*, 1988; Davenas *et al.*, 1987; Endler *et al.*, 1991; Fisher *et al.*,, 1987; Sukul et al., 1986). Furthermore, our studies show that immunological information can be transferred by very high diluted substances, and can have an imprint in the form of antigen memory. The issue of the specificity of this response, though highly suggestive, is not clearly resolved and requires further studies. Especially due to possible cross reactivity between antigens like thyroglobulin and KLH. It would be advantageous to use other, and possibly smaller, antigens to try and resolve this issue. Also, adoptive transfer experiments to show definitively the transfer of immunological memory, will be most critical for better definition of these effects.

Another clearly bothersome aspect of these studies is the variation in the response and reproducibility of the observed effects. The reasons for these variations could probably be the results of a number of variables. Since the system developed by us is very likely to be extremely sensitive to small environmental and experimental changes, such results may not be that surprising (Doucet-Jaboeuf *et al.*, 1984; Carrière *et al.*, 1989). These type of experiments have to be repeated many times in order to get reliable statistical results.

4 References

Bastide, M., Daurat, V., Doucet-Jaboeuf, M., Pelegrin, A., Dorfman, P. (1987) Immunomodulator activity of very low doses of thymulin in mice, *Int.J.Immunother.* **3**, 191-200.

Bastide, M., Doucet-Jaboeuf , M., Daurat, V. (1985) Activity and chronopharmacology of very low doses of physiological immune inducer, *Immunol.Today* **6**, 234-235.

Campion, E.,W. (1993) Why unconventional medicine?, *New England J. Med.* **328**, 282-283.

Carrière,V., Dorfman, P., and Bastide, M. (1989) Evaluation of various factors influencing the action of mouse αβ interferon on the chemiluminescence of mouse peritoneal macrophages, *Annual Rev. of Chronopharmacol*. **5**, 9-12.

Daurat, V., Dorfman, P., and Bastide, M. (1988) Immunomodulatory activity of low doses of αβ interferon in mice, *Biomed. & Pharmacother.* **42**, 197-206.

Davenas, E., Poitevin, B., Benveniste, J. (1987) Effect on mouse peritoneal macrophages of orally administered very high dilutions of silica, *Europ. J.Pharmacol* .**135**, 313-319.

Delbanco, T., L. (1994) Bitter Herbs: Main stream, Magic, and Menace, *Ann. Intern. Med.* **121**, 803-804.

Doucet-Jaboeuf, M., Pelegrin, A., Cot, M.C., Guillemain, J., and Bastide, M. (1984) Seasonal variations in the humoral immune response in mice following administration of thymic hormones, *Annual Rev. of Chronopharmacol* .**1**, 231-234.

Eisenberg, D.M., Kessler, R.C., Foster, C., Norlock, F.E., Calkins, D.R., and Delbanco,T.L. (1993) Unconventional medicine in the United States. Prevalence, costs, and patterns of use, *The New England J. of Medicine* **328**, 246-252.

Endler, P.C., Pongratz, W., Van Wijk, R., Kastberger, G., Haidvogl, M. (1991) Effects of highly diluted succussed thyroxine on metamorphosis of highland frogs, *Berlin J.Res. Hom.* **1**,151-160.
Fisher, P., House, I., Belon, P., and Turner, B. (1987) The influence of the homoeopathic remedy Plumbum metallicum on the excretion kinetics of Lead in rats, *Human Toxicology* **6**, 321-324.
Fisher, P.,and Ward, A. (1994) Complementary medicine in Europe, *B.M.J.* **309**,107-111.
Kewal, J. (1995) Swiss embrace complementary Medicine, *Nature Medicine* **1**, 107.
Skrabanek, P. (1986) Demarcation of the absurd, *The Lancet*, April 26, 960-961.
Sukul, N.C., Bala, S.K., and Bhattacharyya, B. (1986) Prolonged cataleptogenic effects of potentized homoeopathic drugs, *Psychopharmacology* **89**, 338-339.

HEALING CHRONIC WOUNDS PERFORMED ON MOUSE EARS USING SILICA (SiO2) AS A HOMEOPATHIC REMEDY *

A Pharmacological Study of Homeopathic High Dilutions

M. OBERBAUM, Z. WEISMAN, A. KALINKOVICH AND Z. BENTWICH
Ruth Ben-Ari Institute of Clinical Immunology, Kaplan Hospital, Rehovot; The Hebrew University Hadassah Medical School, Jerusalem, Israel

1. Introduction

Homoeopathic medicine frequently uses potencies of silica, especially as a treatment for abscesses, discharging fistulas, chronic ulcers, discharging wounds and keloids and rejection of foreign bodies (Jouanny, 1985; Metzger, 1981). It is also known as a deep-acting constitutional remedy. However, one major advantage of experimenting with potencies of silica is the fact, that it can be applied indicatively without the need to individualize. This characteristic is not common in homeopathy, since individualization is one of the basic requirements of homeopathy. Thus, the property of silica for "anti-chronic inflammation" is exceptional and makes it possible to prescribe it in most cases that suit this definition. It is also worth nothing that unpotentized silica, has a use in conventional biomedical science, as a substance with inhibitory and cytotoxic activity on macrophages (Moseley *et al.*, 1988; Gennari *et al.*, 1987; Bowers *et al.*, 1998; Souvannavong and Adams, 1990). These are the reasons why we have decided to investigate the effect of various dilutions of silica on wound healing. For this purpose we have developed a model of a chronic wound in experimental animals that would simulate the situations in which silica is used in homoeopathy and the results of the studies with this model are hereby reported.

2. Material and Methods

2.1. ANIMALS AND WOUND PERFORMANCE

Female C57Bl/6 mice, 12-16 weeks old, were used in the study. After an adaptation period of 2 weeks, the lobe of one ear of each mouse was perforated with dental wire, weighing 110ϕ,10 mg. The ends of the wire were twisted together to form an "earring". The constant friction of the wire along edges of the wound, resulted in a chronic irritation. Without treatment, the wire caused a hole of variable size, which was a function of the properties of the tissue, the weight of the hanging wire and the

* Communication presented at the 7th GIRI Meeting, November 1993, Montpellier, France.

M. Bastide (ed.), Signals and Images, 191–199.

detrimental effect of the constant irritation caused by the earring. In some cases the wire "cut" with time the ear, and fell off. In the most cases the wound healed and the hole closed around the wire, producing granulation and later scar tissue.

The animals were divided into three or four groups in experiments 1-9 and 10-13, respectively, of ten animals per group (except for experiment no. 1, in which there were four animals per group). The animals were treated with silica dilutions of 5 C, 30 C and saline in experiments 1-9 and an additional dilution, namely, 200 C in experiments 10-13. The animals were given i.p. injections three times a day in experiments 1-2, and oral administration, via drinking water, in experiments 3-13.

2.2. PREPARATION OF DILUTIONS

The silica dilutions were prepared by Dolisos (France) according to the official French Pharmacopoeia. All the dilutions were centesimal dilutions (100^{-n}). The first two steps were performed by trituration of silicium oxide (SiO2) with lactose powder at 1:100 relation (1 C and 2 C), followed by dissolving one part of the triturated powder in 99 parts of 70% alcohol. This dilution (3 C) was vigorously sucussed using a dynamizator. All the subsequent dilutions were prepared using the same technique. In the last two dilutions, saline was used as the diluent, instead of alcohol, these two dilutions were renewed once a week, and kept in a refrigerator at 4° C. The control dilutions were prepared in the same way, "diluting" alcohol in alcohol; the last two control dilutions were done in saline.

2.3. METHOD OF EVALUATION

In experiments 1-3, a subjective evaluation of the hole size was made, on a scale of 0 (closed hole, tight around the wire) to 4 (very large hole). This was done by two independent observers. In the other experiments we added also an objective evaluation of the hole size on the last day of the experiment, using an image analysis system (Galai, Haifa, Israel). The hole was photographed by an electronic camera and the data were transferred to an IBM/AT computer, which calculated the hole size.The hole size of an ear that was "cute" by the wire, was evaluated as if the border of the hole were the edge of the ear.

3. Results

3.1. PRELIMINARY RESULTS

The preliminary experiment was performed on a small number of animals per group, in order to find out whether there would be any reaction when silica is used for a chronic wound. All the mice had been treated for 32 days with three i.p. injections of 0.5 ml of the dilutions, daily. On day 14, the earrings were hanged on the mouse ears.

On the 5th day after the earrings were inserted, a clear difference was observed in the hole size between the ears of treated mice and those of the control group. This difference was maintained until the end of the experiment (day 32). No sign of closure of the holes around the wires in the control group was observed . In the treated group, there was a strong development of granulation tissue, which closed the hole until it disappeared completely (figure 1). An evaluation of the hole size, which was performed with the image analyzer from the photograph of the holes, revealed a significant difference between the groups ($p < 0.01$ in one-way ANOVA test) and between each of the treated groups and the control group ($p < 0.01$ according to the Student t- test).

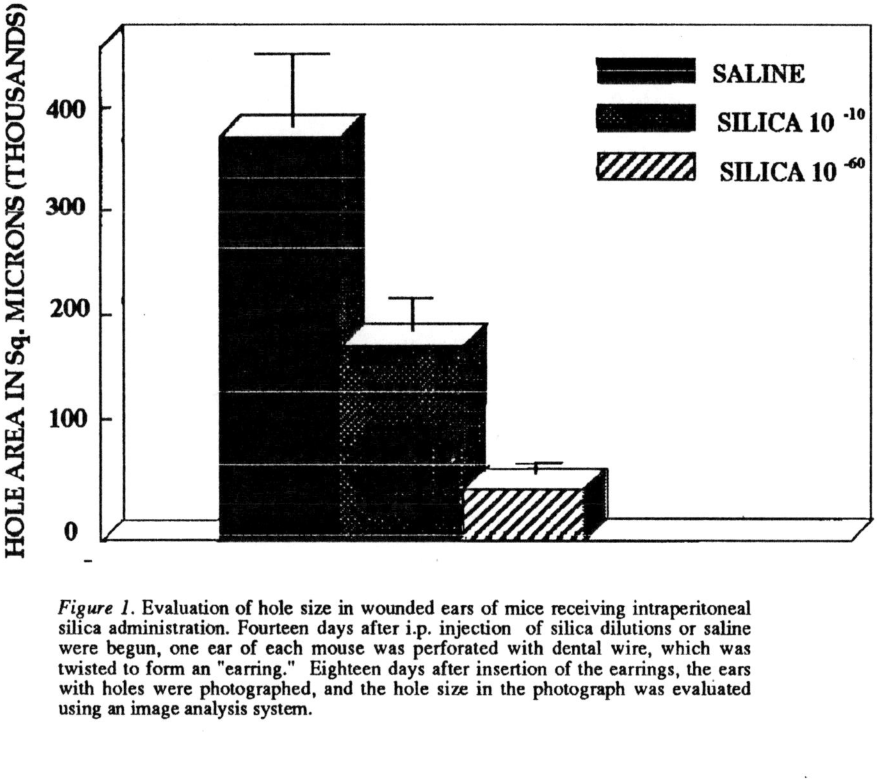

Figure 1. Evaluation of hole size in wounded ears of mice receiving intraperitoneal silica administration. Fourteen days after i.p. injection of silica dilutions or saline were begun, one ear of each mouse was perforated with dental wire, which was twisted to form an "earring." Eighteen days after insertion of the earrings, the ears with holes were photographed, and the hole size in the photograph was evaluated using an image analysis system.

3.2. KINETICS OF THE THERAPEUTIC EFFECT

Encouraged by the dramatic effects reached in the preliminary study, we attempted to evaluate the kinetics of the therapeutic effect. A daily evaluation of the hole size was made, starting from the 3rd day after the start of the experiment, until day 20. Contrary to the first experiment, the dilutions were not injected but were added to the drinking water. In this way, we reduced stress of holding down the animals and injecting them, three times daily. The figure 2 shows the difference in hole size between the treated groups and the control group which progressively increased, reaching its maximum on day 9 ($p < 0.05$ in Kruskal-Wallis one way analysis of variance and in Pearson chi square test). The figure also shows a difference in hole size between the mice treated with the lowest dilution (5 C) and those treated with the higher dilution (30 C). The healing process was stronger at the high dilutions. After day 10, there was no difference between the groups, and the hole size in the control group was equivalent to that in the treated group. Contrary to the first experiment, the beneficial therapeutic effect did not last until the end of the experiment.

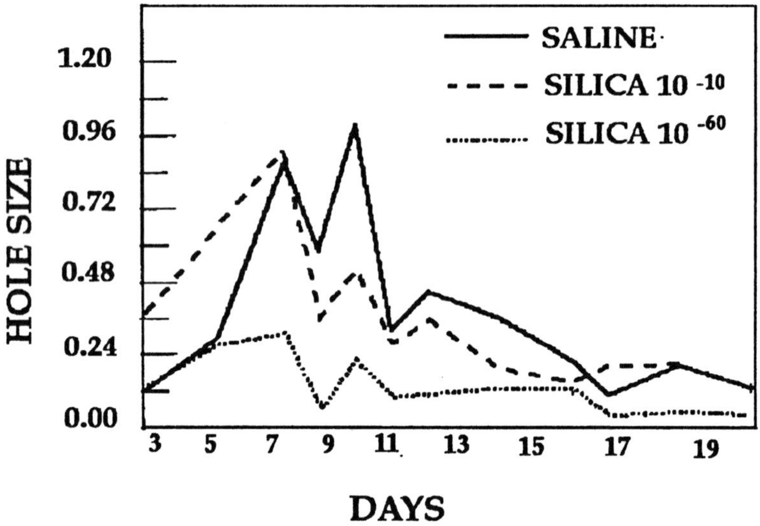

Figure 2. Evaluation of hole sizes in wounded ears of mice receiving oral silica administration. Dental wire was inserted into the ears of mice as described in Figure 1. In parallel, either silica dilutions or saline was added to the drinking water throughout the experiment. The hole size was subjectively evaluated daily, beginning from the third day and continuing until the twentieth day of the experiment.

3.3. THE INFLUENCE OF THE "EARRINGS" ON THE THERAPEUTIC EFFECT

The question, whether there is really a need for the earring, or, in other words, if there is a need for a chronic irritation in order to obtain these results was examined in another experiment in which we perforated an additional hole, in the second ear. The holes were made with an instrument that made holes of a constant size. In these ears, no earrings were inserted. In the attempt to improve the evaluation of the hole size, we added a scanning with an image analysis system in order to measure the hole size.

Table 1 represents a comparison of the subjective evaluation with the objective measurement of the hole sizes, and shows a good correlation between the two. In this experiment also, which was terminated after 5 days, a significant effect of the silica dilutions on wound healing in the ears with earrings was observed. The higher dilutions had a significantly stronger effect on wound healing than did the low dilutions ($p < 0.05$ in one way ANOVA) (Figure 3).

TABLE 1. Comparison between objective and subjective evaluations of hole sizes in the ears of treated and untreated mice

Treatment Group	Method of Evaluation	
	Subjective*	Image Analyzer**
Saline	1.59	37269
$SiO2\ 10^{-10}$	0.5	23644
$SiO2\ 10^{-60}$	0.4	16469

* Mean of the subjective evaluation using a scale of 0 = a hole closed tight around the wire, to 4 = a very large hole.
** Mean hole size (square microns)

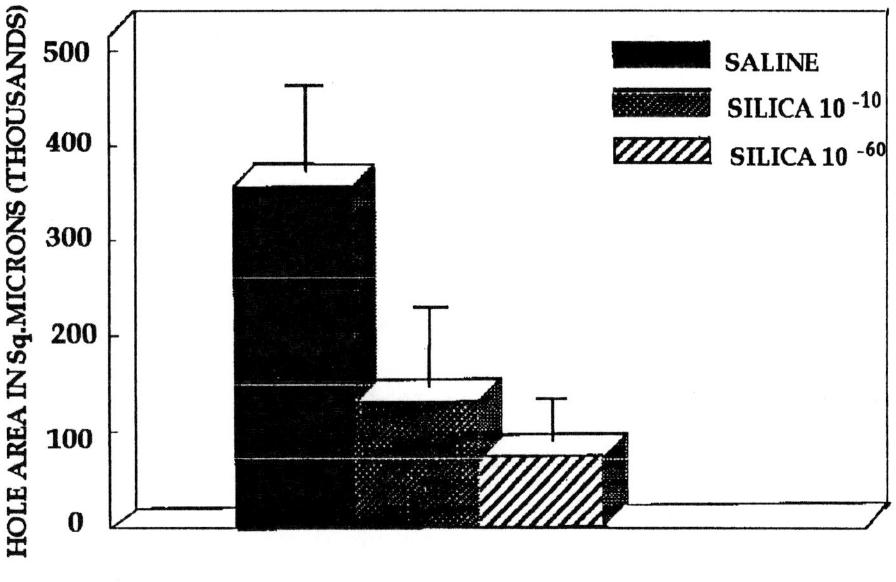

Figure 3. Evaluation of hole size, with and without dental wire, in wounded ears of mice. The protocol was the same as that in Figure 2, and in addition, a hole was made in the other ear without inserting wire earrings. The experiment continued for 5 days, and an objective evaluation of the results was made using an image analysis system. Figure 3 shows the hole size in ears with dental wire earrings.

On the perforated ear without the earring, there was no significant difference in hole size between the treated and the control group (Figure 4).

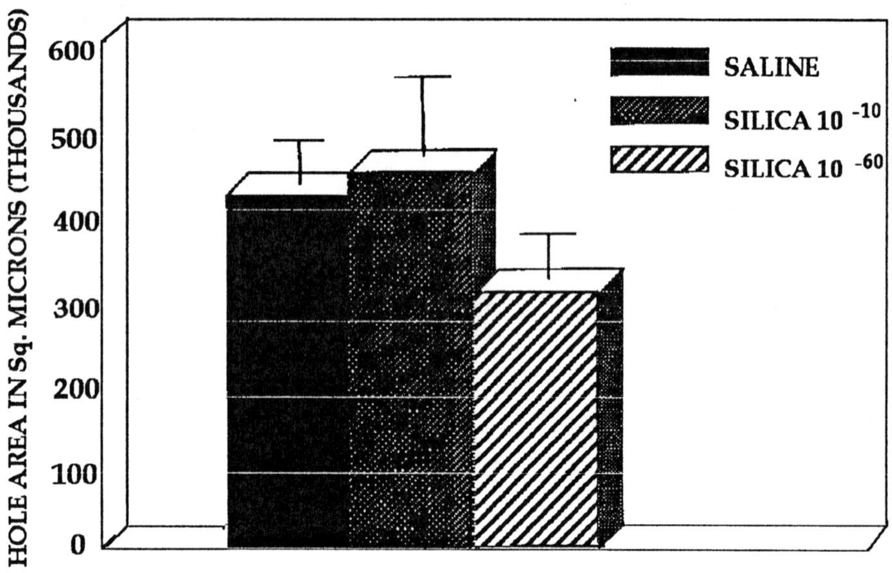

Figure 4. See legend to Figure 3 for protocol. Hole size in ears without dental wire earrings: no significant difference was seen between the treated and untreated groups.

3.4. THE THERAPEUTIC EFFECT OF VERY HIGH DILUTIONS

In order to determine whether the observation that the higher the dilution is, the stronger are its therapeutic properties is consistent, we added another dilution to the experiment, namely a 200 C. There was a similar finding as in the previous experiments, with a stronger effect of the 200 C dilution on the animals treated with it (on day 7 of the experiment): the average area of the holes in the ears of the mice treated with saline was 72799 square microns whereas the average area of the holes in the mice treated with Silica 5 C, 30 C and 200 C was 56195 square microns, 31484 square microns and 28222 square microns, respectively (Figure 5), the difference between hole areas being significant ($p < 0.001$ in one-way ANOVA).

3.5. SUMMARY OF REPETITIONS OF THE EXPERIMENTS

In three of the remaining nine experiments, a clear therapeutic effect of silica dilutions was observed while in the six others, no significant therapeutic effect could be shown; in three of the six the reason was most probably technical: In the first, we tried to hang silver rings in order to obtain a stronger inflammatory effect; the silver caused a very strong irritation and large holes which did not show a tendency to close. In the second experiment, we used very thin dental wire, its weight was not heavy enough to cause sufficient chronic irritation, the wounds remained too small and closed quickly and tightly around the wire. In the third experiment we hung earrings on both ears of each animal in order to see if there was a difference in hole size between ears; there was no significant difference between groups or between the two ears of an individual. In the remaining three experiments, the holes did not develop and closed immediately around the wire; the reason for this was not clear.

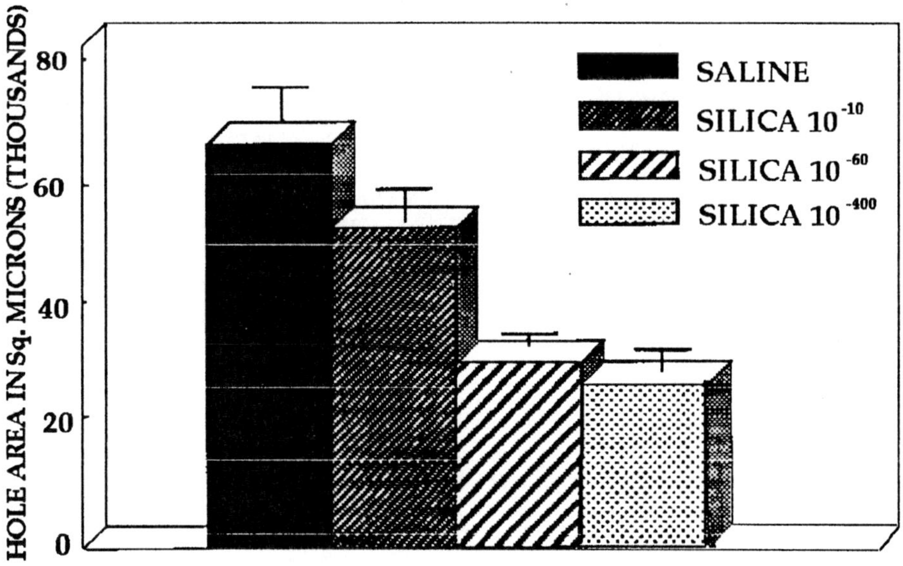

Figure 5. The protocol was similar to the previous experiment except that a group treated with a higher dilution of silica (10^{-400}) was added. The experiment continued for 7 days.

4. Discussion

The present study represents the results of almost 2 years of work. It could be shown, that the healing of chronic irritated wounds, caused by hanging a metal earring from mouse ears, could be enhanced, by treatement with homoeopathic dilutions of silica ($SiO2$). This phenomenon was repeated several times, checked carefully, and was statistically highly significant. The process of wound healing reached its peak after a period of 3-10 days. Subsequently there was an evening up between the treated groups and the control group (experiment b, Figure 2). In that experiment it seems that the wound healing proceeded in two phases (Figure 2): in the first phase it led, most probably due to the chronic irritation and the weight of the earrings, to an enlargement of the holes (the peak in Figure 2 occurred at days 7-9). At the end of that phase the reparative process started, in which the wound started to heal and its size diminished. This phase reached its peak in all groups at the same time (day 10 in Figure 2). According to this observation it seems that the silica was active mainly in the first phase, in halting or slowing down the process of wound enlargement, and less in diminishing the wound size.

It is difficult to assume that these results are due to wrong evaluation methods. In the first two experiments we evaluated the hole size subjectively. In the third experiment we added an objective method and from the fourth experiment on, we used only this objective method, namely evaluation with an image analysis system. It should be noted that the results of this method, which are free of an observer bias proved to be similar to the results of the subjective evaluation (Table 1). It is also unlikely that the effects that were shown are due to "mistakes" in the preparation of the dilutions, since they could be shown in almost all the groups that were treated, as compared with the control groups. These effects were consistent (Figure 2). The mice in the control group had been treated in the same way, kept in similar cages and under the same conditions as the treated animals, and thus these factors could not be the reasons for the significant differences between the groups.

The argument that the "stress condition" of the injected animals could be the reason for the difference between the groups, can hardly be supported since there could not be any reason why the animals that were treated with the active dilution should react differently to the "stress", from the mice that were treated with the control dilution. The argument that there was a "difference" in the way the injections were given between the groups, is also very unlikely since in most experiments there was a clear advantage to the groups that were treated with the active dilutions over the control groups, and also when these were added to the drinking water.

Additional support to the idea that the active dilutions of silica may have a curative effect on healing of chronic wounds, can be obtained from the experiment that showed a therapeutic effect only in the ears that had earrings hung on them. There was no difference in the hole size of the ears that were just "perforated" and did not have earrings hung from them, between the treated groups and the control groups (experiment c, Figures 3 and 4).

Another possibility that was raised was that some molecules of SiO2 remained in the solution due to adhesion to the tube walls. This idea has to be rejected, since in any dilution just 1 % of the previous dilution was used and besides, it is hard to ignore the fact that the higher the dilution was, the stronger the therapeutic effect was ($10^{-400} > 10^{-60} > 10^{-10}$). This observation resembles the effects of homoeopathic remedies in the homoeopathic medicine in which the higher the dilution is, the deeper, stronger and more long-lasting the remedy acts.

There is however a difficulty explaining the "failures" of some experiments: six experiments out of 13 failed. In three of them, the reason was technical: different wire (too thin, silver wires), and an experiment in which we hanged a wire on both ears. In the three other experiments which failed, the reasons for the failure are not clear. It is important to note that the significance of the successful experiments was more than enough to draw the conclusion that the effect was more than a random result. Moreover: in none of the experiments was there a "therapeutic effect" of the control dilution. The issue of the reproducibility of homeopathic studies is certainly one of the big difficulties in homoeopathic research.

An interesting aspect of the study is the "conventional" aspects of the influence of SiO2 on biological systems. As mentioned at the beginning of this paper, there are many scientific publications showing the influence of SiO2 on macrophages, and through them, on the immunological reaction (Moseley *et al.*, 1988; Gennari *et al.*, 1987; Bowers *et al.*, 1998; Souvannavong and Adams, 1990). There are also reports on curative effects of SiO2 on experimental induced autoimmune diseases such as *myasthenia gravis* and *diabetes mellitus* (Kinoshita *et al.*, 1988; Charlton *et al.*, 1988). As far as we know, there has been only one controlled study examining the influence of very low doses of SiO2 on an animal model (Davenas *et al.*, 1987). In that study it was shown that application of 10^{-19}M SiO2 in the drinking water of C57BL/6 mice caused an increase in the activity of the macrophages which were isolated from the treated animals, compared with a control group. In that work, the dilutions still contained some molecules, even though the concentrations used, were several orders of magnitude lower than those used in the conventional studies. In a part of our experiments, (experiments 10-13), we went even further and used dilutions above the Avogadro number. To the best of our knowledge, these dilutions do not contain any molecules of SiO2.

We are aware of the problematic nature of the results described herein. Not only do we claim that the lower the concentration, the stronger the reaction is, but we found "dilutions" that do not contain anything that still have a biological activity. We don't have any good explanation for this phenomenon. There are today several attempts to explain the activity of such "dilutions" (Bernard, 1965; Kumar and Jussal, 1979; Agnastostatos *et al.*, 1991). Almost all of them claim that the "drug information" is transferred to the diluent, and that the diluent turns into the carrier of the "drug information". In other words, if the drug/remedy is the "key" and the diluent is the "lock", so the therapeutic effect is attained with the "lock" and not the "key". Be it as it may, the significance of the results we describe is in showing that such effects do exist and that they cannot be accounted by placebo effect.

References

Anagnostatos, G.S., Vithoulkas, G., Garzonis, P. and Tavouxoglou, C. (1991) A working hypothesis for homoeopathic microdiluted remedies, *The Berlin J. Res.Homoeopathy*, **1**, 141-147.
Bernard, G.P. (1965) Microdose paradox- a new concept, *J. Am Institute of Homoeopathy*, **58**, 205-212.
Bowers, W.E., Ruhoff, M.S., Goodell, E.M. and Stoltenborg, J.K. (1988) The effect of silica treatment on accessory cell-dependent rat T lymphocyte proliferation, *Immunobiology*, **176**, 179-194.
Charlton, B., Bacelj, A. and Mandel, T.E. (1988) Administration of silica particles or anti-Lyt2 antibody prevents beta-cell destruction in NOD mice given cyclophosphamide, *Diabetes*, **37**, 930-935.
Davenas, E., Poitevin, B. and Benveniste, J. (1987) Effect of mouse peritoneal macrophages of orally administered very high dilutions of silica, *Eur. J .Pharmacol.*, **135**, 313-319.
Gennari, M., Bouthillier, Y., Ibanez, O.M. *et al.* (1987) Effect of silica on the genetic regulation of antibody responsiveness, *Ann. Inst. Pasteur Immunol.* **138**, 359-370.
Jouanny, J. (1985) *The Essentials of Homoeopathic Materia Medica*, Laboratoires Boiron Publishers, Ste-Foy-les-Lyon, France. pp.86-91.
Kinoshita, I., Nakamura, T., Satoh, A. *et al.* (1988) Role of the macrophage in the pathogenesis of experimental autoimmune *myasthenia gravis*, *J. Neurol.Sci.* **87**, 49-59.
Kumar, A.and Jussal R.A. (1979) A hypothesis on the nature of homoeopathic potencies, *Br .Hom. J.*, **68**, 197-204.
Metzger, J. (1981) *Gesichtete Homoeopathische Arzneimittellehre*, K.F. Haug Publisher, Heidelberg, pp.1320-1332.
Moseley, P.L., Monick, M. and Hunninghake, G.W. (1988) Divergent effects of silica on lymphocyte proliferation and immunoglobulin production, *J Appl Physiol*, **65**, 350-357.
Souvannavong ,V., Adam, A. (1990) Macrophages from C3H/HeJ mice require an additional step to produce monokines: synergistic effects of silica and poly(I:C) in the release of interleukin 1, *J. Leukoc. Biol.* **48**, 183-192.

CHAPTER III

THERAPEUTICS AND PROVINGS

The therapeutical application of low doses or high dilutions occurs only in homeopathic medicine. Even though previous models could demonstrate the activity or the transfer of non-molecular information, the therapeutic application of "infinitesimal" dilutions depends on the law of similar which is the keyword of homeopathy. The similarity is that of the symptoms observed in a healthy subject receiving a certain remedy (pathogenetic effect or proving) being the same as those observed in a patient needing that same remedy to cure him. As the expression of the symptoms concerns the whole body and mind, the second principle of Homeopathy is the principle of totality. Thanks to Hahnemann's observations, the remedies chosen according to the principle of similarity are more powerful when they are diluted and succussed: the "potentization"(or dynamization) increases the effectiveness of the remedy even though no more molecules are theoretically present. The pathological symptoms may be considered as an attempt at

M. Bastide (ed.), Signals and Images, 201–202.

communication by the patient's body: the symptoms become the process of communication about the pathology, the organism being unable to find itself the mechanisms of regulation necessary to find again the normal state. These symptoms are classically read by the physician either to identify the pathology and to find the corresponding treatment (allopathy), or to identify the remedy able to give the whole symptoms in a healthy subject (homeopathy). The allopathic therapy will consider the body as an object and try to cancel the main symptom by an opposite effect. The homeopathic therapy will consider the body as a whole and will transmit an information on the whole symptoms given by the remedy. The organisation of the symptoms is related specifically to a subject (individualisation). It is unique for each patient who develops pathologies related to specific morbid conditions throughout his life. These symptoms and related pathologic manifestations are not randomly observed but seem to follow an organisation: the complex system could be a model of understanding. The complex system (Le Moigne) was proposed as a model characterized by a "phenomenological hypothesis" and a "teleological hypothesis" which presents a non-traditional determinism based on the final cause or the intention. The patient himself is his own producer of symptoms (or the healthy subject in the pathogenesy or proving) and the diseases he produces represent his personal clinical production (H.Carée). In the model of complex system, the patient (pathological symptoms) or the experimenter (pathogenetic symptoms or proving) operates according to his own particular criteria, evolves in accordance with them and has a teleological project for his own self-realisation. As such, the proving of a remedy (symptoms developped by a healthy subject) should be tested according to rules. Walach *et al.* tested some of them and tried to define a protocol including a placebo. The criteria remain to be considered and the quality of a proving has to correlate with the informative function of a highly diluted successed remedy received by a healthy subject according to his sensitivity; moreover, even in a proving, a memory of the information could continue to exist following administration of the remedy : is a cross-randomized design a possible evaluation for a proving ? Randomized trials (van Erp *et al.*) or veterinary treatments (Grandmontagne *et al.*, Blostin) show how classical rules of statistical evaluation can be employed more or less successfully. However these rules should be proposed for clinical evaluation of homeopathy (Brands).

M.B.

ON MODELLING COMPLEX SYSTEM APPLIED TO HOMEOPATHY.

I. On Modelling Complex system: Wonderful but not Incomprehensible *

J.L. LE MOIGNE
G.R.A.S.C.E. URA CNRS 935, Université d'Aix-Marseille, Centre Forbin
15-19, Allée Claude Forbin, 13627 Aix-en-Provence Cedex 1, France

II. Contribution of Systemic Modelling to the Renewal of Homeopathic Theory Understood as a Complex Phenomenon **

Henri CARRÉE
MD Homeopath - Lecturer at the Faculty of Medecine of Marseilles,
13100 - Aix-en-Provence, France

1. On Modelling Complex system: Wonderful but not Incomprehensible (J.L. Le Moigne)

Is it possible to establish and communicate some rules that could guide the good use of reason in those complex and familiar messes called human affairs (Simon, 1983) ? The project of systemic modelling (SM) should allow any actor who practises it, thus becoming a model builder, to construct some systems of symbols on which he can exercise his reason while transforming them freely, from representation to representation, in turn diplomatic and theatrical[1] ; an exercise that we usually call "reflection" whereby the actor constructs the project in his mind before and in order to illustrate it by his action[2]. It is this conception of SM which legitimates its project of understanding or intelligence of the complexity (Le Moigne, 1985) : perception, itself complex, of phenomena which are themselves perceived as complex and thus irreductible to a finite model, as detailed and specific as possible. Perception not allowed by Analytic Modelling which, since almost two centuries, is quasi monopolistically set up in western scientific culture.

---

* Plenary lecture presented at the 8 th GIRI Meeting, December 1994, Jerusalem, Israel.
** Communication presented at the 8 th GIRI Meeting, December 1994, Jerusalem, Israel.

[1] In reference to the article of J. Ladrière (1989) where the two aspects of the representation concept, the passive (result) and the active (action), are characterized by these two pictures : diplomatic and theatrical.

[2] In reference to K. Marx's parable "The Bee and the Architect".

M. Bastide (ed.), Signals and Images, 203–214.

1.1. ON FORMALISATION AND FORMALISMS OF MODELLING IN GENERAL

This project of SM, progressively re-instituting itself not only as a plausible alternative to analytic modelling, but also as a more ambitious alternative, establishing its capability of modelling intelligence of the non modelable complexity of human affairs, to be operational, needs to make explicit some "formalisms" (or a system of rules) at least as explicit as those proposed by Descartes in his unfinished *Regulae ad directionaem ingenii : Rules for the Direction of Mind*[1] that, for almost three centuries, nevertheless constituted the matricial formalism of analytical modelling. This reference is always cited, but seldom re-read, to the point where it constitutes for many the auto-referential definition of any formalism : *to represent by plain figures* (Rule XIV).

Thus, we have to reject straightaway these analytical definitions of formalisms, and of the formalisation used to establish them, definitions still cited in the contemporary dictionaries of epistemological and logical philosophy. We do this not to reject or dispute their relevance in their own domains of definition, provided that their severe limits are recognized, but to surpass them to obtain a container not mutilating a content which is hoped sufficiently powerful, and which can be nevertheless easily fitted into the semantic framework embroidered by contemporary scientific communication.

Let us admit *formalisation as the cognitive exercise allowing an action (a concrete system) to be transformed into a form (an abstract system)* and let us accept that the opposite operation be understood as the interpretation, these concrete and abstract systems being considered as "systems of signs" (we shall soon say : *systems of symbols*). And let us admit *formalism as a symbolic system of rules carrying out such a process of transformation of experiences into knowledge* (from the ϕ, ϕ universe for physics, to the ψ, ψ universe for psychics, according to P. Valéry's notation). *Formalism is therefore any system of signs resulting from formalisation*, but this definition does not reduce formalisms to some varieties of a formal logic and includes them without any restriction.

This inclusion is very welcome since it incites us to use the millennial experience of formalisation and, in particular, that of the many initiatives of mathematics, arithmetic, geometry and logic ; an inexhaustible patrimony is at our diposal where many apprenticeships of formalisation can be performed and many previous models of formalism rediscovered. For one century, the temptation has of course been permanently *to reduce formalisation to mathematisation, and formalisms to systems of formal logic*[2], so attractive are the successes of theoretical physics becoming mathematical physics to the model builders... in biology, economics or theoretical linguistics. Such successes are sometimes deluding, as was very nicely shown by Simon (1967) considering the emergence of formalism in chemistry during the XIXth Century:

"*The formal languages that have revealed to man the patterns of Nature in one domain are not necessarily appropriate or useful languages for understanding*

---

[1] Most probably written about 1629 in Latin, before the "Discourse on Method" (1637) and unfinished (21 rules instead of the 36 announced) ; this text was published only fifty years after Descartes' death.

[2] According to L. Frey, 1967, "The logicians give to the term formalisation the meaning of elaboration of a formal logical system" (p. 328).

phenomena in other domains. The tools of calculus, so powerful for physics, contributed little to XIXth Century chemistry. Chemistry had to develop its own language : this language had to admit the enormous combinatorial variety of compounds that could be formed from the elements. It had to represent the interactions, many of them non-additive, within and amongst complex molecules... This chemical language... was fundamentally non-numerical, discrete and combinatorial. Its rules of derivation were numerous, eclectic, highly empirical : in all these respects, it contrasted sharply with the language of physics. We must not assume that because a particular formalism has proved powerful in one domain, it will exhibit equal power in another..." (p. 306).

This reflection on the nature of formalisms reveals, above all, the two criteria implicit to any enterprise of formalisation.

1.1.1.*A Criterion of Intellectual Rigour*

Their can be, wrote Simon (1967), formal models which, however rigorous, do not resemble very closely the models using traditional mathematics (p. 304). The moralist, the jurist, the grammarian, the rhetor and, sometimes, the poet, all know this form of ancestral experience. This is, of course, the case as long as rigour is understood as the reasoning process developed in reference to an explicit system of axioms, axioms which are not necessarily those of deduction by non-contradiction. The three axioms constituting the perfect Aristotle syllogism and nowadays, the formalism of deductive logic (seldom explicitly known by most of the integrists of the mathematico-logical formalisation) are a system of formal axioms among many possible others. They are not more "rigorous" than any others, and Aristotle already noticed that they were not imperative to reason. "*What compels us to draw the conclusion of a syllogism" ?* asked Valéry in "Les Cahiers" (1979). "*Nothing in logic compels ; and we don't always draw it*" (p. 320). The essential objective of any formalisation will thus be that of making explicit the axioms or the conventional rules which will guide any exercise of modelling. Logic or "*Art de penser*" of the "Messieurs de Port-Royal" in 1662 (Arnaud and Nicole, 1970) had grouped together, at the end of the XVIIth Century, the axioms of Analytical Modelling (AM) which were borrowed from Aristotle's Analytics and Descartes' Discourse on Method (and therefore the *Rules for the Direction of Mind* published at the beginning of the VIIIth Century). At the same time, some other "Messieurs de Port-Royal" were publishing Pascal's *Thoughts* (1963) ten years after his death, probably with an apologetic goal, but thus giving us indirectly an original axiomatic of Dialectic Modelling which had to wait until Hegel's *science of logic* (1972) to be awarded a cultural and academic status comparable to that of AM. Is it nevertheless necessary to re-read this pascalian axiom which constitutes the first modern alternative to AM axiomatic, and one of the best written ?

"*Since everything, then, is cause and effect, dependent and supporting, mediate and immediate, and all is held together by a natural though imperceptible chain which binds together things most distant and most different, I hold it equally impossible to know the parts without knowing the whole and to know the whole without knowing the parts in details*" (*Thoughts*, no 72).

From Heraclitus to Plato, from Pascal to Hegel, including many others[1], dialectic model builders were able to construct some corpus of axioms allowing them to practise it and on which reasoning could evaluate its own rigour. The monopoly of rigour implicitly claimed by AM in the name of formal logic for two centuries turns out to be contingent ; in establishing its own axiomatic (its topos) during a rich history, dialectic modelling showed the feasibility of the alternatives to AM. SM should be able, in clarifying its axioms of modelling (thus its formalisms), to ensure its own criteria of intellectual rigour and be cleared of the suspicion of laxity that too many wordy, superficial and, sometimes, quasi-mystic discourses on the system approach may have suggested to the traditional academies, thus discouraging many scientific people from the difficult work on epistemological meditation implicit in any interdisciplinary investigation.

1.1..2. *A Feasability Criterion*
An exercise of formalisation has to be an exercise actually and universally feasible. Writing his *Analytics*, Aristotle revealed the feasibility and the possibility of teaching this classical formalism of AM that is the syllogism. Using many roundabout ways (the *tropes*), he accumulated examples, illustrations, analogies and metaphors to assert a feasibility that could have been hidden by the esoterism of syllogism formalism : *it is not BARBARA or BANALIPTON* (two of the fifteen figures of the syllogism proposed by the scholastics) *which constitute reasoning. The mind must not become stilted*, said Pascal in his thoughts on the *"Art de persuader"* (Pascal, 1963).

The feasibility of systemic modelling is supported by a meditation on the countless exercises of modelling of complex phenomena accumulated by mankind. The exercise which would deserve to be attempted is that which will put in evidence the countless contributions of Aristotle's Rhetoric and Topics to contemporary SM. We can limit ourselves here to a more "modern" argument, that of the definition of the general system concept in *"L'Encyclopédie, Dictionnaire raisonné des sciences, des arts et des métiers"* edited by Diderot :

"The general systems, which they more commonly named diagrams, were formed by the sum of all particular systems and, consequently, included all the senses used in the threnody" (1779). The detour by the Encyclopédie to actually construct the concept of system whith the systems of musical notations (among many possible others, from the world system to the financial system of Law), perhaps discloses the very heart of complexity modelling : with systems of symbols, easily and universally recognizable, it turns out that it is possible to represent — or to model — what is ineffable, inexpressible, non modelable in the strict analytical sense of the term ; the symbol systems of musical notation allow us to represent effectively and intelligibly (in the ψ universe), in the form of mediated symbolized knowledge, the experiences which are perceived or are perceptible by an actor who can be recursively the model builder of his own action (the composer). Representation which nobody claims completely depicts the considered phenomenon ; representation which is guileful, teleological and complex but nevertheless intelligible. Representation which does not claim to separate formalism and

[1] I think, for instance, of Y. Barel. See "Sur l'invention de la "Nouvelle Dialectique" selon Y. Barel" (M. Amiot *et al.*, 1993).

interpretation, borrowing from dialectic modelling the familiar rule of recursiveness (which AM excludes since the *Principia Mathematica* (1910) revealed the dangers to its own deductive axiomatic). The symbol, *sign of re-cognition* (*symbolus* in Latin, *sumbolon* in Greek), meaning *action of conjunction* (in Greek *sumballein* = to join, to put together), again becomes the tool of SM, probably since Pierce (1987) characterised it in its instrumental complexity by differentiating it from the two other forms of the sign where it was too often confined : the indice (or index, which defines a causal link) and the icon (which defines an analogical link)[1]. And the system of symbols will allow this instrumentation by putting together the two functions of symbol, *production* (and, therefore, of formalisation) and *interpretation of symbols* : forms which are potentially producers of forms and of sense for and by the actor who "deals" (Newell and Simon, 1972; 1976) with them. Can't he conceive those *More Subtle Numbers* which P. Valéry (who noted them : "N+S" for "Nombres Plus Subtils") dreamed of :

"*I dreamed that — as symbols and complex expressions were created (by a sort of necessity) to satisfy the needs of mathematical physics (a whole series of processing units which allow one to think like Maxwell, etc.). This way it could be attempted to create the ad hoc notions to think the functioning of living matter — feeling — moving — thinking*" (Cahiers, p. 857).

"*... quasi-quantitative notions, non numerable but that can be combined and which summarize the permanent condition of knowledge and that can be enlarged to the perception as well as to the transformation*" (Cahiers, p. 815).

By this way, the two criteria that we intend to use to identify the formalism of systemic modellign are fitted together :

A system of explicit rules (*heuristic search*), the "topos", or an axiomatic, a grammar or syntagms which allows us to evaluate the rigour of the construction by its conformity to the "directions for use" which were made, as a perequisite, explicit and communicable.

A system of symbolisation, the "*tropes*", or graphic and pictorial as well as discursive (languages) paradigms in order to produce sense, the imagination of the potential properties thereby allowing deliberate and intelligible representations by and for the actors of phenomena perceived as complex, in which they intervene, phenomena that they model so that they reason their own behaviour[2].

1.2. THE EPISTEMOLOGICAL STAKES OF SYSTEMIC MODELLING

This conceptual framework (heuristic rules and symbolic systems, topes and tropes, syntagms and paradigms,...), which allows us to define a formalism for modelling should be immersed in an epistemological bath in order to be filled up : if the formalism of systemic formalisation is not that of analytical formalisation, it is not because the framework within which it lies is different, since it is always defined by a set of rules and a system of symbols. It is because the epistemological options on which it is based,

---

[1] This interpretation of C. Pierce's way of thinking was often used. It was presented for the first time, I believe, by G.G. Granger (in J. Piaget, 1976).

[2] This presentation of formalisms by the duality of the "topes and tropes" was suggested by a study of C. Roig (1990) where this duality is presented in a much finer and qualified way than I can here.

and by which such-and-such sets of rules and such-and-such system of symbols are specified, are different.

Formalisms of analytical modelling were so pervaded by positivist epistemologies, naturalistic or realistic, in which they were progressively drawn up during the XIX th Century (to the point of being considered, for a time, the archetypes of "the" scientific method) that, even nowadays, it is difficult to identify them in their methodological autonomy : the frequent reduction of formalisation to mathematisation was the price of this confusion. Calling after 1934 for a "New Scientific Mind", Bachelard (1934) concluded by asking for the restoration of a "*non-cartesian epistemology*". This came true between 1950 and 1970 due to the development of the "new sciences" in the melting pots of cybernetics and structuralism. Epistemologies which are the legacy of the nominalist and dialectical paradigms whose influence on the practice of scientific modelling is at least as ancient as that of the ontological and determinist paradigms on which the contemporary realism and positivism are based. Since J. Piaget suggested in his encyclopoedia *Logique et Connaissance Scientifique* (1967) to recognize them as constructivist epistemologies, it is easier to pinpoint their basic hypothesis and to argue them in order to deduce the few sets of rules which, nowadays, characterize the formalisms of SM.

The basic constitutive hypotheses of the contemporary constructivist epistemologies have already been mentioned (J.L. Le Moigne, 1977, 1991) in most of their formulations (Piaget's Genetic epistemology, Morin's Epistemology of complexity, Simon's Engineering or empirical epistemology (1969), von Glasersfeld or Watzlawick's Radical constructivism, von Foerster's Second order cybernetic epistemology, Bateson's Epistemology of pragmatism,...). Thus, it is possible to limit to a short summary of these hypotheses, "generally accepted" by most researchers who participate today in developing systemics, accepted as the science of systems (and, therefore, *the methods of modelling as and by a system in general*). Let us admit that the various tenets of systemism, understood as an ideology, or "ism", sometimes somewhat imperialist, do not share this agreement ; this discrepancy does not affect our intention, which deals with epistemology, not with ideology, with the scientific discipline, not with the philosophical doctrine.

Indeed, it seems possible to group into three large conceptual options the basic hypotheses of the many constructivisms which the formalisms of SM nowadays refer to.

A "phenomenological" hypothesis : We only perceive operations, namely acts, wrote Valéry (Cahiers). The modelable perception is that of actor's experience intervening in a universe in which he feels active. Valéry makes Leonardo da Vinci say : "*To visualize a tree, one needs to visualize somewhat — or to represent — the background that puts the tree in evidence*". To perceive this tree is to perceive the interaction between the tree and the surroundings, the action produced by the perception of this interaction.

A quasi trivial as it might seem, this phenomenological hypothesis will be, in practice, very anti-intuitive in most contemporary cultures, which are based on a strong ontological hypothesis (this tree exists by itself and it can be analyzed as such, as a sawdust heap, for example) on which the formulations of AM are easily founded.

The phenomenological hypothesis implies almost necessarily an important corollary which will withdraw the constructivisms from the criterion, presumed to put our minds at ease, of objective truth. The "reality" that is experienced by the actor and that he visualizes when perceiving it, is that which is given by this representation. Thus, it is observer and of the observation, that the actor might know. He only reaches for his self-made representations of the singular reality that he experiences (Watzlawick, 1988). It is no longer an objective reality, most probably independent of the observer and which can therefore be imposed as a truth or as an evidence to the many other possible observers.

A teleological hypothesis : as soon as the observer agrees that he only reaches his experience of action, he has to recognize his own project of access to this experience; it is through to *the causa finalis*, the final cause, the intention, that the active experience can be conceived. "*In order to light up the room, he now puts the switch on*". Von Foerster (1991) emphasizes the difference between this representation of action and that proposed by the traditional deterministic hypothesis, while noticing the meaning of the change of conjunction : from "*because*" to "*in order to*" (or from the *efficient cause* to the *final cause*, according to Aristotle). "*Because he put the switch on, the room is now lighted up*", one would say in analytical modelling referring to this dterministic hypothesis (which is also very present in the contemporary scientific cultures). For SM, the teleological hypothesis will constitute in practice a very powerful reference : the famous 1943 paper by Wiener, Rosenblueth and Bigelow *Behaviour, Purpose, Teleology*[1], which restored the status of the teleological hypothesis in the contemporary scientific cultures, made the first formalisms of *cybernetic modelling* culturally conceivable (Klir *et al.*, 1966), and, in this way, those of *systemic modelling* by giving them the concepts of a black box ("*what is it doing ?*" instead of "*how or from what is it made ?*"), of feed-back, and of teleological behaviour.

A hypothesis of "procedural" rationality is defined by contrast to the positivist hypothesis of the **naturality (or "substantivity") of deductive logic.** To reason well is not necessarily to reason according to the perfect syllogistic matrix of deductive logic. The "verifying demonstration" of the hypothetical-deductive reasoning constitutes only one of the ways in the large field of the reproducible and argumentable reasoning that the human mind can conceive. For Aristotle, the reasoning *methods* constituted a more important resource to investigate the complexity of the world than the only syllogistic logic used in his *Analytics*. The reading of *Topics* and of *Rhetoric* shows this in many ways. In 1977, Simon proposed to take account of this infinite capacity of the multiple procedures that the reason can construct and put into action, by differentiating a substantive rationality (or deductive, or perfectly syllogistic, or algorithmic) and a procedural rationality (or deliberative, argumentative, or dialectic, or heuristic[2].) This procedural (and thus communicable) conception of rationality is of course aware of the modesty implied by the absence of any universal criterion of ultimate validation other than the "beliefs" of the actor. It makes nevertheless possible the proceduralisation and, in consequence, the cognitive mastery of these familiar forms of reasoning, which are

---

[1] For this 1943 paper, its translation in French and on the comments which were made, see A. Demailly and J.L. Le Moigne, 1987.

[2] See J.L. Le Moigne, 1993, for a discussion of these concepts.

banned by traditional deductive logic, represented by reflexive and recursive, specular and auto-referential reasoning, reasoning by cause and effect and by trial and error experimentation and, of course, analogue and dialectic reasoning through heuristic searches (Newell and Simon, 1976). The project of reasoning is to produce sense instead of ruining it by giving too much emphasis to the form. The recognition of conceivable and communicable forms, the "procedures", makes possible this reasoned production of sense, so familiar in the exercise of analogy reasoning. Is it possible to substitute the *criterion of projective feasibility* of the procedural reasoning for the *criterion of objective truth* ? : the practitioners of SM reread willingly Vico (1710) to whom *the truth lies in the doing* ("*verare et facere idem esse*") (Vico, 1987).

2. Contribution of systemic Modelling to the renewal of Homeopathic Theory Understood as a Complex Phenomenon (H. Carrée)

2.1. THE HOMEOPATHIC MEDICINE TAKES CARE OF THE SPECIFICITY OF THE SUBJECT AND THE SPECIFICITY OF THE DISEASES

During the course of a lifetime, each human being is liable to present acute or chronic diseases of some kind ; these diseases pertain to the individual patient and are the semeiological expression of his own imbalance.

Is it possible, through the observation of successive and/or concomitant diseases in any given patient, to discern the connections and reciprocal relations existing between the different nosological structures ?

Allopathic medecine interprets and defines diseases in terms of anatomo-clinical entities or distinct nosological structures which evolve independently of each other along separate lines. Homeopathic medecine by means of its all-inclusive (or holistic) conception of medecine, is able to see different diseases as the multiple expressions of a single morbid condition such as each individual discloses or may disclose during his lifetime.

This morbid condition is unique in the case of each patient, if it is seen to be the permanent theme of the sequences (or illnesses) occurring within space and time: in this way it defines the concept of diathesis. All patients present the specific diseases or disorders which are particular to their individual case, and this is never in an arbitrary manner, at randon times or without specific reason.

2.2. THE COMPLEXITY OF THE NOSOLOGICAL STRUCTURES

For example : how are we to interpret or pinpoint the appropriate connections between the different symptoms in the case of a depressed managing director under stress suffering from both a sudden onset of sciatica in the lower left limb and a bad attack of constipation following an excessive intake of wine ? How can we account for the fact that the same managing director also suffers from insomnia between two and three in the morning and from allergic rhinitis every spring, especially in windy weather ? Does a certain logic become apparent when the case is observed clinically, in that if the patient becomes chronically constipated, his allergic rhinitis will also recur whenever he is exposed to a minor draught ?

Understanding the intricacy of nosological structures and explaining the connections between them is complicated. By means of its methodology the modelling of these complex systems removes the sense of complexity and enables us to grasp the concept of a complex system ; it is also the explicit expression of the conjunction between the human phenomenon seen as an indecomposable whole, whose coherence, vital project, unity can be discerned, and the active components of this phenomenon which are able to interreact internally and of which this phenomenon is resulting composition.

Thus, the wealth of clinical symptomatic information, if it is all-inclusive, explains the clinical finality, or rather the concurrence of exogenous and endogenous parameters contiguous to the pathological condition.

Observation demonstrates that in each patient one or several diseases are always organised and presented along the lines of one or more complex pathological systems ; this testifies to the presence of an intrinsic self-organisation which is informative of its own imbalance.

Furthermore, each individual system appears to be ordered according to semeiotic criteria which are classified and codified. How does the homeopathic concept establish the notion of a complex system of semeiotic coding ?

2.3. THE COMPLEXITY OF THE EXPERIMENTAL PATHOGENESIS

Experimental pathogenesis allow for the manifestation of these complex systems.

Experimenting with pharmacologically active substances on healthy individuals in a homeostatic condition has been and still is the means whereby modelling and simulation of complex systems of pathogenesis have been brought about. Such active substances do not act only as the generative agents of experimental pathological systems (or pathogenesis) ; they also reveal a process of prememorisation inherent in the living being, and of complex pathological systems which are latent and moderating ; they also represent the potential clinical combinations or probabilities that each individual is liable to bring into play or produce at some time during his lifetime ; there is therefore a reasonable degree of predictability in diathesis. These active substances also indicate the involvement and the permanent interractivity between the living being and his environment.

The all-inclusive classification of clinical signs and their modes of expression serve to confirm and corroborate the pathogeneses. In this way the pathogeneses prove to be informatory of the way in which a given pathology organises itself and of the symbolics within semeiological semantics ; they are also after high dilution potentialy inductive of a decision of self-regulation in a case of *similimun*.

The homeopathic concept is based on the principle of adequation and the parity between an observed pathological system and the pathogenetic mimesis of it achieved by means of experimentation ; all pathogeneses substantiate the systemic conjunction between different nosological structures and give a clear understanding of the dynamics of their reciprocal relations.

2.4. THE MODELLING OF COMPLEX SYSTEMS CONCURS WITH THE DYNAMICS OF THE HOMEOPATHIC CONCEPT

To come back to the case of the depressed managing director suffering from stress, insomnia, constipation, sciatica and allergic rhinitis, we find ourselves faced with several distinct nosological structures estimated independent of each other. However, by means of experimentation, *Nux Vomica* has been seen to simulate a complex pathogenetic system to which all of these particularly five nosological structures belong and which takes into consideration all the semeiological information and the ways in which symptoms appear, worsen or improve.

Nux Vomica is the ideal systemic remedy corresponding to a patient presenting these elements, which have not made their appearance in an arbitrary manner. Experience teaches us that until the patient has been treated with *Nuw Vomica*, the appropriate substance-remedy, he will always be liable to suffer from one or more of the nosological structures, or even others belonging to the *Nux Vomica* syndrome which have not yet surfaced. Knowing the pathogeneses enables us to make a diathetic evaluation of the psychophysiological condition of the patient and to determine the degree to which he is affected lesionally and functionally.

For a second example : how are we to interpret the case of a mother and her child, each presenting several semeiotically different pathological systems but who share a common denominator in that both belong to the same pathogenous system of a given substance-remedy ? This is in itself a tangible proof of the dynamic reactions of reciprocity both between living beings themselves and between living beings and their milieu.

The patient himself is his own producer, and the diseases he produces represent his personal clinical production ; by means of the substance under experiment, the experimenter produces a pathogenesis which is a clinical construction of his own.

Thinking about modeling concept, we must try to find the particular links to understand the interactions or the intricates between the diseases and the patient himself.

By experimenting with the substance, the experimenter can reproduce the model of a complex system and interpret it *in vivo*. The experimenter is projective, that is, he is in a position to elaborate his project (a disease, in this instance) and elecit its meaning, by executing its programme and outcome.

Disease is both the projection and the projective capacity of the patient, with respect to the clinical formulation of his internal imbalance.

This approach allows us to apply to a patient (or a pathology) and to a substance (or a pathogenesis) the paradigm of general systems. Each human being, immersed in his own environment, operates according to his own particular psychophysiological criteria evolves in accordance with them throughout his life and has a teleological project for his own self-realisation. Each substance-remedy takes into account the environmental conditions, makes the means of operation explicit, enshures the logical transformations and promotes the development of the teleological project.

The canonical model of the organisation finds its exact application in the human being ; man functions and transformes oneself within his environment. He sustains, connects and produces, and at the same time he sustains himself, connects himself, projects himself with a view to certain goals.

The modelling of complex systems concurs with the dynamics of the homeopathic concept, since they both demonstrate the value of the intelligibility and the intelligence of pathological systems ; they ensure the integrity of meaning of the sign and the signifier in the pathology of the human being.

Aknowledgements

J.L. Le Moigne: I wish to thank very warmly Dr Marc Bonzon who has translated in a rather short time this article from French to English and Dr R. Cassada for a critical reading fo the manuscript. I thank also Pr. H. Greppin for his agreement to use a slightly modified part of a larger study published in 1993 by "University of Geneva" in a collective book under his direction ("*Tools for understanding of living systems*").

References

Amiot, M., Brahms, L., and Billard, I., (eds.) (1993) *Systèmes et paradoxes ; autour de la pensée d'Y. Barel*, Seuil Publisher, Paris.
Aristote (1967) *Rhétorique* (Livres I, II, III), Les Belles Lettres Publisher, Paris.
Aristote (1983) *Les Topiques*, Librairie J. Vrin Publisher, Paris.
Aristote (1983, 1987) *Les premiers analytiques* (III) et *Les seconds analytiques* (IV), Librairie Vrin Publisher, Paris.
Arnaud, A.and Nicole P. (1970) *La logique ou l'art de penser* (1662 : La logique de Port-Royal), Flammarion Publisher, Paris.
Bachelard, G. (1934-1940) *Le nouvel esprit scientifique*, PUF Publisher, Paris.
Demailly, A. and Le Moigne, J.L. (1987) Actualité de la téléologie, concept scientifique par excellence, *Revue Internationale de Systémique* **1**, 239-245.
Descartes, R. (1953) *Oeuvres et Lettres*, La Pléïade, Gallimard Publisher, Paris.
Diderot, D. (1779) *L'Encyclopédie, Dictionnaire raisonné des sciences, des arts et des métiers*, vol. 32, Pellet Publisher, Genève.
Frey, L. (1967) Langages logiques et processus intellectuels, in *Les modèles et la formalisation du comportement*, CNRS Publisher, Paris, pp. 327-340.
Hegel, F. (1972, 1976, 1981) *Science de la logique*, 3 vol, Aubier Publisher, Paris.
Klir J. and Valach M. (1966) *Cybernetic Modelling*, Hiliffe Books LTD Publisher, London.
Ladrière, J. (1989) *Représentation et Connaissance*, Encyclopaedia Universalis, Paris.
Le Blond, J.M. (1939-1973) *Logique et méthode chez Aristote*, Librairie J. Vrin Publisher, Paris.
Le Moigne, J.L. (1977, 1990) *La théorie du système général, théorie de la modélisation*, 2ème édition augmentée ; 3ème édition augmentée, PUF Publisher, Paris.
Le Moigne, J.L. (1985) The intelligence of complexity, in *The Science and Praxis of Complexity*, United Nations University Publisher, Tokyo, pp. 35-61.
Le Moigne, J.L. (1991) Sur les fondements épistémologiques de la science de la cognition, in E. Andreewsky (ed.), *Systémique et Cognition*, Dunod Publisher, Paris, pp. 11-49.
Le Moigne, J.L. (1993) Sur la capacité de la raison à discerner Rationalité Substantive et Rationalité Procédurale, in J.C. Passeron, L.A. Gérard-Varet (eds), *Calculer et raisonner, les usages du principe de rationalité dans les sciences sociales*, EHESS Publisher, Paris.
Marx, K. (1965) *Oeuvres - Economie*, 2 vol. La Pléiade, Gallimard Publisher, Paris.
Morin, E. (1977) *La Méthode, La nature de la nature*, vol. 1 Seuil Publisher, Paris.
Morin, E. (1980) *La Méthode, La vie de la vie*, vol. 2 Seuil Publisher, Paris.
Morin, E. (1986) *La Méthode, La Connaissance de la Connaissance*, vol. 3, Seuil Publisher, Paris.
Morin, E. (1991) *La Méthode, Les idées, leur habitat, leur vie, leurs moeurs, leurs organisations*, vol. 4, Seuil Publisher, Paris.
Newell, A. and Simon, H.A. (1972) *Human Problem Solving*, Prentice Hall Inc., Englewood-Cliffs Publisher, New Jersey.
Newell, A. and Simon, H.A. (1976) *Computer science as empirical inquiry : symbols and search*, Communication of the ACM **19**, 113-126.
Pascal, B. (1963) *Oeuvres complètes*, Seuil Publisher, Paris (See English translation in Great Books of the Western World, Vol. 33, Encyclopaedia Britannica, Inc., Chicago, 1952).
Pierce, C.S. (1987) *Textes fondamentaux de sémiotique*, Méridien Publisher, Paris.
Piaget, J. (1967) *Logique et connaissance scientifique*, La Pléiade, Gallimard Publisher, Paris.
Piaget, J. (1976) Les sciences sociales avant et après Jean Piaget (Hommage publié à l'occasion du 80ème anniversaire de Jean Piaget et précédé de son autobiographie) (Busino G., ed.). *Revue Européenne de Sciences Sociales*, tome XIV, nos 38-39, Droz Publisher, Genève.
Rey, Y.and Prieur, B. (1991) *Système, éthiques, perspectives en thérapie familiale*. ESF Publisher, Paris.

Roig, C. (1990) Rhétorique et analyse socio-politique, in G. Duprat (ed.), *Connaissance du politique.* PUF Publisher, Paris, pp. 96-133.
Simon, H.A. (1967) The use of information processing languages in psychology, in P. Fraysse (ed.), *Les modèles et la formalisation du comportement*, CNRS Publisher, Paris, pp. 303-317.
Simon, H.A. (1977a) Does scientific discovery have a logic ? (1973), in H.A. Simon, *Models of Discovery*, D. Reidel Publisher, Dordrecht, Holland.
Simon, H.A. (1977b) *Models of Discovery*, D. Reidel Publisher, Dordrecht, Holland.
Simon, H.A. (1969, 1981) *The Sciences of the Artificial*, MIT Press Publisher, Cambridge, Mass.
Simon, H.A. (1982) From substantive to procedural rationality, 1976, in H.A. Simon (éd.), *Models of Bounded Rationality*, Vol. 2, MIT Press Publisher, Cambridge Mass, pp. 424-443.
Simon, H.A. (1983) *Reason in Human Affairs.*, Stanford University Press Publisher, Ca.
Valéry, P. (1979) *Cahiers 1894-1945.* 2 vol. La Pléiade, Gallimard Publisher, Paris.
Valéry, P. (1972) *Oeuvres complètes*, 2 vol. La Pléiade, Gallimard Publisher, Paris.
Vico, G.B. (1987) *De la très ancienne philosophie des peuples italiques* (1710), Trans Europ Express Publisher, 32120 Monvezin, France.
Von Foerster, H. (1991) Ethique et cybernétique de second ordre, in Y. Rey, B. Prieur (eds), *Système, Ethiques, Perspectives en thérapie familiale.* ESF Publisher, Paris. (See English translation : Ethics and Second Order Cybernetics, in *Cybernetics and Human Knowing*. vol. 1, n° 1, pp. 9-19, 1992).
Watzlawick, P. (1988) *L'invention de la réalité - Contribution au constructivisme*, Seuil Publisher, Paris.

EFFECTS OF BELLADONNA 12 CH AND 30 CH IN HEALTHY VOLUNTEERS *

A Multiple, Single-Case Experiment in Randomization Design

H. WALACH, S. HIEBER AND E. ERNST-HIEBER
University of Freiburg, Department of Psychology, D - 79085 Freiburg

1. Introduction

The practice of homoeopathy lies on two principles: the law of similars and the potentization of homoeopathic remedies. The law of similars states that likes be cured by likes: "*Similia similibus curentur*" (Walach, 1986). This means for example that certain types of feavers which exhibit symptoms like an intoxication with deadly nightshade can be cured by the homoeopathic preparation *Belladonna*, which is prepared from this very same plant. While this principle has been known for a long time, it was the German doctor and pharmacist Samuel Hahnemann who made practical use of it, by giving drug substances to volunteers, noting down the symptoms they experienced, and using the very same symptoms as indication of this remedy in the case of a disease. His volunteers experienced sometimes quite toxic reactions. Therefore, Hahnemann diluted the remedial substances stepwise, by adding 99 drops of alcohol to one drop of substance and vigourously shaking it. This procedure he called "potentization" or "dynamisation". For every dilution process he used a new glass vial. He experimented with very high dilutions not knowing that he had way beyond transgressed Avogadro's number which states that 6.0231023 molecules are contained in one mole of any solution. By virtue of the homoeopathic potentization procedure, no molecules are to be expected in preparations beyond and above 12 CH or 24 XH, 12 stepwise agitated dilutions in the ratio 1:100, or 24 stepwise, agitated dilutions in the ratio 1:10. Hahnemann however, not knowing about Avogadro's number, very frequently used potentizations as high as 30 CH or higher, and in the 6th edition of his Organon (Hahnemann 1979) he even prescribed 30 CH as a standard potency for curative and experimental procedures alike.

The law of similars was in Hahnemann's time a tradition with no direct practical implications. But by testing the available drug substances, Hahnemann found a working basis. Since then homoeopaths have "proven" new remedial substances by giving substances in both crude and diluted form to healthy volunteers and noting down their symptoms. These symptoms form the homoeopathic materia medica. Rarely however it has been tried to find out whether symptoms supposedly produced by these substances in healthy volunteers are in fact distinctive symptoms, or whether the symptoms experienced by the provers are due to the accidental situation of observation and other unspecific factors, and are therefore to be considered as placebo symptoms.

* Communication presented at the 8th GIRI Meeting, December 1994, Jerusalem, Israel.

M. Bastide (ed.), Signals and Images, 215–226.

In a previous study (Walach, 1993) it was shown that, although on a single case basis differences could be seen between placebo and verum in a double blind crossover trial, on a group level there was no difference between placebo and verum symptoms. This was due to the fact that some persons clearly showed more symptoms under the test substance, *Belladonna* 30 CH, and some provers showed clearly more symptoms under placebo. Since this trial was not designed as a single-case trial, no statistics could be provided in order to test for single subjective significant effects. We therefore decided to design a study which would allow to draw conclusions on a single case basis.

We chose a randomization design proposed by Edgington (1987), which is a true single-case experiment, and can be evaluated statistically on a single-case basis. In this design treatment periods are randomized such that the single subject undergoes a randomized series of treatments not knowing which treatment is to be applied at what time period. A full report of the study is available (Ernst-Hieber and Hieber, 1995).

2. Method

We randomized 4 weeks of treatment and 4 weeks of placebo per person, preceeded by an observational run in week.

This allows for $\frac{8!}{4!\,.\,4!}$ permutations.

For each test subject one of these 70 permutations was chosen at random. Test substances were plant extracts of *Belladonna* prepared in 12 CH (group I) and 30 CH (group II). The test substances were freshly prepared by the German manufacturer Deutsche Homöopathie Union (DHU), Karlsruhe, according to the German pharmacopoe (Homöopathisches Arzneibuch, HAB). The homoeopathic dilution process yields a substance containing 10^{-24} and 10^{-60} parts of the original material, dilutions commonly used in homoeopathy for experimental and curative purposes. The test substances were dispensed as size 3 globules. Placebo consisted of the same batch of sugar globules medicated with the same amount of unsuccussed alcohol used for the potentization procedure of *Belladonna*. Placebo and verum substances were indistinguishable. One container per week was dispensed, labeled with the serial number of the week.

A permutation table containing all possible 70 permutations was sent to the department of biostatistics of the University of Ulm (Prof. Wilhelm Gaus) who, using a random number generator, selected independently 40 sequences of possible permutations. These sequences were sent to the manufacturer who packed the test substances according to the instructions, which were fixed in advance in a protocol. The code was held by the manufacturer. Sealed envelopes containing a copy of the code were sent to doctors who could be consulted in the case of an emergency or in case of an illness which might be traceable to the test substances. The code was not handed out until all data had been collected and a printout of the data with an electronic copy had been deposited at the department of biomedical statistics, University of Ulm. Thus, the study was strictly double-blind.

Data collection was done using a diary which presented the subjects with predefined symptoms derived from the homoeopathic materia medica in order to query subjects on *Belladonna* as well as non-*Belladonna* symptoms. Apart from general well-being subjects were askedto report on any changes and if so, what type of changes these were. (e.g. did their well-being change for better or worse, did their well-being change suddenly or slowly, during the day or during the night, was it aggravated or improved by

heat or cold, what types of changes did they notice and where did the changes occur. A sample copy of the diary can be obtained from the authors.)
Using the key symptoms and modalities of *Belladonna*, it was decided in advance which kind of responses would be counted as *Belladonna*, and non-*Belladonna* symptoms, respectively. There were an equal number of possible *Belladonna* or non-*Belladonna* responses. *Belladonna* and non-*Belladonna* symptoms are given in table 1

We aimed at unselected subjects unfamiliar with the theory and nature of homoeopathy. Advertisements were distributed in several computer newsnets, placed in public places, and handed out as leaflets. Subjects were in good health, in no need of medication, under no medical treatment or supervision, not being heavy users of alcohol, caffeine or recreational drugs. Other criteria which might be handled as exclusion criteria in traditional homoeopathic remedy provings were noted down in a questionnaire, but were not applied, as their validity is not experimentally confirmed. These included unstable lifestyle, anticipation of major life-events, of stress and change in daily habits. Subjects were also screened using the Freiburg Complaint List (FBL) (Fahrenberg, 1995) and the Freiburg Personality Inventory (FPI) (Fahrenberg, Hampel & Selg 1989), two widely used and validated instruments. They were informed about the possibility of experiencing transitory symptoms according to homoeopathic theory, and of the essentially harmless nature of the experiment, and also that they would receive a preparation which was pharmaceutically inactive as it would not contain any molecule of an active substance, but which nonetheless was claimed to be able to produce transitory symptoms. They all gave written informed consent. They were instructed to take 2 times 3 globules of the respective test substances on Mondays, throw away the rest of the container, and to use a new container each week, according to the serial number.

TABLE 1. *Belladonna* and non-*Belladonna* symptoms asked for in the diary

Belladonna symptoms	non-*Belladonna* symptoms (chosen at random)
modalities	
better warm	better cold
worse cold	worse warm
sudden change	slow change
during night	during day
type of changes	
redness	paleness
swelling	weakness
heat	cold
pain	itching
pulsating	strong thirst
dryness	lameness
pressure/fullness	disturbed sleep
restlessness	trembling
localisations	
head	breast
neck and throat	stomach
skin	back
limbs	

The subjects were provided by mail with a full set of test substances, diaries and written instructions, how to conduct the data collection, and how to proceed with the test substances. These included some precautions, e.g not to place the test substances under any electromagnetically active device, not to expose them to exceptional heat or cold and to store them in a safe and neutral place.

Statistical evaluation was done in the following way: for each week a Belladonna score was calculated, summing up all the *Belladonna* symptoms which had been experienced in this week. The non-Belladonna scores were not used for calculation. The differences of mean *Belladonna* scores were calculated between the four *Belladonna* and the four placebo weeks. This calculation was repeated for all 70 possible permutations, using an author -written Basic program which performed the permutations and calculations. Then the proportion of differences was determined which was greater or equal than the empirically obtained one. This value devided by 70 yielded the real probability that this distribution of *Belladonna* scores could have been obtained by chance. We conducted one-sided tests (i.e. more Belladonna symptoms with Belladonna). This procedure is clearly described by Edgington (1987). Visual analysis was done using the graph of the data as shown in the figures below. The general logic of single-case randomization experiments opens the possibility of conducting independent experiments with individual subjects. We hypothesized that, if there were any effects present in the potentized homoeopathic dilutions, then at least one subject out of a group of 20 people should show a reaction. This design does not produce multiple tests out of one experiment but is a series of independent experiments.

3. Results

Seventeen subjects volunteered to be enrolled in group I (*Belladonna* 12 CH). Three out of these 17 subjects dropped out and did not continue the experiment. The final group consisted of 14 subjects. In group II (*Belladonna* 30 CH) 16 subjects volunteered to participate, 5 dropped out leaving a sample of 11 subjects. Dropouts were not related to the test substances, but were motivated by lack of interest or bad compliance. Demographic data are presented in table 2.

In group I (*Belladonna* 12 CH) one subject showed changes on visual analysis, which were expected, i.e. more Belladonna symptoms with *Belladonna*. This result is not significant, however. One subject showed changes which were unexpected, namely more *Belladonna* symptoms with placebo ($p=0.057$). A similar result was obtained in group II (*Belladonna* 30 CH): one experiment showed a significant tendency, with more *Belladonna* symptoms with *Belladonna* ($p=0.071$), two more experiments showed striking effects on visual analysis, yet yielded no significant statistics. Data and graphs are presented in figures 1-6.

The figures present the data in the following way: The first week always was observation with placebo for run-in purposes which was not counted. Then follow the 8 experimental weeks in random order of placebo and *Belladonna* weeks, indicated by the dotted line (hills: placebo, troughs: *Belladonna*). The number of *Belladonna* symptoms is given in black bars, and the number of non-*Belladonna* symptoms in white bars. The left abscissa gives the absolute frequency of symptoms, the right abscissa gives the rating of general well-being on a 10 cm visual analogue scale. The data presents the number of *Belladonna* symptoms per week.

As can be seen, a variety of pattern arises:

TABLE 2: Demographic data

Group	I (Belladonna 12 CH)	II (Belladonna 30 CH)
Marital status		
unmarried	10	12
married	6 *	3**
Sex		
female	8	9
male	8*	7
Age		
mean (range)	35 (27-52)	33 (24-39)
Education		
A-level	13	13
secondary education	3*	2
primary education	0	1

* only 16 subjects returned the questionaire.
** only 15 subjects returned the questionaire.

We observed any type of reaction from low level responders presenting nearly no symptoms to highly responsive subjects presenting a large number of symptoms. Not only Belladonna symptoms were reported but also a large number of non-*Belladonna* symptoms. The experiment of subject number 16 (figure 1) showed 4 Belladonna symptoms in the last *Belladonna* week and no symptoms at all in all the other weeks and only 2 non-Belladonna symptoms in the *Belladonna* week. However, the randomization test was not significant.

Subject 8 experienced a great number of Belladonna and a smaller number of non-*Belladonna* symptoms during the weeks 5 and 6, that is 2 weeks after having taken *Belladonna* 12CH. The *Belladonna* symptoms were pain in the throat and related symptoms. A smaller cluster of symptoms can be seen in week 7 (*Belladonna*) and in week 8 (placebo). There were clearly more *Belladonna* symptoms under placebo (p=0.057). It should be noted that not only were more symptoms reported with placebo but more speicific *Belladonna* symptoms, as the non-*Belladonna* symptoms were not included in our evaluation.

Subject number 24 (*Belladonna* 30 CH, figure 3) showed an interesting pattern, albeit not significant: symtpoms were exhibited in three of the four *Belladonna* weeks, always on the day following administration.

The experiment with *Belladonna* 30 CH (figure 4, subject no 31) shows a statistical tendency (p=0.071) and a clear picture: we see altogether 14 *Belladonna* symptoms only in the *Belladonna* weeks with no *Belladonna* symptoms in the placebo weeks as well as a number of non-*Belladonna* symptoms in the *Belladonna* weeks which were not counted.

There were also a couple of other interesting observations which however did not yield significant results. Subject number 30 showed visually impressive results which for some reason did not reflect in a significant statistical effect (figure 5).

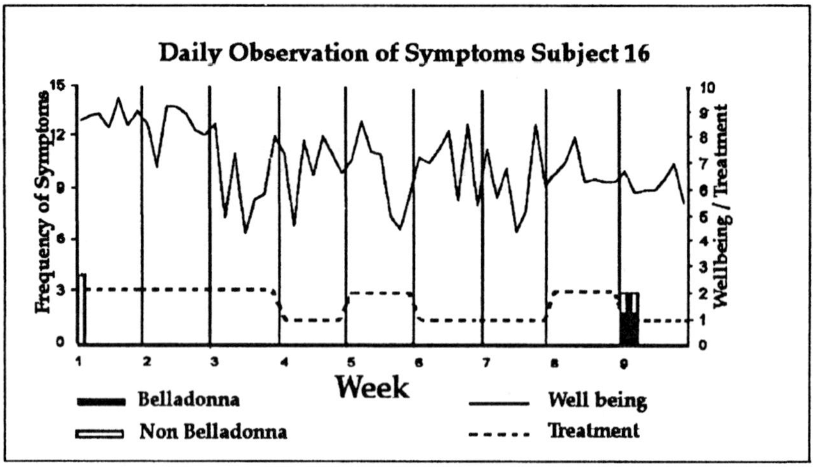

Run-in P P V P V V P V

Figure 1. Striking in visual anlaysis but not statistically significant, more Belladonna symptoms with Belladonna 12 CH.

Placebo (Score per week)	0 0 0 0
Verum (Score per week)	0 0 0 4
Probabilitity	.5

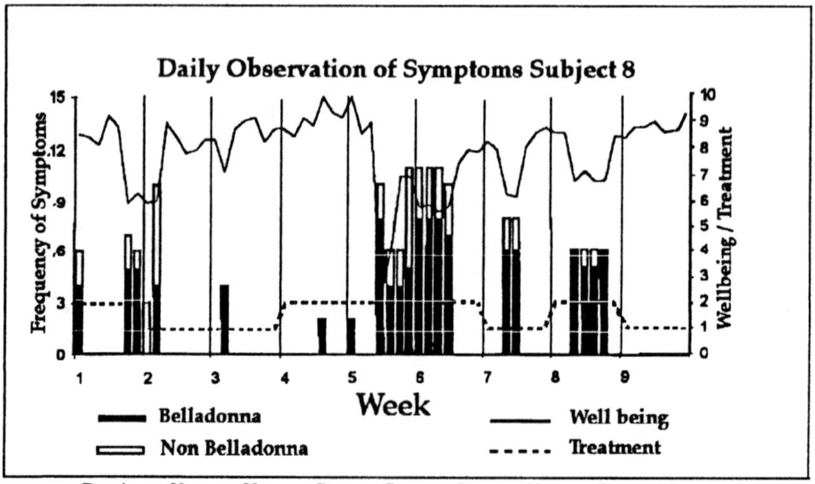

Run-in V V P P P V P V

Figure 2. Significant randomization test, more Belladonna symptoms with placebo (Belladonna 12 CH).

Placebo (Score per week)	2 23 31 22
Verum (Score per week)	4 4 12 0
Probabilitity	.057

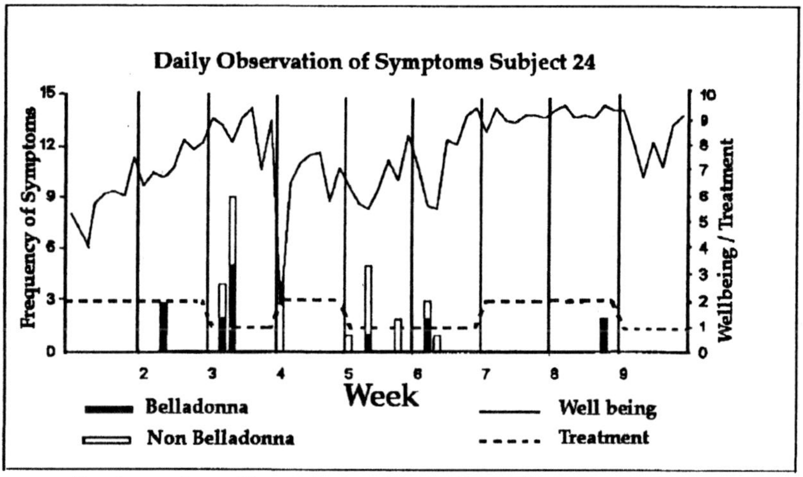

Run-in P V P V V P P V

Figure 3. Striking in visual analysis; more Belladonna symptoms with Belladonna 30 CH. Belladonna symptoms with Belladonna 30CH showed on the day following administration.

Placebo (Score per week)	3 0 0 2	
Verum (Score per week)	7 1 2 0	
Probabilitity		.357

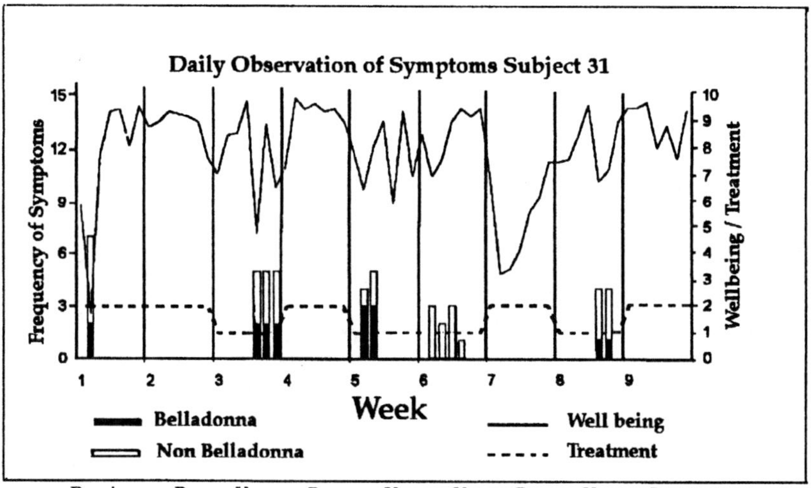

Run-in P V P V V P V P

Figure 4. Significant randomization test, striking in visual analysis; more Belladonna symptoms with Belladonna 30 CH.

Placebo (Score per week)	0 0 0 0	
Verum (Score per week)	6 6 0 2	
Probabilitity		.071

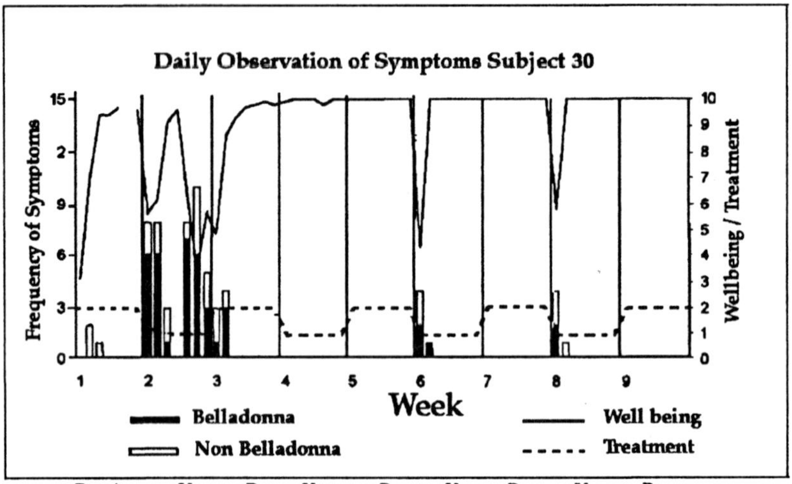

Run-in V P V P V P V P

Figure 5. More Belladonna-symptoms in visual analysis, no significant randomization test with Belldonna 30 CH.

Placebo (Score per week)	4 0 0 0
Verum (Score per week)	29 0 3 2
Probabilitity	.242

One can clearly see that *Belladonna* symptoms show up more frequently with *Belladonna* 30 CH (35 altogether) than with placebo (3 altogether). What can also be seen is a clear drop of the well-being data on Mondays exactly when the test substance had been taken, accompanied by the experience of *Belladonna* symptoms as well as some non-*Belladonna* symptoms. Intuitively one would expect a significance of the result which, however, did not show. We do not know, why this is so, but we suspect that this is a property of the permutation procedure which underlies the rationale of this kind of statistical testing.

A vital presupposition of this kind of design is the absence of carryover effects. For if carryover effects should be present, the statisctical procedure looses power and a differentiation of the seperate experimental units, i.e. experimental weeks, is more difficult. We indeed saw some carryover effects which of course can't be proven as such.

The data of subject number 19 (figure 6, *Belladonna* 12 CH) show a large cluster of symptoms beginning in the seventh week and going through to the end of the experiment. It is pain in the throat which is a *Belladonna* symptom but which can't be in fact attributed to *Belladonna* as it carries through the two following placebo weeks. The same was seen in the subject number 21 (data and graph not shown, *Belladonna* 30 CH): pain in the throat carrying through to the next placebo week.

It is generally assumed that homoeopathic provers should be healthy in a homoeopathic sense i.e. they should show no symptoms in their normal state of health so as to be able to discern any change in their general well-being. We tried to operationalize this demand by using the Freiburg Complaint List, which is a very sensitive list of subjective complaints in any area of the body. A low scoring in the

FBL means that a subject generally reports no or very few complaints on a phenomenological basis. We correlated FBL scores and the number of symptoms but found no clear correlation. Also, we saw some contradictory patterns i.e. subjects with very low FBL scores producing a lot of symptoms as well as producing very little symptoms and vice versa. We also saw subjects with very low FBL scores and still showing symptoms only under placebo or in a very unsystematic way. If there is any effect present in the data, it is unrelated to prior scores of either the Freiburg Complaint List or the Freiburg Personality Inventory scores.

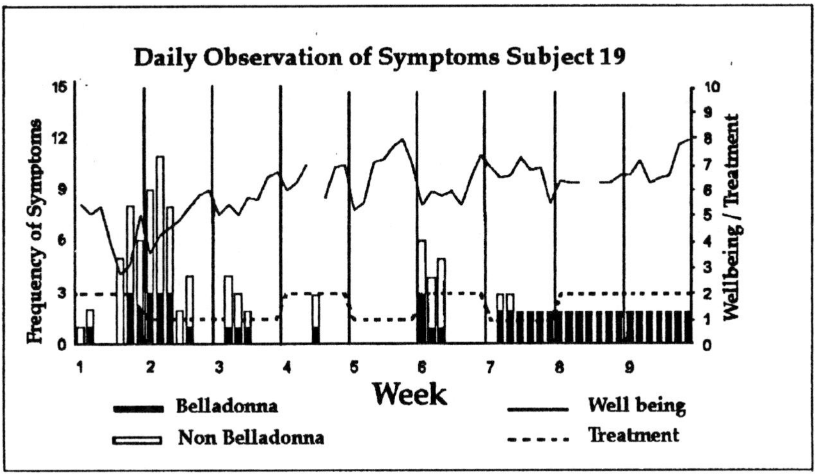

Run-in V V P V P V P P

Figure 6. Hints for a carry-over effect with Belladonna 12 CH

Placebo (Score per week)	1 5 14 14
Verum (Score per week)	10 3 0 12
Probabilitity	.3

We therefore conclude that *Belladonna* 12 CH and 30 CH does have effects different from placebo. The effects however are very small: one out of 25 experiments is significant, three more out of 25 experiments are striking in visual analysis but not statistically significant. There are unexplained paradoxical effects: one out of 25 experiments show more *Belladonna* symptoms with placebo. We can not exclude the presence of carryover effects in this kind of experimental pathogenesis study and therefore recommend not to use this kind of randomization design as a general approach to experiments following the homoeopathic rationale of a remedy proving.

4. Discussion

We have shown, using single-case randomization experimental design, that homoeopathic preparations (*Belladonna* 12 CH, 30 CH) produce effects different from placebo. One should observe that this can not be argued away by saying that so many experiments had been carried out for detecting the effect. As all the experiments were independent, the significant results are not artefacts of multiple testing, but true experimental results. But the fact that only 1 out of 25 test persons showed a significant effect and three more showed visually detectable effects means that the effects are very small. This, however, was expected: therefore the hypothesis was formulated that one out of the group of 20 would show a significant effect. The phenomenon, therefore, is not reproduceable in everybody and under any circumstances, but can be seen only infrequently in certain persons. Our data do not give a clear hint as to what kind of preconditions should be met. We examined the demographic data and screening information, but did not find any clear pattern or significant correlations. The type of result we found closely resembles the results obtained by Weiss *et al.* (1980) who asked whether food additives in soft drinks would trigger allergic reactions in children. As they, like we, expected a very small effect, they conducted a multiple single-case experiment with 22 children. Most of the children did not show any signs of allergic reactions, but one of the children showed a very clear and significant allergic reaction to the food additives. The authors therefore concluded that the effect was present but was very small. Still they issued a recommendation to the authorities to acknowledge food additives as a potentially allergogenic component of soft drinks, which is recognized as clinically relevant. The same is true for our results: the effect is very small and at the same time it seems to be there. As a randomization experimental procedure provides for intraindividual control, the effects can not be attributed to failure in randomizing groups of people as could have been in small parallel group designs. Of course, it could have been the case that *Belladonna* symptoms appeared under *Belladonna* just by chance. This is true, and the chance that this might have happened is given by the true p value. It is roughly 2/100. It also cannot be argued that our data collecting procedure generated the data. We also saw a large, approximately equal, number of non-Belladonna-symptoms, which we did not count. It does not seem to us that the experimental results can be explained away.

At the same time we obtained paradoxical results: clearly more *Belladonna* symptoms with placebo. The same fact was also observed in the predecessor study (Walach 1993), where 11 out of 45 subjects showed inverse reactions, that is, less *Belladonna* symptoms with *Belladonna* and more *Belladonna*-symptoms with placebo. Since this experiment had been conducted as a traditional cross-over design, and as a group experiment, no single-case statistics were possible. We hypothesized that the paradoxical effects might have been curative effects, that is, the test substance *Belladonna* cured symptoms which had been there anyway, and which had shown under placebo. The present experiment also was designed to test this hypothesis. We did not find data corroborating this hypothesis. Only two other subjects who experienced more *Belladonna* symptoms with placebo indicated that the effect was possibly due to a curative effect, and these two subjects did not show significantly more symptoms. That is, the person who experienced more *Belladonna* symptoms with placebo cannot be attributed to a possibly curative effect of *Belladonna*. In order to appreciate what this means, one should bear in mind that the target of our study was the number of *Belladonna* symptoms and not the number of symptoms altogether. Thus these results mean that one subject reacted with placebo, as if he or she had taken *Belladonna*. This is as yet unexplained and indicate that "placebo" is a highly complex and poorly understood intervention.

One might argue that this is just what one would expect by chance: sometimes more *Belladonna* symptoms with *Belladonna*, sometimes more with placebo, sometimes none. One should bear in mind, however, that we designed our experiment in a way that we would be able for a single case to answer the question: Does *Belladonna* produce more *Belladonna* symptoms than placebo? This was statistically true in one out of 25 cases and visually in three more out of 25 cases. One has also to consider that our methodology was a rather crude one: we did not look for all possible symptoms, neither did we probe for specific or individual symptoms. Rather did we screen for very crude general symptoms of *Belladonna*. Therefore, our results are not definite proofs, but hints that this venue of research should be followed on with more careful and possibly more sensitive methodology.

Our study shows two shortcomings from a methodological and external validity point of view: first, the randomization test procedure requires random permutation of sequences. This again assumes that carryover effects must be excluded. This presupposition may not always be met. Although *Belladonna* normally is supposed to have shortlived action in homoeopathy, this does not exclude the possibility of carryover effects. Because of this, the design is not ideal for testing this phenomenon and it should not be taken as a prototype of experimental studies in this field. A design, which is also a single-case randomization experiment but which randomizes the intervention point, would be more appropriate. Another shortcoming is the fact that we operated with fixed dosages. Many homoeopaths recommend flexible doses, adjusted to individual reaction. We felt that this could not be accomodated within the framework of a randomization design, so we dropped this requirement and instead opted for a small single dose distributed over one day. In this way we hoped to provide a stimulus which would be true to homoeopathic theory and to the requirements of experimental design. Future studies, however, should take this into account and find a design which can allow for flexible dosing.

In summary: the high dilution effect of *Belladonna* 12 CH and 30 CH seems to exist, albeit very weak. This warrants further investigation, preferably in a design, where carryover effects do not play any important role. Traditional exclusion criteria, such as perfect health or stimulant free life style, should be considered with caution, as in our study they did not contribute to explaining the effects. The effect is there, as well as the paradox: significantly more Belladonna-symptoms with *Belladonna*, as well as with placebo. The small effect size calls for confirmation in a study with larger numbers and more sensitive methodology.

Acknowledgement:

This study was supported by the Robert-Bosch-Foundation, Stuttgart. The authors gratefully acknowledge the help of Prof. W. Gaus, Ulm, in preparing the randomization table, of Frank Wieland, formerly DHU, and Dr. Marianne Heger, DHU Karlsruhe, for preparing the test substances, and of Dr. Patrick Onghena, Leeuven, for a critical reading of a draft version of this manuscript and helpful criticism of a former version of the statistical evaluation.

5. References

Edgington, E. (1987) *Randomization Tests*. Dekker Publisher, New York.

Ernst-Hieber, E. and Hieber, S. (1995) *Wirkt eine homöopathische Hochpotenz anders als ein Placebo?. Randomisierte, doppelblinde multiple Einzelfallstudie*. Hippokrates Publisher, Stuttgart.

Fahrenberg, J. (1994) *Die Freiburger Beschwerdeliste FBL. Form FBL-G und revidierte Form FBL-R*. Hogrefe Publisher, Göttingen

Fahrenberg, J., Hampel, R. and Selg, H. (1989) Das Freiburger Persönlichkeitsinventar FPI. Revidierte Fassung FPI-R und teilweise geänderte Fassung FPI-A. Hofgrefe Publisher, Göttingen. fifth Ed.

Hahnemann, S. (1979) *Organon der Heilkunst. sixth Ed.*,. R. Haehl (ed.), Reprint - Hippokrates Publisher, Stuttgart.

Walach, H. (1986) *Homöopathie als Basistherapie. Plädoyer für die wissenschaftliche Ernsthaftigkeit der Homöopathie.* Haug Publisher, Heidelberg.

Walach, H. (1993) Does a highly diluted homeopathic drug act as a placebo in healthy volunteers? Experimental study of Belladonna 30 CH in double-blind crossover design - a pilot study, *J. Psychosom. Res.* **37**, 851-869.

Weiss, B., Williams, J.H., Margen, S., Abrams, B., Caan, B., Citron, L.J., Cox, C., McKibben, J., Ogar, D. and Schultz, S. (1980) Behavioral responses to artificial food colors, *Science* **207**, 1487-1489.

THE VALUE OF USING HOMEOPATHY AT THE END OF GESTATION IN THE DOG AND PIG *

Report on two clinical trials

Y. GRANDMONTAGNE[1] AND A. RIAUCOURT[2]
[1] *LTP-Dolisos - BP 20 - 51051 Reims Cedex France*
[2] *Filiaporc-Dynal, 69, rue H. Le Vezouet BP 471 - 22 Loudéac Cedex France*

1. Introduction

In the last few years, there has been a trend towards increased use of homeopathy in animals. Treatment of this type has several advantages: the low cost of the treatment, efficacy against a variety of disorders, the absence of side effects... In farm animals (cattle, pigs, poultry ....) the improvement of production potential and the absence of residues following such treatment constitute two further major advantages.

In the specific context of birth, homeopathy can contribute to the welfare of the dam: these effects are apparent from the delivery conditions (quality, duration ...) and also to the growth potential and survival of the offspring. For some years, various authors have tested homeopathic dilutions around the time of delivery. Most of these studies have been carried out in farm animals because the intensive husbandry conditions which pertain are particularly suitable for experimental purposes, notably in making it possible to constitute various treatment groups within the same facility which are subject to the same conditions. Seifert (1987) has been able to demonstrate the value of rational homeopathic treatments in Mammitis-Metritis-Agalactia syndrome (particularly obvious efficacy with regard to the survival of piglets at weaning. Gavaret (1989) has tested the prenatal treatment of sows with Sulfur 9 CH and has demonstrated the beneficial effects on piglet growth. The clinical trials reported here were carried out, in one case in bitches and in the other in sows. Both studies evaluate homeopathic treatments administered to the dam during delivery.

2. Study in Bitches

2.1. PROTOCOL

2.1.1. *Aim of the Study*

The aim here was to evaluate the effect of a homeopathic medication on the delivery process in bitches.

---

* Communication presented at the 8th GIRI Meeting, December 1994, Jerusalem, Israel.

M. Bastide (ed.), Signals and Images, 227–235.

The aim here was to evaluate the effect of a homeopathic medication on the delivery process in bitches.

2.1.2. *Method*
The trial was a multicenter, placebo-controlled, double-blind trial and the treatments were randomly assigned to the individual dogs.

2.1.3. *Subjects eligible for inclusion*
Bitches confirmed to be pregnant, presenting with no pathological disorder or malformation which could hinder delivery by the natural process.

2.1.4. *Initial assessment*
The initial assessment of each subject was reported on the individual case report form:

-Data on each subject included: breed, age, presumed date of mating, expected date of delivery (=D0) and, if appropriate, dates of any previous deliveries.

-Previous history of any reproductive disorders.

2.1.5. *Treatments*
Many homeopathic remedies can be used in order to prevent any risk of dystocia in gravid females; these are prescribed on the basis of the homeopathic signs shown by each animals and observed by the prescriber. In the context of systematic prescription, the remedies had to be effective in all individuals; in this case they were prescribed, not on the basis of the various signs presented by the individual, but in this study, on the basis of the state of pregnancy common to all the bitches to be treated.

For the purposes of this clinical trial, we selected two "classical" remedies: Arnica and Caulophyllum.
Arnica: this is the principal homeopathic remedy used to treat physical injury, contusions, the effects of physical overwork and repeated exertion.
It also promotes the venous and capillary circulation. Arnica is recommended both during the final stages of pregnancy and during delivery.
Caulophyllum: this is the principal homeopathic remedy used to treat dystocia related to rigidity of the uterine cervix. It promotes dilatation at the moment of delivery; it is also reported to reduce uterine pain during gestation.

Each veterinarian received 12 numbered treatment kits, one per animal for the expected number of animals to be treated during the trial.
Presentation: each individual treatment was presented in the form of a kit consisting of 2 bottles: A and B.
1 individual kit = 1 bottle A + 1 bottle B.
Each kit was numbered.

2.1.6. *Beginning of Treatments*
The Veterinarian instructed the owner of the bitch to start the treatment two weeks before the expected date of delivery (D-15). A deviation by up to 5 days was acceptable.

2.1.7. *Dosage/administration regimen*
The treatment consisted of daily alternate administration of either 20 drops (about 1 ml) of the contents of each of the two bottles (regardless of the bodyweight of the animal) until delivery actually took place.

i.e.	20 drops from bottle A	on day 1
	20 drops from bottle B	on day 2
	20 drops from bottle A	on day 3
	20 drops from bottle B	on day 4
	20 drops from bottle A	on day 5
	20 drops from bottle B	on day 6

and so on, until the animal had pupped.
The drops could be mixed with the feed.
Concomitant medical treatment was permitted, as long as this was reported on the individual case report form -Appendix 1).

2.1.8. *Assesment Criteria*
The following quantitative criteria were taken in account during the trial:
- number of procedures (manual, medical or medicinal) carried out during delivery,
- Duration of the two successive stages of delivery:
from the onset of labor to the birth of the first pup;
from the birth of the first pup to the birth of the last pup.
- Number of live born pups, still born pups and the number of pups alive at 3 days, which makes it possible to evaluate any injuries attributable to birth and the viability of the pups.

2.2. STUDY PROCEDURE/FINDINGS

The trial was a placebo-controlled trial carried out under double blind in several veterinary clinics between March 1993 and June 1994 in the context of the work of the C.N.V.S.P.A. (Conférence Nationale des Vétérinaires Spécialisés en Petits Animaux - National Conference of Veterinarians specializing in Small Species) by the G.E.B. (Groupe d'Etude en Biothérapies - Biotherapy Study Group).
Twenty-two veterinarians took part in the trial and provided 73 interpretable cases.

2.2.1. *Number of Procedures*
The small number of procedures performed made it impossible to carry out any statistical analysis of the data.

2.2.2. *Duration of Delivery*
The differences observed occurred during the labor phase i.e. the time between the breaking of the waters and the expulsion of the first pup. The mean duration of this phase was longer in the placebo group than in the homeopathy-treated group (verum group): 3.12 h vs. 1.82 h ($p < 8\%$) - see Figure 1.

2.3. RESULTS

The results are expressed in figure 1.
Evaluation of the viability of the pups at birth: this parameter was assessed from the mean ratio of the total number of pups per litter (% live births/total births). The mean value of this ratio was 96.1% in the verum group versus 92.4% in the placebo group; a difference which was significant to the 10% threshold.

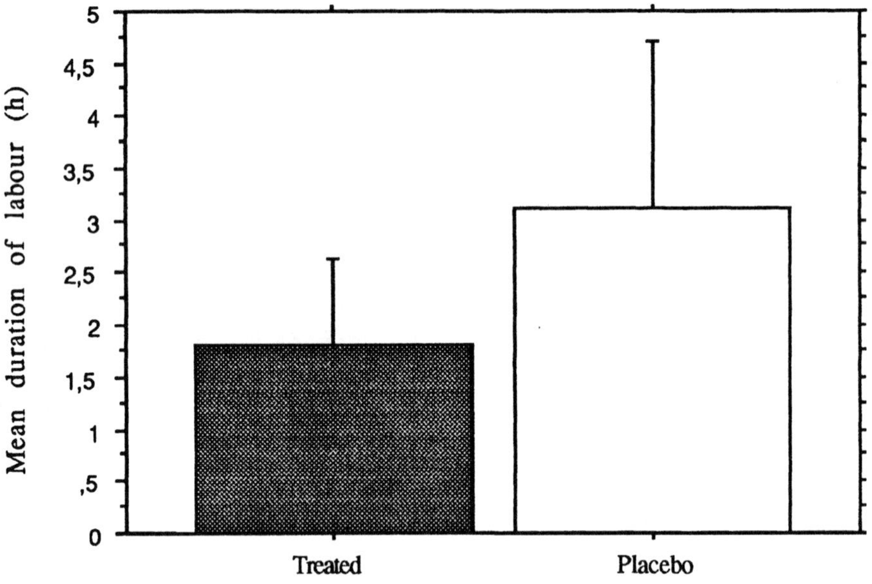

Figure 1. Onset of the labour

3. Study in sows

3.1. PROTOCOL

3.1.1. *Aim of the study*
The aim of this study was to evaluate the effect on the farrowing process and the survival of the piglets of treating gestating sows with a homeopathic complex.

3.1.2. *Method*
The trial was an open trial versus a reference treatment. The treatments were randomly allocated by a process involving equilibration for litter rank and expected date of delivery.

3.1.3. *Subjects eligible for inclusion*
Only sows carrying their second to sixth litter were eligible for inclusion. Those expecting their first litter were not eligible due to the lack of information about their farrowing capacity and the wide range of variability of their behavior. Litters after the

sixth were not included because they are not typical of husbandry stock (only good sows are kept). Sows which had presented with any particular problems during previous farrowing(s) (MMA syndrome, abnormally prolonged farrowing) were also excluded from the trial.

3.1.4. *Treatment*
The treatment chosen was a complex formulation containing 3 constituents which had been in routine use in several animal husbandry facilities for several months:

Caulophylum	30 CH
Sepia	6 DH
Cocculus	6 DH

Caulophyllum thalictroides: a plant which contains methyl cystisine (Guermonprez *et al.*, 1985). This is a hormonal substance, which explains why this substance has a polarity which promotes opening of the uterine cervix.
Caulophyllum has been subjected to numerous clinical trials in the sow which have demonstrated its role in facilitating farrowing (Wolter, 1980) and its indirect beneficial effect in improving prolificity criteria, reducing the number of piglets stillborn (Day, 1984) and increasing the litter weight (Casini *et al.*, 1990). In experiments carried out in female rats, *Caulophyllum* has demonstrated and estrogen-like activity, resulting in increased ovarian and uterine organ weights (Seifert, 1987) In women, a homeopathic complex containing *Caulophyllum* reduced the duration of delivery and the number of cases of dystocia (Arnal, 1983; Coudert, 1981).

Sepia is prepared from squid ink. *Sepia* (Zissu and Guillaume, 1986) contains derivatives of amino acids, such as DOPA and serotonin-like substances. Its action consists of two phases: a short excitation phase (venous and vasomotor erythema, venous overload) followed by a prolonged phase of depression of the elastico-connective and nervous systems as well as of the cutaneous and mucosal orifice linings. The main areas affected are the pelvic pole and the hepatoportal pole. Administered concomitantly with *Caulophyllum*, *Sepia* is traditionally used in women in preparing for childbirth.

Cocculus indicus: obtained from the fruits of a Menispermacea, *Cocculus* contains alkaloids and picrotoxin, which induces hyperparasympathicotony. *Cocculus* has already been tested in pigs (three actions investigated: cannibalism, mortality during transportation and during farrowing); some idea of its potential usefulness has already emerged (Riaucourt, 1991).

Five milliliters of the preparation in liquid form was administered into the feeding trough by means of a dosing gun. This was repeated daily for the 5 days preceding the scheduled date of farrowing.

3.1.5. *Concomitant Treatment*
Such treatments were authorized and administered in accordance with the usual stock practice of the farmer, and administered routinely to both groups.

3.1.6. *Assessment Criteria*
Qualitative criteria:

"Farrowing assessed by the stockman

- duration (quick, normal, slow)
- demeanor of the sow during farrowing (nervous, normal, calm)

Prolificity criteria

- number of live births

- number of still births
- number of mummified fetuses
- number alive at 48 h

(adoptions are carried out only after 48 hours)
- percentage of live births still alive at 48h.

3.1.7. *Monitoring the trial*

The information about each sow was reported on an individual case report form.

3.2. TRIAL PROCEDURE / RESULTS

- The trial was carried out between December 1991 and April 1992 in a farrowing/fattening unit in the Côtes d'Armor region (of Brittany, France), with 7 successive cohorts, each of about 25 sows every 3 weeks.

From each cohort, up to 2 groups of 10 sows were selected by means of random selection equilibrated with regard to the number of previous litters, so that these 2 groups of sows could farrow in the same farrowing unit.

- All the animals in both groups received the routine treatments administered in the unit:
  - on day 114: Planate" (morning) + Suanovil" (afternoon)
  - at the end of farrowing: Ocytocin (30 IU)
  - the day after farrowing: combination of antibiotic + anti-inflammatory (Ampimycin Dex").

Animals included: 111 animals were included.

Animals excluded during the trial: 10 multiparous sows, either because they were not pregnant or because they became ill before farrowing.

101 sows completed the trial in entirety.

3.2.1. *Preliminary Analysis*

The results are presented in Table 1.

The distribution of the litter ranks in each group was fully equilibrated.

TABLE 1: Comparison of the distribution of litter ranks in the two groups (K^2 test: NS)

	Control Group	Treated Group
2nd litter	14	12
3rd litter	8	15
4th litter	8	6
5th litter	8	9
6th litter	8	8
TOTAL	51	50

3.2.2. *Analysis of the Criteria Investigated*

- **qualitative criteria**

No difference was detected between the two groups with regard to the demeanor of the sow or the duration of farrowing.

- quantitative criteria

No difference was found between the groups with regard to the number of live births, still births, mummified fetuses or live offspring at 48 hours. The number of stillborn piglets was however, clearly higher in the control group (20 still births versus 10 in the treated group).

TABLE 2: Summary of the study data

	Control Group	Treated Group	Statistical analysis
Number of Subjects (sows)	51	50	
Qualitative Criteria			
Duration of Farrowing			
- quick	10	10	NS
- normal	39	38	NS
- long	2	2	NS
Demeanor of the sow during farrowing			
- nervous	6	3	NS
- normal	29	35	NS
- calm	16	12	NS
Prolificity * *Criteria*			
Number of live birth	10.1± 2•5	10.1 ± 2	NS
Number of still birth	0.4 ± 1.1	0.2 ± 0.5	NS (p < 7.6%)
Number of mummified	0.1± 0.3	0.1± 0.3	NS
Number alive at 48h	9.1 ± 2.2	9.5 ± 1.9	NS
Number alive at 48h / number live birth	91.9% ± 10.5	95.1% ± 6.3	p < 4%

* Mean ± SD

The percentage of piglets alive at 48 h in terms of the number of live births showed a ***significant difference, to the threshold of 4%,*** in favor of the treated group. ***Farrowing process:*** in contrast with the findings reported by Wolter (1980), no difference was found between the two groups. The difficulty in carrying out evaluations of this type is related to the nature of the parameters selected (qualitative criteria marked by some degree of subjectivity) and to the timing of farrowing (this often occurs at night, which does not always make it possible for the observer to be present): the percentage of sows showing "normal" demeanor or farrowing described as "normal" may therefore have been overestimated.

Prolificity parameters

The intergroup comparison of the number of live born piglets did not reveal any difference:

the equivalent number of mummified piglets in these two groups can easily be accounted for by the fact that mummification process occurs earlier than the period of the test treatment.

- the facilitation of the passage (distention of the cervix) was not confirmed statistically, on the criterion "number of still births", despite the clearly smaller number of in the treated group: 10 still births versus 20 in the control group. The "reduced distress" during farrowing is however reflected in the improved survival of the piglets in the treated group: 95.1% survival rate versus 91.9% in the control group (which is equivalent to mortality rates of 4.9% and 8.1% respectively during the first 48 hours). Another analysis of the survival rate, this time in terms of the total number of births (= live births + still births + mummified) gives values of 92.8% in the treated group versus 87.4% survival rate (difference of more than 5 points, significant to the 7% threshold).

The two tests described, carried out on two very different types of animal (dog and pig) consisted in both cases of a comparison of a homeopathic treatment administered at the end of gestation with a placebo. In both cases significant differences between the treated/placebo groups were noted: both studies demonstrated a beneficial effect of the homeopathic treatment on the viability of the offspring at birth (number of live births/total number of births) in the dog and (number alive at 48 hours/number of live births) in the sow. In the sow, a beneficial effect of the homeopathic treatment on the duration of farrowing was also observed.

These findings probably reflect the known effect of *Caulophyllum* on the dilatation of the uterine cervix. *Caulophyllum* is included in the formulations used in both studies, but is used at different concentrations. These findings confirm those reported by Day (1984). They also show the financial value of administering some homeopathic preparations in combination with conventional "allopathic" treatments.

4. References

Arnal, M.N.(1986) ***Préparation à l'accouchement par Homéopathie: Expérimentation en double insu versus Placebo,*** Thesis Medicine Doctor, René Descartes University, Paris, France.

Casini, L., Del Francia, F. and Bosi, P. (1990), Studio dell'effetto del rimedio omeopatico *Caulophyllum* somministrato alle scrofe nell'ultimo periodo di gravidenza, *Rivista di Suinicoltura* **10**, 43-45.

Coudert M. (1981), ***Etude expérimentale de l'action de Caulophyllum dans le faux travail et la dystocie de démarrage,*** Thesis Medicine Doctor , Limoges University, Limoges, France.

Day, C. (1984), Control of stillbirths in pigs using homeopathy, *Vet.Record* **114**, 216.

Dorfman, P., Lasserre, M.N.and Tétau, M. (1987), Préparation à l'accouchement par Homéopathie, *Cahiers Biothérapie* **94**, 77-81.

Gavaret, T. (1989), *Traitement homéopathique prénatal en élevage à haute productivité,* Thesis Veterinary Doctor, Maison-Alfort Veterinary School, France.

Guermonprez, M., Pinkas, M. and Tork, M. (1985), *Matière Médicale Homéopathique*, Doin Publisher, Paris.

Riaucourt, A. (1991), *Cocculus* en élevage industriel porcin, *Dynamis* **5**, 6.

Seifert, U. (1987), *Die Behandlung des Mastitis, Metritis, Agalaktie (MMA) Syndroms der Muttersausen mit homöopathischen Arzneimitteln im Vegleich mit einer praxisüblichen allopatischen Therapie*, Thesis Veterinary Doctor, Free University, Berlin, Germany.

Sunil, K., Anil Kumar, S. and Chandrasekhar, K. (1981), Effects of *Caulophyllum* on the uteri and ovaires of adult rats, *Brit.HomJ.* **70**, 135-138.

Voisin, H. (1976), *Matière Médicale du Praticien Homéopathe; Thérapeutique et Répertoire Homéopathiques du Praticien* (2eme ed), Maloine Publisher, Paris.

Wolter, H. (1980), Wirksamkeitsnachweis von Caulophyllum D 30 bei der Wehenschwäche des Schweines im doppelten Blindversuch, in K.H. Gebhardt (ed), *Beweisbare Homöopathie*, Haug Publisher, Heidelberg, pp 63-69.

Zissu, R. and Guillaume, M. (1986), *Fiches de Matière Médicale Homéopathique*, Doin Publisher, Paris.

ARSENICUM ALBUM AND NEUROTOXIC POISONING IN DOGS AND CATS *

An example of Homeopathic Treatment in Animals

R. BLOSTIN
Dr Veterinary, 42 rue F. Poulenc 77430 Champagne sur Seine, France

1. Introduction

In his practice, the vet has to treat in emergency dogs and cats with acute severe poisoning by neurotoxic agents,without knowing exactly the toxic agent and the doses absorbed. For more than 6 years, we have been using in a systematic and exclusive way, *Arsenicum album* with remarkable success.

2. The basic observation (Blostin, 1990)

Jennie, a female Doberman of 18 months, has just absorbed one hour ago a bait containing arsenic, despite of the notices pointing out the presence of sausages containing this poison. Jennie has vomited a part of this, half an hour ago.
Symptoms observed:
Physical and mental hyperexcitability. Jennie runs about in all directions, howling as though she is going to die. She only stops moving for a few seconds at a time looking for shade. Her movements are jerky. When she is at rest all her muscles seem ready to react to the slightest tactile or auditory stimulus and this is expressed by a local or general uncontroled reaction and by howling. Jennie appears to feel cold despite the surrounding heat. Jennie presents a very fluid hypersyalorrhea, accompanied by watery rectal emissions (cf urine) emitted in little involontary jets and apparently painful.
Treatment undertaken:
Faced with this typical clinical scene reminding one of both the observations of the toxicologists and those of the *Materia Medica*, and because of the toxic agent, I decide to undertake a treatment with *Arsenicum album*.

I choose to administer highly diluted doses directed at the nervous system and I choose 15 C because it was available in my surgery. A phial of 2 ml of *Arsenicum album* in 15 C is given at 14h30, then at 14h45 a perfusion (Sodium Chlorate 0.9%) and an injection of a dose of *Arsenicum album* in 15 C.

At 3 o'clock the discharges stop.

At 15h15 and 15h45 two more doses are given intraveinously.

* Communication presented at the 7th GIRI Meeting, November 1993, Montpellier, France.

M. Bastide (ed.), Signals and Images, 237–243.

At 4 o'clock the howling disappears and the movements became pratically normal. The perfusion is stopped temporally and Jennie galops about in the garden. Slight muscular hyperexcitability still persists in the presence of pronounced tactile stimuli.

At 16h15 perfusion is started again to compensate for the liquid losses estimated at 500 ml.

At 16h15 another oral dose is given.

At 5 o'clock Jennie returns home. Another dose, still in 15 C is given at 7 o'clock and Jennie asks for an evening meal, swallows it and keep it down.

A dose is given the following day and for four days thereafter. No sign of poisoning is observed as from the morning following the day the poison was taken.

3. Generalisation

Jennie's case is an easy one, and in this respect it is rare, because the poison was known and the symptoms presented by the dog were like " a text-book case " . *Arsenicum album* was the " similar " and the *Similimum* in this poisoning.

I live in a region where we come across many cases of this kind of poisoning, but after this "text-book case", I was faced with cases of poisoned animals where I did not know the poison and where the symptoms, variable from one to another, did not call a priori for the immediate prescription of *Arsenicum album*. In certain cases the neurotoxic action is more central (predominance of convulsion), in other cases more peripheral (predominance of gastro-intestinal spasms for example).

However I noted that *in every case there were symptoms irrespective of the individual or the toxic agent.*

These commun symptoms are:

1 - muscular hyperexcitability
2 - aggravation through visual, auditory and tactile stimuli (light, noise, touch)
3 - anguish

These symptoms are described in the *Materia Medica* of *Arsenicum album* and were present in the case of Jennie. So I decided to use *Arsenicum album*, as in Jennie 's case, even if this remedy was not a priori the *Similimum*. The results were immediately spectacular, provided one respected the modalities of aggravation i.e. isolating the animal by eliminating the maximum number of sensory stimuli. ***These three symptoms became for us the three indispensable elements for the prescription of Arsenicum album.***

They appear at degrees which vary according to the poisoning and the nature of the poison, the quantities swallowed, the reactivity of the individual. For each symptom there exists a scale of variability according to the poisoning:

as for symptoms 1 and 2 one may observe either a very slight hyperexcitability brought on by hitting the surgery table very hard or violent epileptic-like attacks at the slightest touch;

as for anguish, it appears either simply in the animal's eyes or by its characteristic howling or mewing, full of anguish and communicating anguish to the observer.

It is to be noted that in the case of serious poisoning, the animal arrives howling or mewing; in even more severe cases, these cries appear only in the course of the treatment as a step towards recovery (recovery of consciousness). It is as though the animal was "calling for his arsenic" and taking it calms him within a quarter of an hour.

4. Scheme of treatment

First and foremost, it is extremely important whatever the degree of poisoning to respect the modalities of aggravation and therefore to isolate the patient by eliminating the maximum number of stimuli. From a practical point of view, we install the animal in a cage (not too big in order to limit movement), covered by a thick blanket (to reduce the intensity of auditory and visual stimuli), the whole thing isolated in a dark room away from every day activity (for the same reasons as before).

The scheme suggested below concerns the case of serious poisoning and shows the possibilities of treatment according to different stages observed in the course of hospitalisation; for less serious poisoning the method is taken from the stage observed. It is as though there existed a scale of seriousness for each case of poisoning with characteristic symptomatic stages ; a scale which the animal seems to climb like a staircase in the course of the treatment by going through the preceding stages.

First phase: epileptic-like attacks repeated at the slightest stimulus.

A dose of *Arsenicum album* in 15 C is given every half hour for two hours, then every hour until the following "noisy" stage. Gradually the sensitivity to stimuli declines; the stimuli must be more and more violent to bring on an attack. *Arsenicum album* is at first administered orally then if need be in perfusion as soon as this is possible.

Second phase: the noisy phase; anguished howling or mewing.

The animal demands his arsenic; a dose of *Arsenicum album* in 15 C calms it within a quarter of an hour, and is administered upon demand.This phase generally lasts from a half an hour to 3 hours.

Third phase: the recovery phase.

The animal is calm; only a slight muscular hyperexcitability to stimuli persists; the latter must be gradually more violent. This phase usually lasts from 12 to 48 hours; *Arsenicum album* in 15C is given upon demand approximately every 3 hours at the beginning, then spaced out. We intersperse doses of 5 C (1 or 3). Finally a dose of *Arsenicum album* in 15 C is given after a week.

Of course this classification in three stages is schematic and didactic rather than rigorous. In fact one should try to administer *Arsenicum album* in 15 C upon demand, as always in homeopathy, i. e. as soon as one thinks that the previous dose has exhausted its effect. When the animals are brought to us in a comatose state the improvement often takes the form of the sudden and spectacular appearance of the convulsive phase; it is not a question of an aggravation but of a natural development in the course of the treatment.

5. Results

Over 6 years we have had to treat more than fifty cases of neurotoxic poisoning in dogs and cats, including - when we have been able to check the facts - cases of poisoning by arsenic, metaldehyde or organophoshorus compounds. These animals received *Arsenicum album* as their only remedy . The results are expressed in Table 1.

TABLE 1: Results Obtained in 50 Dogs and Cats

- 4 deaths out of 50 dogs or cats
- 80% returns home within 12 hours after the beginning of the treatment
- 100% within 24 hours
- No secondary effect were observed.
With classical allopathic treatment, ± 50% death is classically observed.

We should note however that in 3 cases - and particularly in the case of a cat which was treated for 2 poisonings at an interval of one year - we were informed of temporary relapses, taking the form of loss of balance or trembling which lasted several hours; relapses occuring every fortnight or 3 weeks with symptoms which gradually became weaker. This fact is interesting and must certainly correspond to a cyclical discharge of the neurotoxic substance by the storing organs (liver, kidney). We should note that a 3 week cycle is classically described in the proving of *Arsenicum album.*

When we can know the toxic agent, treatments with Arsenicum album are very satisfactory in intoxication by Arsenic, Strychnine, Organophosphorus compounds, carbamates, Chloralose, but certainly weaker in pure methaldehyde intoxication.

A bibliographical study confirms these clinical results (C.N.I.T.V. Notes, 1985; Chaoui, 1988; Goodman and Gillman, 1985) : muscular hyperexcitability, aggravation through visual, auditory and tactile stimuli (light, noise, touch), anguish are together described by the toxicologists in intoxication by Arsenic, Strychnine, Organophosphorus compounds, carbamates, Permethtine, Chloralose; no aggravation by stimuli are described in intoxication by methaldehyde in dog and cats.

After dogs and cats, we tried to adapt this scheme of treatment to a mouse model using a parathion intoxication before *Arsenicum album* treatment. The results confirm the aggravation by stimuli, in spite of gentle handling for gavage. Unfortunately this aggravation was fatal as the mice died right at the beginning of the treatment. They died before the necessary repetition of the doses of *Arsenicum album.* This protocol was not adapted for mice. No statistical conclusion are possible for the use of *Arsenicum album* in mice.

6. Discussion and Conclusion

The treatment of acute cases of poisoning by neurotoxic substances in dogs and cats by *Arsenicum album*, following the procedure described above, gives spectacular positive results far better than those obtained with allopathic treatments and today several vets use them. It is interesting for animal and human health.
But some fundamental questions may be asked:

- What is the relationship between a prescription based on the "Similia Law" used by the homoeopaths and the pharmacological reality of the receptors ?

-Why does *Arsenicum album* in 15 C act efficiently irrespective of the individual and the toxic agent ? how and where does *Arsenicum album* work?

At the level of traditional homeopathy, apart from 2 or 3 cases - including that of Jennie - where *Arsenicum album* was "the Similar" or the *Similimum*, more often than not it is only a simile. The search for a *Similimum* in the course of these poisonings seems to us pointless since the use of the simile *Arsenicum album* is so reliable. And a good thing too !. For to find the *Similimum* of an animal, which in the throes of a convulsion, in the presence of its owners who may themselves be quite justifiably in a state of panic, would appear to us to be quite an achievement.

The problem would remain if we decided to use treatment by " the Similar", since it is so difficult practically to identify the poison chemically.

However, while observing these animals certain remedies occurred to us. Were they the *Similimum*? We did not try them ! Thus the dog, which as soon as it was touched started an attack of convulsions accompanied by loss of consciousness, coldness and cyanosis of the mucous membrane, simulating death and which made us think of hydrocyanic acid. Would hydrocyanic acid have worked ? . We do not know. *Arsenicum album* certainly did work.

But this systematic use of *Arsenicum album* whatever the poison and whatever the individual, leads the homeopath to make certain comments since it appears at first sight to belong much more to the domain of allopathy than to that of homeopathy . Indeed, at first sight, you cannot find in this use of *Arsenicum album*, in this " recipe ", the 2 basic precepts of homeopathy i.e. ***the law of totality*** and ***the law of the Similimum.*** Again at first sight, this therapeutic recipe is homeopathic only in so far as it uses a diluted remedy : ***the law of infinitesimality***. So only one out of the three laws is respected and yet this treatment works, why ? .

We shall in fact demonstrate that for us these three laws are indeed respected and that this really is homeopathy. It all began by observing the case of Jennie for whom *Arsenicum album* was without dispute her *Similimum* in the course of the poisoning. Subsequently by observing other cases of poisoning we saw that three symptoms (Table 2) existed which were common to all these poisonings and this led us to use *Arsenicum album* as in Jennie 's case.

TABLE 2: Indispensable symptoms for *Arsenicum album* indication

1 - muscular hyperexcitability
2 - aggravation by all the sensory stimuli
3 - anguish

They form a very solid triangular base of homeopathic symptoms according to Hering. Indeed we have one general symptom, one modality and one mental symptom.

Arsenicum album is only prescribed when these three symptoms appear together. They should be present in the course of the evolution of the poisoning .

At this stage in our reasoning we are very far from allopathy; in fact we are back again right in the heart of homeopathy, in the heart of those constant and fascinating considerations on the collection of symptoms.

We use infinitesimality, we have just seen that we also respect the law of totality. What about of the law of the *Similimum* ?***In fact, for us, Arsenicum album may be considered also as a Similimum.*** It all depends on the level of organisation at which we view the situation and the importance of this level for the collection of symptoms.

Whatever the poisons may be, there is at the biochemical level a common mechanism, or different mechanisms, which involve the presence of the three common symptoms described above in these poisonings. It is at this level that I believe that *Arsenicum album* acts and certainly at the synaptic level. This remain to be confirmed. By the same process, or by different processes, *Arsenicum album* involves the return to normality of the transmission of nervous impulses. If we consider this biochimical level, it is common to this type of poisoning whatever the individual.
At the biochemical level the basic mechanisms involved in the action of these poisons are sufficiently strong to efface individual reactions.

Besides they must act more on the storing and eliminating capacities depending on the specific enzyme equipment of each individual; it's may be why we are obliged to repeat the doses (= repeat the signal). The neurotoxic poisonings affect neurobiological processes common to all individuals; in view of this idea, the individual reaction perhaps becomes of secondary importance in the search for symptoms and the symptoms which we use in order to prescribe *Arsenicum album* are directly linked with this biochemical level. ***At this level of organisation Arsenicum album can in our opinion be considered as a Similimum.***
We certainly must follow this way in order to continue our research.

- Why is it necessary to repeat the doses before improvment?

After Hahnemann (See Hahnemann, Baillière 1986), several homoeopaths like Rouy (1989), recommend to repeat the dilutions in acute desease. In our treatment, we must repeat the doses before improvement. It's the same when we have to treat dogs or cats for a poisonous snake bite with *Apis* or *Vipera*. When we are sure of the similitude in acute cases, we must not be afraid to repeat the doses and certainly not to stop the treatment . The homoeopath will say that he has to give enough information for the organism to react; the pharmacologist could say that we have to sature some receptors....

- Originality of this kind of treatment ?

Arsenicum album is a well-known remedy in homoeopathy for the treatment of food-poisoning (Duprat, 1985; Kent, 1982; Voisin, 1986), but not for acute neurotoxic intoxication in dogs and cats. We just can find *Arsenicum album* in a complexe remedy "Arsenicum album comp. " proposed by Pascoe Laboratory (Germany) (Pascoe, 1991), as a drainage remedy after a chronic exposition to insecticides.

To end with two more ideas:

-compared with a prescription using "the similar", the use of *Arsenicum album* seems to us more effective both theoretically and practically because:

1- in case of a severe poisoning it is difficult to discover the poison quickly

2- *Arsenicum album* is a substance which is already present and active in the organism; methaldehyde and organophosphorus compounds are normally absent.

-as for the procedure itself it can certainly be modified or even improved upon. Other centesimal or Korsakovian dilutions may be used. So far we have not attempted to do this . "You don't change a team when it is winning all its matches ".

7. References

Blostin, R. (1990) *Arsenicum album* and neurotoxic poisoning in dogs, *Proceedings of the Second International Congress for Veterinary Homoeopathy* , I.A.V.H , Zutphen, The Netherlands.
C.N.I.T.V., (1985), *Notes de Toxicologie Vétérinaire*, 3, Ecole Nationale Vétérinaire de Lyon.
Chaoui, A. (1988), *Influence de certains facteurs physiques et chimiques sur l'activité de dilutions infinitésimales d'Arsenic*, Thèse de Doctorat de Sciences Pharmaceutiques, Lille 2, France.
Duprat, H.(1985), *Traité de Matière Médicale Homéopathique*, 3 édition, J.B. Baillière Publisher, Paris.
Goodman, A. and Gillman D.H. (1985), *The Pharmacological Bases of Therapeutics*, Macmillan Publisher, London.
Hahnemann, S.(1986), *Doctrine Homéopathique*, J.B. Baillière Publisher, Paris.
Kent, J.T (1982), *Final General Repertory*, 2 édition, N.H.Publisher, New Delhi.
PASCOE Laboratory (1991), *Nosoden-Complexe*, D-6300 Giessen, Germany, 1991
Rouy, P. (1989), *Enseignements sur la Thérapeutique Homéopathique* , tome 1 et 2 , Similia Publisher, Paris.
Voisin H.(1986), *Matière Médicale du Praticien Homéopathe*, 2 édition Maloine Publisher, Paris.

MALARIA AND HOMEOPATHIC REMEDIES IN GHANA *

An Open Study and a Double-blind Randomized Clinical Trial

V.M.A. VAN ERP [1] AND M. BRANDS [2]
[1] *BSc, Vrije Universiteit Amsterdam*
[2] *MD, Homeopaths without Borders Netherlands*

1. Introduction

Malaria is a disease caused by four parasites: *Plasmodium falciparum*, *P vivax*, *P. ovale* and *P. malariae*. Each of them has its own morphological signs, period of incubation, clinical picture and sensitivity to medical drugs. (Meuleman, 1989). All the parasites are transmitted by different species of the female *Anopheles* mosquito (Zahar, 1984).

General symptoms are intermittent fever, headache, general malaise, bone- and waist pain. The diagnosis is confirmed by clinical presentation and a thick blood smear. Malaria tropica, caused by *P. falciparum* has as possible complications cerebral malaria and renal insufficiency (Peters, 1985), due to parasite infestation of the capillaries in the brain and the kidneys; these complications can be fatal. In many countries malaria is endemic; people are not clinically ill from malaria, but many of them have chronic complaints like weakness and loss of concentration. They have usually a positive thick blood smear; this was the case in the area where this study was done.

Standard treatment in Ghana according to WHO-protocols is chloroquine, 25mg/kg bodyweight, in resistant cases quinine in combination with antibiotics like tetracycline (Afro technical papers, 1992). Other drugs exist for chloroquine resistant malaria, like halofantrine (Peters, 1987)

Malaria is a major health problem in most developing countries (World Health statistics, 1992). Resistence against chloroquine, a rather cheap and easy available treatment, has been assessed at several places (Notten, 1992). Also resistence against other drugs is reported. These circumstances make malaria a difficult disease to cure. Trials to develop a vaccine have not yet been successful on a large scale (Valero *et al.*, 1993).

The clinic where the study was done, is in a rural area, where the large majority of the population has endemic malaria. In this clinic a teaching programme of "Homeopaths without Borders" is given at regular intervals since 1993, to train the staff in treating common diseases as diarrhea, feverish conditions, trauma and skin

* Communication presented at the 8th GIRI Meeting, December 1994, Jerusalem, Israel.

M. Bastide (ed.), Signals and Images, 245–252.
1997 *Kluwer Academic Publishers. Printed in the Netherlands.*

suppurations.

The study was performed to see whether a substantial number of patients would react on homeopathic treatment. Until now there was only casuistice evidence of malaria patients being treated, described in the homeopathic literature. This is the first controlled study of the efficacy of homeopathy on malaria.

As every patient can have different accompanying signs to his classical symptoms of malaria, the treatment is individualized according to the inter-individual differences. The homeopathic diagnosis consists actually of determining the susceptibility of a given person for a homeopathic remedy, at the basis of a similar set of symptoms of that person to cases being cured in the past. Although a standard treatment might not be possible on theoretical grounds, an "epidemic approach" is however useful (Hahnemann, 1986). This means that in epidemics as diarrhea and endemic diseases as malaria, a limited number of remedies might be indicated for 70 to 80% of all cases. In diarrhea this limitation is already shown by studies in Nicaragua (Jacobs *et al.*,1994). This might be also important for the future transfer of knowledge to e.g. health assitants in clinics in developing countries about the homeopathic treatment of malaria. The cases are described in the literature, of which is given now a brief account.

In the *"Materia Medica Pura"*, Hahnemann's documentation (1988) of the symptoms arising from provings, drug pictures are described which resemble malaria. In the clinical literature descriptions are found of treatment of malaria with homeopathic remedies. Already in the beginning of this century, in Allen's *"Therapeutics of fever"* (1983) and Farrington's *"Comparative Materia Medica"* (1989) (both reprinted) clinical cases are described with the main remedies used. Other casuistic evidence for individual prescribing of homeopathic remedies in malaria was shown afterwards. (Puhlmann, 1920; Fenner, 1925). However until now no controlled group study was done, comparing standard malaria treatment (chloroquine) with homeopathy.
Research comparing chloroquine with homeopathic treatment seems to be relevant because it could offer an additional perspective in treating malaria.

2. Material and Methods

The main research-questions were: "Would homeopathic treatment work better than chloroquine?" and: "Which homeopathic remedies can be mainly used, when prescribed individually?"

The research took place from october until december 1993 at the Shekhinah clinic in Tamale, in the Northern Region of Ghana. The open study was done during the wet season, the double blind at the end of it. The patient selection was done by the nurse in charge.

2.1. INCLUSION AND EXCLUSION PARAMETERS

Patients were included in the research if the patient had some of the following clinical parameters:

- fever (T > 37.5 ^{o}C rectal) eventually remittent fever
- chills
- general headache
- headache above the eyes
- waist-pain (paravertebral pain, lumbar region)
- bone-pain
- anemia: Hb women < 7,4 mmol/l; men < 8,0 mmol/l
- eventually splenomegaly
- other complaints: abdominal pains; dizziness, palpitations
- parasitology: a positive thick blood smear

The usual parasitological parameter: a positive thick blood smear (Meuleman, 1989), was problematic as a valuable parameter for evaluation, as in a prestudy screening, almost all subjects had a positive smear, if they had clinical signs of malaria or not. This is often the case in endemic malaria (WHO, 1992). For the evaluation of the efficacy of the therapy therefore, this is not used. The central outcome measurement thus is defined in terms of the clinical symptoms, listed above.
Exclusion criteria were: children younger than ten years and pregnant women. Patients treated by the research protocol with no signs of improvement, were counted as negative results and got chloroquine or quinine. (WHO, 1991).

2.2. OBJECTIVES OF THE TREATMENT

The malaria-treatment can have different objectives: i) to finish an attack; ii) to achieve complete cure; iii) to improve a patient's resistence against a next attack, during a certain period

2.1.1. *To Finish an Attack*

- an improvement in at least **three** of the six clinical symptoms: chills, headache above the eyes, general headache, bone-pain, waist-pain.
- a decrease in temperature with 0.5° C rectal

2.1.2. *A Cure is Defined as*

- being free of fever episodes during five days after treatment
- a rise in Hb of 0.3 mmol/l
- a negative thick blood smear

2.1.3. *Improvement Evaluation.*

The improvement of patient's resistance against a next attack can be measured if the frequency of malaria-attacks in one year has been reduced.

The duration of the research period was not sufficient for evaluating the follow-up after six or twelve months; moreover, a positive thick blood does not seem to be related to the presence of clinical symptoms, pathognomonic for an attack of malaria. Therefore this study was meant only for evaluating the possible difference in effect on an attack of malaria by chloroquine and homeopathic treatment, based on an assessment of clinical symptoms.

2.3. CHOICE OF THE TREATMENT

The homeopathic remedy was given after repertorization with Kent (1974), using the following rubrics

-time of onset of chills
-part of the body where the chills begin and extend from
-periodicity of chills and/or fever
-succession of stages of chill, heat and perspiration
-thirst/thirstlessness/quantity of thirst/time of maximal aggravation
-location of headache
-kind of headache
-nausea/vomiting
-stool/diarrea

It is emphasized that these symptoms were used for the diagnosis of the remedy-picture, and that the evaluation of the treatment was done with the clinical symptoms mentioned in the definition of finishing an attack.
Some examples of remedies are given (Allen, 1983; Farrington, 1989; Gaucher *et al.*, 1993):

*** arsenicum album**
Often used in quotidian, tertian and other types of malarial intermittent fever with marked periodicity.
Symptoms

- irregular chills preceded by yawning and stretching
- hot stage is intense, longlasting and dry burning
- stages not clearly defined

Characteristic signs

- drinks frequently but little at a time
- drinking causes nausea and vomiting
- heat is as if hot water is running through blood vessels

*** natrium muriaticum**
Every type of fever with a characteristic paroxysm of fever in the morning.
Symptoms

- chills beginning in the fingers and toes
- long severe heat with excessive weakness
- stupifaction and unconsciousness

Characteristic signs

- drinks often and much at a time
- bursting headache
- water tastes putrid

*** pulsatilla**
Intermittents of irregular type with irregular stages which are apt to overlap each other; Accompanied by digestive irregularities and stomach disorders.
Symptoms

- chills begins in the hands and feet

- general heat sometimes with coldness of single parts

Characteristic signs

- no thirst

2.4. CLINICAL TRIALS

In order to assess the feasibility of the clinical trial to be undertaken, the research was started with an open study. This had a duration of four weeks. All patients received homeopathic treatment. Depending on the frequency of attacks previously of the treatment the patient returned for check-up in one or three weeks. The division for follow-up was as such that if the patient had an attack daily he returned after one week, while if the patient had weekly relapses he returned after three weeks.

After the positive results of the open study, it seemed relevant to try to confirm these preliminary results in a double blind study, using the double-dummy method.

Every patient was assigned at random to one of two groups. One group received placebo-homeopathic treatment and verum-chloroquine tablets. The other group received verum-homeopathic treatment and placebo-chloroquine tablets. The number of tablets were five for chloroquine (verum or placebo) and one homeopathic grain of C200-potency (verum or placebo). So every patient got the same number of drugs, and with the same visual aspect and taste. For chloroquine the standard dosis was used: 25mg/kg bodyweight. None of the patients was treated by both chloroquine and homeopathic remedie nor only by placebos. All patients entered the study after informed consent.

3. Results

In the open study 92 patients were treated: 17 patients did not return for the follow up. Moreover, 68 out of the 75 (90,7%) remaining patients had at their follow-up check improvement of complaints.

TABLE 1. Frequency of utilized homeopathic remedies in > 5% of cases open/double blind study

Arsenicum	10.9 / 9.5
China	7.6 / 5.4
Eupatorium P	5.4 /
Natrum Mur	10.9 / 6.8
Nux Vomica	5.4 / 14.9
Pulsatilla	10.9 / 17.6
Rhus tox	5.4 /
Sulphur	8.7 / 17.6
total %	61.4 / 71.8

The following homeopathic remedies have been mainly used: Arsenicum album, China, Eupatoriatum perfoliatum, Natrum muriaticum, Nux vomica, Pulsatilla, Rhus toxicodendron and Sulphur.

In the double blind study 74 patients were treated: 41 in group I (= homeopathy verum and chloroquine placebo) and 33 in group II (= homeopathy placebo and chloroquine verum). 11 patients of group I and 8 patients of group II didn't return on their follow-up visit. Twenty five out of 30 (83.3%) patients in group I showed improvement of at least three symptoms of the listed symptoms. The 95% reliability-interval is 65.3-94.4%. In group II were 18 patients out of 25 (72%), who showed an improvement. The 95% reliability-interval is 50.6%-87.9%. The difference between the homeopathic treatment and the chloroquine treatment is 11% in favor of the homeopathic-treatment. The X^2-test is 1.03 with a p-value of 0.31.(non-significant)

In case of all the dropped out would have had no improvement the percentages would be 61.0% (25/41) and 54.6% (18/33) instead of 83.3% and 72%. The X^2-test is 0.31 with a p-value of 0.57. In case all the dropped out would have had improvement the percentages would be 87.8% (36/41) and 78.8% (26/33). The X^2-test is 1.09 with a p-value of 0.29. In both cases there would have been no statistically significant difference between group I and II.

TABLE 2. Results of the Double-Blind Study

	group I (%)	group II (%)	total (%)42 (56.8)
average age*	36.0	37.1	36.6
dropouts	11 (26.8)	8 (24.2)	19 (25.7)
improvement	25	18	43
improvement in % (95% CI)**	83.3% (65.3-4.4%)	72% (50.6-87.9%)	78.2%
no improvement	5	7	12
no improvement in % (95% CI)**	16.7% (5.7-34.7%)	28% (12.1-49.4%)	21.8%

* the age is an estimate made by the nurse, because the exact age is not known in general
** X^2-test is 1.03 and p = 0.31

4. Discussion

Kent (1974), Fenner (1925) and Farrington (1989) listed 45 homeopathic remedies; most frequently were used: *Arsenicum album*, *Bryonia*, *Ipecacuanha*, *Eucalyptus globulus*, *Natrum muriaticum* and *Veratrum album*. In Puhlmanns Handbuch (1920) the following remedies were added: *Nux vomica*, *Pulsatilla*, *Carbo vegetabilis*, *Ignatia*, *Digitalis* and *Rhus toxicodendron*. Recent databases as Reference Works (Macintosh) list

113 remedies; most of these are rarely indicated, at least they are not of much use in a situation where many patients are to be treated daily, like in the rainy season.

In this research the following remedies have been used: *Arsenicum album*, *China*, *Eupatoriatum perfoliatum*, *Natrum muriaticum*, *Pulsatilla*, *Rhus toxicodendron*, *Nux vomica* and *Sulphur* (each of them has been used in > 5% of the cases - see table 1).

The difference with the literature can be possibly explained by historical differences. Considering the experience that in two studies, which have been undertaken in the same area and by the same person, it would be more likely that this is an indication of what is a main feature of homeopathy: the adaptation of the therapy to different individuals as well to groups of patients. Moreover, the rainy season can account for the differences between the open and the double blind study. Two remedies used only in the open study during the wet season, *Eupatorium perfoliatum* and *Rhus toxicodendron*, are known for acting especially in humid climate circumstances.

The percentage of 90.7% in the open study and 83.3% in the double blind study can be attributed to the homeopathic treatment, but it is also possible considering the remittent character of the disease that these results are not caused by the treatment, but are partly coinciding with the natural course of the disease; however, an attack usually lasts for several days, while most of the patients recovered within the usual period. This would need to be assessed with a daily report scale; the compliance of the patients needed for that, would require more staff than was available given the small budget. To draw definite conclusions about the efficacy of the treatment, a study with a longer follow-up period would be necessary to assess the number of disease-episodes comparing the chloroquine group with the homeopathic group.

The criterion of at least three symptoms for a clinical improvement can be used in an area where a positive blood smear has no correlation with the clinical picture, and if only the treatment of the acute attack is the object of the study.

5. Conclusion

The question: "Would homeopathic treatment work better than chloroquine?" cannot be answered from this research. The only conclusion that can be drawn is that homeopathy has an effect, comparable with and slightly (non-significantly) better than chloroquine. The effect of chloroquine might be difficult to calibrate as the level of resistance against chloroquine is not known in the population studied.

The second question "Which homeopathic remedies can be mainly used?" can be answered as follows: *Arsenicum album*, *China*, *Natrum muriaticum*, *Nux vomica*, *Pulsatilla* and *Sulphur*. Interesting is the difference with the remedies used in the literature. This confirms the general rule that homeopathy is a matter of individualizing the treatment.

Given these results, and the potential interest for health care, it would be advisable to repeat this research with larger samples, combined with making thick blood smears before and after treatment, and thus be able to evaluate also the second and third objective of malaria-treatment, i.e to achieve a complete cure, and to improve the resistance against next attacks, respectively.

The following parameters can be taken: clinical symptoms, Hb, temperature, parasitology and a follow up after six months, assessing disease episodes and use of medication in these episodes.

Acknowledgments.

We thank dr D Abdulai and his staff (Shekinah Clinic, Tamale), dr Deville (VU statistics department), Hannah Kostelijk (VU medical biology), prof I Wolffers (VU Social medicine), and VSM Geneesmiddelen BV (Alkmaar), for their cooperation in this project.

6. References

Allen, H.C. (1983) *The therapeutics of fevers*, World Homeopathic Links Publishers, New Delhi.
Björkman, A. *et al*. (1991) Susceptibility of P. falciparum to quinine in Liberia, WHO bulletin, **69**.
Boericke, W. (1976) *Materia Medica*, Harjeet and Co Publishers, New-Delhi.
Farrington, E.A. (1989) *Comparative Materia Medica*, B.Jain Publisher, New Delhi.
Fenner, G. (1925) Tropenkrankheiten, Schwabe Publisher, Leipzig.
Gaucher, C. *et al*. (1993) Cholera and homeopathic medicine, *Brit. Hom. J.* **82**, 155-163.
Guidelines for the diagnosis and treatment of malaria in Africa, Afro technical papers 22/1/1992.
Hahnemann, S. (1988), *Materia Medica Pura*, B. Jain Publishers, New Dehli.
Hahnemann, S. (1986), *Organon of Medicine*, Gollancz Publisher, London.
Jacobs, J. (1994), Treatment of acute childhood diarrhea with homeopathic medicine; a randomized clinical trial in Nicaragua, *Pediatrics*, **93**, 719-725.
Kent, J.T. (1974) *Repertory of the homoeopathic Materia Medica*, Harjeet and Co Publisher, New-Delhi.
Lucas, A.O. (1984) *Applied field research in malaria in Africa*, **62**-suppl., W.H.O.Bulletin.
Lehrbuch der Homoöpathischen Therapie (1906) Schwabe Publisher, Leipzig
Meuleman, E.A. (1989) Medische parasitologie en entomologie, V.U. Publisher, Amsterdam.
Notten, J. (1992) Malariaparasiet wordt snel resistent tegen 'wondermiddel' chloroquine", Internationale samenwerking 7/10/10-1992.
Peters, W. (1985) The treatment of severe falciparum malaria, *Brit Med J*, **291**, 1146-1147.
Peters, W.(1987) Chemotherapy of rodent malaria. XLV. Reversel of chloro-quine resistence in rodent and human plasmodium by anti-histaminic agents. Ann. Tropic. Med. Parasitol., **84**, 541-551.
Puhlmanns (1920) Handbuch der Homoöpathischen Praxis, T.Hangstebeck (Ed.) , Schwabe Publisher, Leipzig.
Valero, M. *et al* (1993) Vaccination with SPf66, a chemically synthetised vaccine, against *P. falciparum* malaria in Colombia, *Lancet*, **341**, 705-710.
World Malaria situation 1990, Division of control of tropical diseases, World Health statistics, 45/2,3/1992.
Zahar, A.R. (1984) Vector bionomics in the epidemiology and control of malaria, part I and II, W.H.O.

AIDS AND INTEGRAL MEDICINE *

A Controlled Time-serial Study

M. BRANDS
Lauriergracht 47, 1016 RG Amsterdam

1. Introduction

This research is a protocol for a pilot study in the effects of integral medicine on patients with symptomatic HIV-infection and Aids. Three treatment modalities are tested: homeopathy, detoxification according to naturopathy, and psychosynthesis.

This study has three aspects:

1. The therapies: Integral Medicine
Aids is a disease affecting different levels of the human organism, so a multifactorial therapy targeting these levels, is an option to be considered and evaluated seriously; this approach is operationalized as a form of "integral medicine" (for definition see "Therapies") (Aakster *et al.*, 1989).

2. The methodology: Time-Serial Design
A multifactorial therapy has to be evaluated with many parameters at different levels of the human organism; thus, a design has to contain the capacity of assessing many parameters simultaneously. Moreover, due to the heterogeneity of the complaints of the patients under study, group means are not easily assessed. The "controlled time-serial design" is appropriate for this situation, see under "Methods" (Neter *et al.*, 1982; Barlow and Hersen, 1984; Ottenbacher, 1986; Wagenaar, 1990).

3. The patients with symptomatic HIV-infection and AIDS
Many patients with symptomatic HIV-infection and Aids have chosen one of the therapies studied here; they can be categorized under "integral medicine". Many of these patients improve in health condition, unexpected according to official prognosis. Almost all these data stem from casuistic descriptions (see also "Health Policy Relevance").

2. Therapies

The therapies belong to the field of integral medicine. Integral medicine can be defined

---

* Communication presented at the 7th GIRI Meeting, November 1993, Montpellier, France.

M. Bastide (ed.), Signals and Images, 253–261.

as: "integration of different therapies, with as common criteria: i) efficacy, to be tested with validated scientific methods; ii) economy (low-cost): availability for all; iii) autonomy-inducing for patient and therapist; iv) psycho-physiological approach". The theoretical basis of integral medicine lies in systems theory, which sees the human organism as a set of *interdependent* subsystems. Changes in one subsystem induce subsequent changes in another one. Therefore, therapies influencing the immune system might do so by addressing different subsystems, i.e. following different pathways through the human organism; they might even have a synergistic effect.

Three therapies are selected, as these are frequently used by Aids patients (Van Dam *et al.*, 1992; De Boer, 1992). Until now no synergistic effect has been documented in a clinical study. The objective of the three methods is to stimulate the immune activity along different pathways.

1. Homeopathy claims to do so by directly affecting the regenerative capacities of the organism by giving potentized i.e. dynamized remedies to the patient (Weissman *et al.*, 1992). Recently, biological research (Van Wijk *et al.*, 1994) has shown that high dilutions of a toxin stimulate "protector protein" or heat-shock-protein (HSP)-synthesis in response after the same substance in a toxic dose has been administered to cell cultures.

2. Psychosynthesis mobilizes, when using imagination techniques, the phantasies and imaginations about illness and then releases old psychological painful experiences, thereby lowering the general stress level of the organism. According to the theory of psycho-immunology this enhances the immune capacity.

3. Drainage/detoxification, based on naturopathy, eliminates by specific herbal and dietary regimens, old remains of disease metabolites and of drugs from liver, kidneys and intestines. Furthermore, shortages of vitamins, minerals and trace elements are supplied, and the nutritional status is improved (Kunzendorf and Sheikh, 1990).

3. Methodology

In Aids the immune system is the center of gravity of pathological processes. The immune-system is susceptible to the influence of many factors; so in a clinical effect study in Aids treatment with the objective to reinforce this system, this multifactorial character has to be dealt with at a methodological level. The different medical interventions to be evaluated here are not easily substituted by a placebo. Moreover, samples of patients are usually heterogeneous of composition. These three problems: multifactorial processes, absence of available placebos and heterogeneity of samples can have consequences for the design.

Especially the heterogeneity of the patients has led us to the application of the "time-serial design". This is derived from the "cohort model", - in contrast to the "case-control model", the basis of the randomized placebo-controlled trial (Barlow and Hersen, 1984).

The time-serial model calibrates the patient as its own control during the course of time. For measuring subsequent changes in one person many dependent variables are necessary, and a close relationship between therapy and its effect, independent of the course of time, has to be shown. The "time-serial" model is formed by a sequence of baseline and treatment periods. In the baseline phase the variables of the patient are measured only, the treatment phase is used for giving the therapy and measuring its effect on the same variables of the patient. The trends in the changes of effects are evaluated with the proper statistical techniques (see below, Statistics).

The conditions for this approach are:

- many parameters (dependent variables) are to be acquired at the same time, obtained from questionnaires, medical history, clinical and biochemical examination, all (except the biochemical) scored at specifically prepared ordinal scales (mainly Visual Analogue Scales).
- the scales have to be scored a sufficient number of times to apply statistics. The minimum number of moments of measurement is three within one evaluation period (baseline or treatment phase). In this design the dependent variables measuring the three therapies are clustered in three groups.

4. Approach and Planning of the Study

Research questions:

1. is this approach feasible for measuring clinical changes following Aids-therapies?
2. is a measurable effect of the therapies present?

4.1. HYPOTHESIS

There is a measurable effect in the health status:

1. psychosynthetic imaginative therapy decreases psychological distress, fear, anxiety and depression
2. homeopathy enhances the general energy level, and mental and physical health
3. detoxification diminishes symptoms and signs of intoxication and dysbiosis, decreases the shortage of vitamins, trace elements and minerals, and improves the nutritional status.

4.2. INTERVENTIONS

The patients are treated by homeopathic and naturopathic practitioners, as well by psychosynthesis therapists; all have experience with the treatment of HIV-related disease and Aids.

During both baseline and treatment periods, questionnaires are filled in by the patients weekly, measuring their health with the dependent variables which are relevant for the perspective of psychotherapy, homeopathy and naturopathy. Biochemical parameters are scored simultaneously (like lymphocytes T4 and T8-counts), not weekly, but according to the existing follow-up protocols of hospital outpatient clinics.

Number of patients for pilot study: 40 persons;

Ethical approval will be obtained from the ethical commission of the Faculty of Medicine of either the Free University of Amsterdam or the University of

Amsterdam.

4.3. INCLUSION CRITERIA

Adult (20-55yrs) HIV-positive patients with symptoms or with AIDS, irrespective their initial CD-4 count, as the changes in CD-4 counts are to be assessed individually. A specific minimal CD-4 level is not required, as very low levels would exclude patients who might benefit of the therapy in general well-being without an increase of CD-4 counts, as is the case in many patients under treatment regimens. The correlation between CD-4 counts and other measurements will be assessed in this study.

Men and women are both included, as possible inter-individual differences are to be calibrated per person. They enter after informed consent. No other chronic therapies are to be used during the study for an unbiased interpretation of the results: other therapies will confound the results of the therapies investigated in this study, as all of them might act upon the immune-status of the patients. Only for acute dangerous infections (e.g. PCP, toxoplasmosis) standard treatment will be available; their use will be registered as an extra dependent variable in the evaluation of the effect of the treatments under study.

Recruitment: at the beginning of the study the patients are ambulantly treated, and referred by general practitioners, internists or coming on their own initiative. Hospital admissions are in principle not a reason to stop the participation, if the treatment can be continued.

4.4. EXCLUSION CRITERIA

HIV-positive patients with neither symptoms nor Aids. Admission before or during the study in a hospital is not a reason per se to exclude the patient from the study, but the number of therapeutic interventions other than the ones tested specifically in the study will be scored descriptively. Patients who cannot continue the treatment for any reason will be excluded from the study.

4.5. OPERATIONAL DEFINITION OF IMPROVEMENT OF HEALTH STATUS

- A positive change in the trends of the sum of scores, at the Visual Analogue Scales of the three clusters of questionnaires as mentioned above, after the statistical corrections have been performed (see "Statistics").
- A positive change in the trends of the sum of scores, assessed during the course of the treatment phase, can be obtained by:

  an increase in the sum of scores at the "positive" questions i.e questions asking about a favorable feature of the patient

  a decrease in the sum of scores for "negative "questions i.e questions asking about an unfavorable feature of the patient (e.g. complaints).
- The scores are measured in millimeters at the Visual Analogue Scales.
- The correlations between the different scales will be calculated.

4.6. MEASUREMENTS OF EFFECT

4.6.1 *Psychology Questionnaire Cluster, as Evaluation of the Psycho-synthesis Therapy.*

- Weekly scored: Profile Of Mood Scale: dimensions: depression, fear, anger, weakness, force - Rotterdams Complaints List.
- Scored every 4 weeks: General Health Questionnaire .
- At the start and the end of the pilot, the Stressful Life Events test is scored.

4.6.2. *Homeopathy Questionnaire*

This is based at "Kent's medical history", the basic of homeopathic history taking since 1900; the "Kent" contains most questions used to determine the correct remedy, and to evaluate clinical changes during treatment, the items appropriate for quantitative measurement on ordinal scales, are selected. General, mental and physical signs and symptoms are included and all are formulated as such that they can be scored at ordinal Visual Analogue Scales. (At the moment of consultation, the medical record of the patient is scored systematically by the homeopathic researcher, in order to determine the correct remedy. This record however is not appropriate for non-parametric analysis, as it is a nominal scale).

4.6.3. *Detoxification Questionnaire*

- History scale; a nominal scale, (open questions), regarding the toxic load of the excretion organs: liver, kidneys, and intestines.
- Food and intoxication intake; a nomial scale; open questions, to be scored weekly
- Signs of failure of excretory functions; an ordinal scale, to be scored weekly, Visual Analogue Scales. Only this scale will be analysed statistically; the other two are used for eventual correlation assessment.

4.6.4. *Biochemical and Immunological Variables*

They include CD4 and CD8 counts, kreatinine, ureum, liver functions (Gamma GT, alkaline phosphatase, SGOT/SGPT), ferrum, transferrine. They are used to test for immune system-, kidney- and intestine function.

4.7. DESIGN

This project is planned twofold, as a *pilot study* and a *main study.*
The current study is the pilot only, but to show the total logic, the integral design is explained, the main study having to be implemented later.

4.7.1. *The pilot study*

In the pilot the design consists of a baseline period (A) of random length varying between three and five weeks, followed by a period of six weeks, wherein the three therapies are given as single (B, C or D) or as a combination (B+C+D), and the effects are assessed.

Four groups of each 7 to 10 patients are formed, one group receives homeopathy as therapy (B), the second drainage (C), the third psychoimmunity training (D), and the fourth receives all three treatments (B+C+D).

The designs are: group 1: A - B; group 2: A - C; group 3: A - D; group 4: A - B+C+D.

4.7.2. *The main study*

In the main study, an alternating baseline-treatment phase design is used (Barlow and Hersen, 1984), with all treatments given to all patients, so seven periods, four baseline and three treatment periods in between them, are included. The order of the treatments is randomized for each patient. An example of a design for one patient is: A - B - A - D- A - C - A.

Each phase lasts six weeks, so the total duration of the main study will be 42 weeks. Therefore, standard antiviral treatment if taken by a patient upon enrollment in the study, can be continued. The results of the groups of users and non-users of antiviral treatment will be compared after the study.

4.8. STATISTICS

Data analysis in time-series experiments requires several steps (e.g., Wagenaar, 1990). In time-series the effect of different treatment forms can be assessed. Before differential effects between these treatment forms can be analysed statistically, one has to be certain that no general trends are inherent in the data. Whenever there is such a trend within the data, later points can be predicted on the basis of the former, i.e. the data are "serially dependent". Since traditional statistical tests rely on the assumption that (the errors scores of) the data points are independent, trends must be either removed before proceeding with further analysis (e.g. by using first order differences), or modelled and controlled for (e.g. by time series analysis).

In this study, lag sequential analysis is used in order to detect serial dependency within the data points of the individual time series (Ottenbacher, 1986; Wagenaar, 1990). If a secular trend is manifest within the data derived from a specific patient, first order differences or moving average techniques are used in order to remove serial dependency. If this procedure is not succesful when applied for the first time, it will be repeated until significant trends will be removed. Subsequently, differences in efficacy will be calculated by using non-parametric tests.

The statistical test to be used is the Friedman two-way-analysis of variance, as this test is applicated for repeated measurements (Neter *et al.*, 1982). And in this design, three measurements per patient per intervention-phase is scored. (weekly testing, see under Design)

4.9. FEASABILITY

As one of the prime goals of this pilot is to establish the feasability of such a protocol, the following is said on this subject.

- The correlation between the different scales of measurements has to be established;

- The compliance of the patients can be a limiting factor; this will be assessed by the description of the number of cancellations, the number of telephones for changing dates etc. The motivation of the therapists, an important item in a multicenter, out-patient setting, will be scored as well. A separate scoring list will be used for this item.
- If there will be not enough participants to perform appropriate statistics on the data, the study must be stopped; the exact number for this decision will be determined then by the methodological specialists in the Board of Advisers.

The results are to be published in a scientific journal of usual standards, be they favorable or unfavorable in outcome.

5. Health Policy Relevance

5.1. PRIORITY OF CLINICAL STUDIES

Many symptomatic HIV-infected and Aids patients use therapies, not taught at university medical schools in the Netherlands. The scientific basis of these therapies is hard to assess within the medical concept taught at universities, based on Newtonian physics and molecular chemistry. To solve the theoretical controversies in medical science will demand much fundamental and scientific-philosophical research. Clinical effect studies seem for the time being more relevant, be it alone to have a more solid base for information to our patients when they choose their therapies.
So assessing the results of these therapies has a social and clinical relevance.

5.2.CREDIBILITY AND INFORMATION

A growing interest of patients means higher demands on the credibility of the treatments used. Moreover, in choosing their therapies both they and their doctors must be well-informed about results and possible side-effects. Rather than ignoring these strategies, they are better well-evaluated according to scientific standards. Moreover, health policies are served with a well-documented spectrum of therapies, rather then make-believe stories about wonder-cures. On the other hand, university trained physicians might discover some benefits of treatments they are not acquainted with by professional post-graduate training.

5.3. PREVENTIVE ASPECTS

In the case of homeopathy, naturopathy and psychosynthesis, with as yet no reported side-effects on the immunestatus of patients, a general increase of health will prevent opportunistic infections, therefore preventing personal suffering and hospital admissions. The results of the patients treated in London with homeopathy (M Strange) are, even only casuistically reported, promising. Many seropositives and Aids patients in this sample stay healthy for several years time without the use of any other therapies then naturopathy, homeopathy and psychotherapy.
A beneficial effect on health care expenditures is plausible, but should be evaluated

carefully. Also substitutional effects are to be considered in this respect.

6. Organization

The patients will be checked up for medical tests in a Hospital specialized in the treatment of Aids.
Network of practitioners:
- in Amsterdam the group of participating homeopaths, cooperates with some homeopaths in London, UK, with eight years of experience with the treatment of Aids patients.
- cooperated is with Belgian colleagues Marichal and Jenaer; they carried out a study with Aids patients in Zaire; homeopathy had a positive effect on weight loss, fever episodes, infection, diarrhea, mortality in a two years follow-up; 28 treated patients were compared with 11 controls
- naturopathy in the form of drainage/detoxification is used by different therapists in Holland, UK, and the US. Many Aidspatients use these therapies, often in combination with homeopathy. For naturopathy, a strategy for Aids patients has been worked out, and is used with homeopathy as a combination therapy for greater effectivity
- psychosynthesis and imaginative therapy is done by the Institute Psyche in Amsterdam, and used as a combination therapy with homeopathy for Aids patients; this Institute has an experience for years in treating cancer patients.

7. References.

Aakster, C., van Wijk, R. and van Dijk P.(1989) *Integrale geneeskunde*, Wolters-Nordhoff Publisher, Groningen.
Assagioli, R (1977) *Psychosynthesis*, Hobbs Publisher, New York.
Ballieux, R. E. (1991) Impact of mental stress on the immune response. *J.Clin. Periodontol.* **18**, 427-30.
Barlow, D. H. and Hersen M. (1984) *Single Case Experimental Design*. Pergamon Publisher, N.Y. Oxford.
Bastide M. *et al.* (1987) Immunomodulator activity of very low doses of thymulin in mice, *Int J Immunother* **3**, 191-200.
Bastide, M. *et al.*(1985) Activity and chronopharmacology of very low doses of physiological immune inducers, *Immunol Today* **6**, 234-5.
Berckenbosch, F. *et al.* (1991) Neuroendocrine and immunological mechanisms in stress-induced immunomodulation, *J. Steroid. Biochem. Molec. Biol*. **40**, 639-47.
Boer, J de (1992) Quality of Life study in Aids patients, *Dissertation*, VU Publisher, Amsterdam.
Cal, J.C. , Cambar, J. *et al.*(1988) Chronobiological approach of mercury-induced toxicity and the protective effect of high dilutions of mercury against mercury-induced nephrotoxicity, *Liver, Cells & Drugs*, **164**, 481-485.
Cambar, J. (1992) Effect of highly diluted arsenic on arsenic-induced nephrotoxicity in mice, 3rd Int Congress Homeopathes sans Frontières, Montpellier.
Chandra, R.K. (1983) Numerical and functional deficiency in T helper cells in protein-energy malnutrition, *Clin. Exp. Immunol.* **51**, 126-32.
Chandra, R.K. (1992) Effect of vitamin and trace-element supplementation on immune responses and infection in elderly subjects, *Lancet* **340**, 1124-27.
Dam, F. van, de Boer, J. *et al.* (1992) Use of remedies and alternative therapies by patients with Aids (CDC IV), WHO Collaborating Center on Quality of Life, & Dutch Aids Treatment Group, AMC, Amsterdam, International Aids conference VIII, PoB 3402
Dijk, P. van (1991) *Huisarts en integrale geneeskunde*, La Rivière en Voorhoeve Publisher, Hampsen.
Jacobs, J., *et al.* (1994) Treatment of Acute Childhood Diarrhea with Homeopathic medicine: a randomized clinical trial in Nicaragua, *Pediatrics*, **93**, 719-725.

Kerkhof, M.P.N. van de, Maassen, T. and van Rooijen, P. (1992) *Zorg voor Zorg Hulpverlening aan HIV-geinfecteerde homosexuele mannen*, Schorer Stichting Publisher, Amsterdam
Kleijnen, J., Knipschild, P.and ter Riet, G. (1991) Clinical trials of Homeopathy, *British Medical Journal* **302**, 316-23.
Kunzendorf, R.G. and Sheikh, A.A. (ed). (1990) *Psychophysiology of mental imagery: theory, research and application*, Baywood Publisher, New York.
Marichal, B.J., Jenaer, M.C. (1990) Immunotherapy at infinitesimal doses. Int. congress on Ultralow doses, Univ. of Bordeaux
Mulder, N. (1994) Psycho-immunology and HIV-infection, PhD Dissertation, Rotterdam University, The Netherlands.
Neter, J *et al.* (1982) *Applied Statistics*, Allyn and Bacon Publisher, Boston.
Ottenbacher, K. (1986) *Evaluating Clinical Change*, Williams and Wilkins Publisher, Baltimore.
Reilly D.T. *et al.*, (1988) Is Homeopathy a placebo response? Controlled trial of homeopathic potency with pollen in hay fever, *Lancet* , **8512**,881-885.
Smith, E.M. *et al.*, (1986) Corticotropin releasing factor induction of leukocyte derived immunoreactive ACTH and endorphins, *Nature*, **321**,881-6.
Strange, M. (1988) Aids and homeopathy, *Brit.Hom. J.* **8**, 301-305.
Taal, J. (1991) Van angst naar kracht: de waarde van innerlijke beelden bij kanker, *The Practitioner*, **18**, 1083-1088.
Temoshok, L. (1988) Psychoimmunology and Aids,in T.P. Bridge *et al.*, (eds), *Psychological, Neuropsychiatric and Substance Abuse Aspects of Aids*, Raven Press Publisher, New York.
Wagenaar, R. (1990) Functional Recovery after Stroke, *Dissertation*, VU Publisher, Amsterdam.
Weissman, Z. *et al.* (1992), Immunomodulation of the specific immune response to KLH by high dilutions of antigen, *6th Conf of Intern Research Group on Very Low Doses and High Dilution Effects (GIRI)*, Munich.
Wijk, R. van *et al.*(1994) *Cultured mammalian cells in homeopathy research*, University of Utrecht Publisher, Utrecht.

CHAPTER IV

EPISTEMOLOGIC APPROACH

A.LAGACHE
Notes on Conceptual Basis of Science
A.LAGACHE
What is Information ?

The three principles of homeopathy (the similia law, totality and the pharmacological effect of succussed high dilutions) are nonsense for mainstream science which is based upon the logic of objects according to the mechanistic paradigm. This paradigm has been so fruitful that we think it is the only law of rationality. However, other facts necessitate the creation of a rational new way of thinking whose roots can be found in previous centuries (Lagache). The functionning of the living body is one of these facts. The living organism cannot be considered as an object. It is continuously modified as a function of time. It is in a continual and irreversible learning process. As Lagache proposes, the living organism (body and mind) is a complex structure able to exchange information with the external world. Information also circulates in the internal world allowing exchanges at every level: the living organism is an informed-informing structure. Just as the mechanistic paradigm seizes the material interactions between objects and the symbolic paradigm assumes linguistic facts, we need a paradigm that allows the understanding of living structures: living beings communicate with their world in a non verbal way, whether on a somatic or psychological level. This paradigm takes place within the framework of the logic of analogy. Communication will be mediated by "semantic objects". Their application can be used to interprete the therapeutic effects of homeopathy as well as succussed high dilutions (infinitesimal). The dilutions will play the role of semantic objects bringing information to the body.

So, "what is information" (Lagache) ? The semantic object makes sense for the living organism: the circulation of this type of signifier has its own original laws. Each piece of information is not an object even if it has a carrier. For example, the information of the remedy is carried by the

M. Bastide (ed.), Signals and Images, 263–264.

potentized dilution (the transmission of the information being due to the electromagnetic carrier obtained by the succussion of the solvent of the remedy as shown by Endler). The sense is non-local; it concerns the whole organism designated as the receiver. This receiver then creates the meaning and is able to modify its behaviour. The simplest mode of representation for living structures is a passive mimesis. But when the living organism receives this information brought by the semantic object (the succussed high dilution) , it receives it not as a material object but as information about this object which calls for a processing and active regulation by the whole organism. This is an operation of active mimesis. In order to be treated, this semantic object must fulfill certain conditions. It has to be alleviated, diluted, so as not to worsen the symptoms through too strong a presence. It has to be related to the receiver's referent.

The allopathic therapy will consider the body as an object and the pathology has a causality. The therapeutics must cancel the main symptoms by an opposite effect and treat the cause. The homeopathic therapy will consider the body as an informed-informing structure, and the symptoms in an complex organisation: it will transmit an artificial and alleviated information on the whole symptoms given by the remedy. Then, the body is able to recognize and treat the "semantic object" which allows the negation of the symptoms recognized as an erroneous adaptation. The more similar the symptoms of the patient and the remedy, the more effective the remedy. The medical device has to re-inform the patient and makes his symptoms move on towards a higher level of integration. The action of the remedies consists of a dynamic analogy between pieces of information as proposed by Lagache .

One can rationally propose that homeopathy belongs to the world of communication. Starting from this point, scientific experimentations have to be based upon the possibility for the receiver to understand and treat information. In each model, the question of the receiver's referent has to be asked. When the potentized dilution is used, ones must define the framework of information. For example, high dilutions of a poison are able to induce a learning process in a living system but only to fight specifically against this poison (law of identity) as an "expected danger". The endogenous molecules such as bursin, thyroxin, etc.. are automatically recognized according to the genoma. The similia law gives the framework of information by analogical communication when homeopathic remedies are used in animals or in humans.

M.B.

NOTES ON THE CONCEPTUAL BASIS OF SCIENCE *

Looking for the idea of mediation, and its tradition.

A. LAGACHE
Lycée Carnot, 145 boulevard Malesherbes
75017 Paris - France

"The science of Nature is not the thinking impulse which gets straight away at the world of objects, but also the environment in the midst of which mind gains knowledge of himself."
Ernst Cassirer, 1951.

During the second part of the Twentieth Century, some scientists have lost the habit of thinking about the conceptual basis of science : experimental methods, as an evidence, seemed to fill in every need of understanding phenomena ; and if some points remained dark, more experiments by themselves would provide with progress. So much that the deep epistemological questioning of the first part of the century looked like intellectual games for many people.

But Science needs always a conceptual basis ; and there is no knowledge without any actual paradigm, any conceptual set which makes it able to work. We notice it when a dominant paradigm reaches its conceptual limits : as for example a biology lacking a theory of life. Arising paradoxes is the sign that we reach logical boundaries, with regard to a previous way of thinking. Like, for example, the previous situation of homeopathy, and the contrast between its positive curative effect and the lack of interpretation.

1. The Two Monisms

The current situation of science is more peculiar than it seems. It is, in fact, much stranger than we care to think. We inherited the methods of physics and chemistry, which are good in themselves as methods, and have taught us a lot about regulated use of reasoning and experimenting.

Nevertheless, in the same batch lied the overpowering arrogance of positivism : everything is matter, according to the very restricted definition of matter provided by the mechanistic paradigm. Consequently, it was thought that every problem was

---

* Communication presented at the 8th GIRI Meeting, December 1994, Jerusalem, Israel.

M. Bastide (ed.), Signals and Images, 265–278.

the mechanistic paradigm. Consequently, it was thought that every problem was solvable by reducing its terms into simple material elements. Matter rules over everything : that is the narrow materialism claimed by positivism.

On the other hand, many phenomena actually escape from the explicative capacities of a mechanistic paradigm. Human sciences in general are based on the concept of an isolated mind, according to the reigning myth of an independant awareness and will.

Mind and matter are two isolated kings, each within its own kingdom. They ignore each other so as to be certain of being the only true king, each controlling its own totalitarian state. This fact is also maintained and hidden by means of the academic division between disciplines, which protects from a fatal confronting! Therefore, the current conceptual basis of science is not a dualism, but a contradictory juxtaposition of *two monisms*, each one in its own camp.

For example, the way we deal with insanity can represent this situation of unconnected monisms. For about a hundred and fifty years, two notions of insanity have been fighting each other, each one with its own piece of truth, in endless and vain quarrelling. We can spend centuries wondering whether insanity is a somatic disease or a psychological problem : the reality of insanity is outside this split. Thus we can also spend centuries combining chemical and psychological treatments (which is of course better than nothing, and a duty of emergency), making very little progress, unless we find out a third way. The place from where someone is led to distort the relationship between mind and body, between the being and reality, is a mediation : something like a meaningful sensitivity or a corporal sense.

Anyway, the juxtaposition of these two monisms, having probably been fruitful in some period of science, leads now to a dead end and to infertile arguments between sets.

2. What Would a Real Dualism Consist of ?

It is hard to conceive... History, habits of thinking and language resist... Four centuries of science and some millenia of philosophy have forged our tools, our words and our concepts ; always seeming to prove that the prevailing restricted materialism is right, in the same way as any almighty-minded idealism.

Let us not forget that there are a lot of people looking for another way, looking for a conceptual basis that will describe human - or at least living - reality. This reality not being a tragic and irreducible opposition between physical and semantic facts, but actually the dynamic linkage between them.

Yet we can understand that a real dualism, a dynamic one, is different from a simple juxtaposition of two terms.

Two logical pitfalls must be avoided there :

- The whole working of a two-level system cannot be a juxtaposition or a sum. It must be combinatory. Something more than analytical.
- We often keep mistaking the two terms of a contradiction for a dualism. But the aristotelian contradiction asserts only one thing, and the other term is only the negation of the former one, that is an empty category.

For example, if we say that matter is everything which is not mind, and mind everything which is not matter, we don't assert their coexisting but each one separately as non-coexisting. Then, the irreducible difference between mind and matter behaviours remains this unaknowledged and unfortunate defect of human nature that we were told about...

A real dualism means that the different levels of reality are an integral part of nature, working together in the dynamic process of life. The one cannot be reduced to the other, without missing the very aim of understanding life. Thus, why do we need to evolve a real dualism ? We need something which links mind and body, or in general physical and semantic facts, whereas maintaining the distance and difference between them : that is a *mediation*..

The notion of mediation suffers from being neglected. It is always considered like a kind of second-ranked term, facing the magnificent firmness of substance. However the only way to understand a real living dualism is to consider that mediation creates links, movements and ways between the different things, wherefore providing each one with its own identity.

Where it is dualism, there are things which have to deal with each other, without loosing one's own nature. We can find a metaphor of this logical form in anthropology. Let us remind of a piece of the American Indian mythology, as related by Levi-Strauss, (1991) : how have the Earth and the Sky been created ? Because of the *mist*.. Fogs, hazes and mists are the very facts which lead to conceive Earth and Sky with the difference and the bond between them. Consequently, we must think that mist creates Earth and Sky, within their dynamic contact and distance.

Let us take another nearer example : music is a genuine and creative mediation between body and mind. It does not belong neither to the body alone, nor to the mind on its own. But it binds them with satisfaction, while it requires specific skillness from each one. Since it provides the best way to enlighten on the "good-distance-of-communication" between mind and body, you can, one step further, understand that music creates mind and body within a negotiable relationship.

This general scheme of thinking is to be tried. The mediation supplies not only bonds between different things, but also deals with distance and difference between them and maintains it. Then, the mediation is not a kind of average or an intermediate mixed up element, but it has the double function of linking and separating.

Anthropology can again give us an example : in the traditional culture of Australian people, a *walkabout song* defines the individual identity, and is the person's own property. The *walkabout song* describes also an itinerary : the piece of route that a man is able to travel up and to create around him ; his own symbolic existence is linked with this ability to understand the land. He can go all over the continent by exchanging information about a further route with other *walkabout people*. So this symbolic knowledge gives the world its existence and warrants its permanence. This *walkabout song* creates the landscape and sets up terrestrial elements on its sides.The consistency of the internal world and the link with environment share the same signifier.

Therefore, mediation must be taken as the creative element which makes possible that different levels of reality exist according to their own identity, while working together in nature. The mediation logically takes place before the objects which arise on its sides. It is a way to understand creative organisations and temporal regulations.

3. The Balance of Life

In this prospect, we can look for two types of mediation, according to the side we start from. Some psychological facts find their concrete expression through the body ; it belongs to psychoanalysis and freudian concepts to give an account of it. On the other hand, some concrete events can be directly meaningful for the body, as we observe it usually for example in aesthetics.

Then my assumption has been : the living being is an informed-informing structure, a network of relationship between the internal set and the surroundings. Consequently, some biological elements have to be managed not as material things, but as *semantic objects*. A semantic object fulfills the conditions of a mediation : it is a concrete element, far away from any linguistic or symbolic term; but it is treated by the body as a piece of information.

Then, logic of exchanging information can fit ; unlike exchanging objects, where one loses and gains, the exchange of information creates a new situation, different from a sum. It is a dynamic process, through which the partners change and build themselves by dealing with information : the very mediation creates what it mediatizes. Reasoning in terms of mechanistic elements is not enough to understand life, but if we assume that :

- first, living structures imply mechanical organization combined with semantic facts ;
- second, mediations are the dynamic elements which make living structures working as a whole, then we might go ahead.

Let us take an example about the evolution of species. There is a former animal organization with an external skeleton, like *sea urchins* ; this external skeleton is a mechanical filter and protection against environment, at least a part of it ; it implies a radial shape, refered to one axis in space. When we look at the evolution towards *vertebrates* , a different solution is set up : the internal skeleton has two axis, it introduces new motions in space and laterality ; at the same time the first form of an immune system appears ; it seems that the mechanical function of self-protecting had transfered its part to a semantic one, creating a new balance.

In the same way, the genetic code could be considered as a semantic system ; not as a language, because it is not symbolic, but concrete. Actually, reading a poem, nobody is surprised that the poem chooses some words and doesn't use the whole possible words of the dictionary ; but the efficiency of this poem depends up on the wider possibilities of the whole language. The "useless" part of grammar and dictionary makes possible the viability of this useful poem. Is it a coincidence that the "non-coding" part of the genetic code seems to be in the same relation regarding the coding part ? Some specific expression normally requires a wider register of expressive possibilities.

Furthermore, each word of the poem is related with every other, within a general and mutual system of meaningful compatibility. As showed Jakobson (1963), the effective combination is linked to an upper contextual unity. At the end, it defines a *style*. The same idea arises from the notion of "Bauplan" suggested by Ernst Mayer : for example, "flesh-eating animals never have horns". There is no utilitarian justification of that, except that it can be the expression of the consistency of carnivorous general genetic organization. Yet the dogma is : one gene, one protein. But nobody would deny that, although each word in the poem has its own meaning as isolated, the very understanding of the text overtakes this first level. So one gene codes for one protein...but it doesn't exclude that they might do something else together, when we don't stare at them. Likewise, the wedding feather's finery of the bird of paradise is useless in terms of adaptation, as long as we consider adaptation in terms of struggling for energy. But who knows with what it is compatible, at the level of the "Bauplan" ? The "unmotivated" or "gratuitous" character of life's creativity disappears when we consider a semantic order binding elements. Utilitarianism always prefers short cuts. But at each level, limited means of life realize a maximum of combination, so as to constitue a whole coherence of individuals, and it implies some digressions. Hence the genetic code seems to integrate complex mechanical systems into semantic management. There is not so much an evolution of species (although it has a chronology) than a variation on the same semantic theme of organization.

So the hypothesis of semantic functions can enlight some questions in this kind of topics.

4. The Paradigm of Signifiers

Homeopathy and immunology provided a beautiful experimental area for this hypothesis, and, with Madeleine Bastide, since 1988, we have settled the *Paradigm of signifiers* : if some elements of the immune system, or of the homeopathic remedy, are taken as pieces of information, we can describe the process of communication according to the logic of concrete signifiers. (Designation by analogy, paradoxical negation and dynamic analogy are the main operations ; theory of passive mimesis and active mimesis provides the explanation of the different possible ways of information in the receiver)(Lagache, 1988).

The paradigm of signifiers is an epistemological device, based on the logic of analogical communication. It supplies a possibility to understand in what ways the body deals with information, and, to a certain extend, to predict some features of experiment. The work of Madeleine Bastide has given firm experimental basis to this logic of communication, as applied to biological phenomena.(Bastide *et al.*, 1995).

Yet, when we evolved and proposed the paradigm of signifiers, it seemed to be an entirely new and unusual idea, a devilishly new idea from nowhere! And I am afraid it still looks like it! Nevertheless, seeing that experimental work, I must rely on my own idea.

And if it is a good idea, it is not so new as it seems ; because I am convinced that we never invent something out of nowhere. There is no thought without a tradition. Beyond the chasm between materialistic positivism and abstract idealism, there is a tradition in which we can recognize the research of humanism : a view in which living beings can appear diverse and complete, individual and connected with others and environment, temporal and changing to remain the same. According to Spinoza, it is the tradition in which

"We are part of the fullness of Nature, whose measure we share."(Spinoza, 1677).

But what interest in searching about this tradition ? The very progress of sciences has delivered many pieces of truth, which now lie like a disordered jigsaw. We still don't know what makes a being live, evolve, carry on with following its own finality. We need synthesis.

We need also to end in all kinds of reducing thoughts : it has shown its limits. At last we need to connect our scientific knowledge with our ethical concerns : this cultural gap between them cannot last for ever. If there is an order of life, linking matter with sense in a natural way, then ethic begins with the cell.

5. Revisiting History

The tradition of a meaningful sensitivity and a corporal sense has been hidden, or put on the fringe, because we often conceive history from a present point of view ; or furthermore from the angle of winners in history : those who are right for the moment.

Therefore, history can be told like a long road leading to the present successful materialism, throwing away every sideways idea in the dust-bin of man's awkwardness.

Nevertheless, we can find again the thin wire of a philosophy which takes into account the whole human being facts in their combination, and with their meditations. Through that, I would not pretend to rewrite history : it would be presomptuous to pretend to be exhaustive, and many ways are passable in the wide fields of past. I will just suggest another course, marking it out with some milestones. Then we should realize that the advance of science has no fatality.

6. Beginning with the Seventeenth Century: an Open Paradigm

Yet it is arbitrary to choose a beginning, because you still can find some forerunners...But the Seventeenth Century has given to the word "science" the meaning it has, more or less, for us : Galileo Galilei formulated the main points of experimental method, and Descartes the rules of a rational use of mind ; then science as a method took about the consistency it has now.

But the aim and interests of scientists were not the same as ours.

It seemed obvious that this fantastic new intellectual set might be used to understand mankind concerns ; most certainly dealing with a right concept of nature and natural determinism, but including the very facts of sensitivity, life and thought, as natural too.

7. Descartes and the Link between Body and Mind

Descartes was, on some way, not so much cartesian than we are used to being! At least, we are mistaking when we consider him as the mere example of a mechanistic, and we forget, there, both his metaphysics and his focusing preoccupation : that is the real *dualism* of the two substances, and consequently the problem of *linking body and soul.* His conception of a matter - extent substance - which was understandable by means of mechanics, never reduced nor overshadowed the position of the thinking substance ; they were both actual, both parts of nature, both to be scrutinized. Both are ruled by regular and intelligible causes, although still unknown.

That is the reason for Descartes's focusing on the question of relationship between body and soul : he understood clearly that, if we want to have a real enlightenment about nature, we might answer this question. And he dedicated many time and many texts to deal with this question. Let us take some sample of it:*"Since it is made of its own stuff, the soul is not accomodated in the body like a pilot in his ship. On the contrary, their union builds up a "being in itself", a third substance."*(Descartes, 1637).

The body itself can be taken as an automaton, in a methodical way, to study it in its parts. But it is a whole :*"It is one and, in some way, indivisible, because of the layout of its organs ; they so much relate each to the other that, when each one is taken away, the whole body becomes faulty"*(Descartes, 1649). Indeed the sensible soul cannot be deduced from matter. But it is "joined and united" with body, *"in order to have feelings and thirsts like ours, and consequently to make up a true human"*(Descartes, 1637).

Therefore no knowledge of nature could be possible without practising a real dualism. Body or soul are *"unfulfilled and incomplete substances"*(Descartes, 1642), if separately applied to human being. From then on, imagination, sensation and *"dark thought"* of the body are providing with mediations. However these mediations never mix up the terms of dualism, but combine them and make them work together.

For example, about the sensation of burning pain when touching an ember : *"There is no reason to convince myself that there is, into the fire, something looking like this heat, or like this pain ; but I am right in thinking that there is something in the fire - whatever it is - which causes for me these feelings of heat and pain"*(Descartes, 1642). This argument defeats any monism : levels of body and soul are different, but not strangers : reality is their very acting each on other.

Then we can have a glance at this open paradigm of the Seventeenth Century.*"By saying "nature", in a general meaning, I don't imply anything else than God himself, or rather the measure and arrangement that God established in created things"*(Descartes, 1642).

Yet it is expressed with the religious vocabulary, which was in this time the only available one. But we can understand that for Descartes, the open paradigm of science might involve a set of coherent and compatible views of nature. The strength of the cartesian philosophy consists of this operating dualism, assuming distributions, joints and mediations. Once everything was still possible, and especially a concrete but human knowledge.

8. Spinoza: Body and Soul Share the Order of Reason

It would be very long to recall the whole theory of relationship between mind and matter, in the works of Spinoza : it is a cornerstone and involves his whole philosophy. Let us just take a small testimony of it.

" As far as we are not racked by ailments which were contrary to our nature, so far we have the power of arranging and stringing the ailments of body according to the measure of reason.

Demonstration :

Bad ailments, that are contrary to our nature, are injurious only when they prevent the soul from understanding. "(Spinoza, 1677).

This shared measure between body and mind actually looks like an anticipation of psychoanalysis or homeopathy : body and mind share their darkness and their light. Disease is this breaking in the agreement between reason's order in body and reason's order in soul. Health is their mutual lighting. Thus, it would be illogical that body remained an unknown and darkened piece of matter :

" There is a way for the soul, and it belongs to its nature to be able to conceive the essence of the body, from an eternal point of view."(Spinoza, 1677). It belongs to human cleverness to discover the unity and diversity of nature, and it is essentially able to do it.

9. The Aesthetic Knowledge: the Baroque

During the same period but under the sweet light of Italy, the same knowledge about unity, that is mediation within diversity, had been represented by baroque architecture and sculpture. It conveys the same idea of reality : when stones were quivering with life. There the representation of nature and men is concrete, but its materialism is not based on what we intend by matter ; matter is going to move, to include time and to be meaningful for our perceptions. Let us ask Wölfflin to illustrate it : *"The widened form, as treated by the Baroque, is related to a new conception of* ***matter****. I mean : the idea-like-matter whose life and internal behaviour let the constructive elements express themselves.*"(Wölfflin, 1888).

Each period has the architecture it deserves! We can notice that the message of Baroque remains ambiguous in our culture ; it has given rise to many and many contradictory interpretations and comments until now; it shows that we still don't have the real ability to think in terms of connected multiplicity.

10. An Open Paradigm of Science

That is what had been the concern of the very rationalist Seventeenth Century. They certainly did not rely very much on body's influences, because it remained the access to passion and sin ; but although mistrusting it, they subtly describe the mediation between physical and mental events : it is the very power of *imagination* to surprise the mind with concrete images. Yet imagination was considered as the *"master of error and falseness"*(Pascal, 1670) ; but by the same way, it began to win interest.

The power of images, in fact, is the shady power of idols, which are meaningful for mind as well as for body, through the same signifiers.

Even in the most clear part of science, that is in the works of Newton, we have overshadowed some details. Newton himself noticed, about the gravitation, that attraction could not be "*a real physical force*" : the mechanical postulate of matter implies that two material elements only interact by external and negative ways ; that is contradictory with any kind of action from a distance. It is *"incompatible with the absolute reality of space and the property of place."*(Koyré, 1957). Newton noticed the *"foreign element"*, extraneous to the mechanistic paradigm and called the universal gravitation *"a spiritual force"*, because at that time it was the only available word for it. It was at least an honest admission of ignorance, and the clue to a logical limit, that we forgot to realize.

This problem of interpretation will arise again in the Twentieth Century, when confronting quantum mechanics with general relativity : which one is to be taken first, thing or interaction ? And then, what is the nature of interaction ?

All these elements, chosen here and there in the Seventeenth Century, are the signs of an open paradigm : from different places and various points of view, they evolved science as a complete purpose of enlightening nature by means of reason ; and each delicate point of junction between different levels of reality had been holding attention, as requiring the best of awareness, and the most steady reasoning. But there is a point to emphasize : the aim of unity doesn't reduce diversity (to understand that, we must stop to conceive unity as based on identity) ; unity must arise from the right mediations between real and sound differences.

Then, considering their teachings, we can ask the question in their own words : why would "gods or nature" have settled up their laws and organisation over a chasm ?

11. The Eighteen Century: the Touchstone of Science

So the challenge was a wide-ranging rationality, encompassing life and matter. Of course the Age of Enlightenment took it up, and probably deserves its name because of this prospect. The new methods of observation focused the reflection on living beings;

it conveyed new interests, creating "natural history" and what will soon be called "psychology".

The first clear idea of a determinism ruling man's behaviour as well as natural phenomena was formulated ; but if it was an *order of life* to find, obviously it had something to do with ethical order. And the scientific knowledge would have looked absurd if not involving an ethical progress for mankind. Otherwise, the "historical method", applied in natural sciences, plainly implied evolving and changing objects all along time.

It can be shown, for example, by the interest holded on the question of "sensitivity" (that is what the previous Century called "soul"): this ability of the body to *feel,* which is necessarily linked with the ability of the mind to *know*. Philosophers and scientists fervently looked for the doorway between the two levels.

12. Kant: the "Transcendental Scheme of Imagination"provides a Mediation

Kant is known, of course, for the part he gives to the *"pure reason"* : intellect and sensitivity are two separated and different levels. But a rather neglected part of his philosophy is devoted to explaining what imagination and sensitivity bring to mental abilities. Imagination becomes active. The image gives a synthesis and creates unity with the diverse determinations of sensitivity.

Then, how to jump over the gap, towards a pure and intellectual meaning of ideas ? The "transcendental scheme of imagination"(Kant, 1781) is especially forged in order to explain in what way mind uses images : it is a go-between, a kind of vessel, not sensible nor sensitive, which allows mind to give meaning to images. From image to scheme, we go from a passive alteration of sensitivity (which can be explained in terms of material effects) to an active understanding which builds sense.

Anyway, it remains a gap ; a very thin and accute dividing line, a jump over a nothingness that Kant doesn't fill in, and which afterwards will become an ocean. But Kant asked the question, and tried to establish the continuity that philosophy of Nature, before him, had made essential.

13. The Problem of Molyneux

However, this theory of Kant is an answer to the very famous, old and forgotten "problem of Molyneux". "In spite of its abundance and diversity, the detailed research forever leads to a theoretical and basic problem, where all its lines converge. It concerns this question, first asked in the *Optics* by Molyneux, which straight away aroused the greatest philosophical interest"(Cassirer, 1951).

The question concerns perception : if we want to define an empirical perception from a scientific point of view, we must say that a perception is the material modification of a somatic device.

Then, where are the springs of sense ?

How can we enlight the conjunction between somatic facts and mind's activity ?

*"Are the experiences, that we had in one of our sensorial areas, able to set up **another area**, qualitatively different and with another specific structure ?"*(Cassirer, 1951).

The question is not as ambiguous as it seems : we know that visible and tactile perceptions are quite different stuffs. So the problem is: the combination, the junction of two perceptions makes something else than a perception plus a perception. Something new is created, in the convergence of perception, which is the irreversible spring of sense: it comes into a meaning.

In fact, the very passage between passive modification of somatic devices and recognizing meaning is a typical mediation : if one isolated perception can be reduced to material events, the meeting, the confrontation between two perceptions produces a new source, a qualitatively new event : the spring of significance. Perceptions are in themselves different : they don't mix up, they maintain the distance. But they deal with difference : the organisation of brain arouses a new kind of facts from their confrontation, that is the primitive form of thought.

By analogy, we can recognize in the problem of Molyneux a typical form of asking that we meet nowadays in science. For example : how does the organism discriminate between self and self-difference, by means of the same presentation of the Major Histocompatibility Complex?

This inspiration of the Eighteenth Century forever goes round this intuition : life is something else than an addition of material elements. Its most simple manifestations show constantly this special ability : *"the connection between two living molecules is quite different from the contiguity of two inert masses"*(Diderot, 1830), and this new level of organisation is the source of semantic facts.

So work the mediations : the upper level of organisation is always provided by the confrontation of elements in a previous level ; regulation or criticism work in this way : dealing with differences, they mediatize them.

14. Lessing and the Aesthetic Signifiers

Significance can be rooted deeper : in the perception itself, which is a genuine experience of sensitivity and contributes to form the meaning of the event. In his beautiful *Laocoon* (1766), about aesthetics, Lessing answered in an original way to the problem of Molyneux. At this time, the influence of poetry implies that painting must be "a lyrical poetry", and not an art on itself.

No, says Lessing : the own means of each art, the concrete vehicle is a part of significance. We see paintings, and react to colours and forms; we read poetry, or listen to it. Each artistic effect is linked with its means. (As a common fact, we can notice that if a novel is screened, it must be created again, afresh, according to the proper demandings of cinema, and if not, it is a failure!)

It means that the corporal experience of art is constitutive of the meaning and cannot be removed from the sense. Besides, up to my opinion, the difference between an "aesthetic" experience and a "normal" one is totally artificial : which experience is totally free from reacting to beauty or ugliness ? All these works show that, to understand life, we must combinate material organization with semantic order. The very living reaction of a living being is always something more than mechanical.

What Kant, Molyneux and Lessing demonstrated about internal world and perception is also true and relevant about dealing with environment. They explained to us that the internal power of organization links the material function (to be modified by perception) with the semantic one (to give meaning to this modification).

In the same way, Diderot gave a complete definition of evolution of species, according to the double motion of constraint and creativity :
*" A primitive conformation is altered or improved both by **necessity** and by **usual functionning**."*(Diderot, 1830).

So it is possible at this time to evolve together positive and negative interactions of living beings with their environment : negative interactions as darwinian ones, the constraint of necessity which leads to natural selection ; and on the other hand, ability of living structures to learn from experience, the positive interactions or lamarckian ones.

The quarrelling between Darwin and Lamarck theories will split this balance between the two notions. The break, the exclusion arose because we excluded information from real phenomena, by choosing a reduced matter as the only criterion of reality. For a mind of the Eighteenth Century, the balance of life is obviously an original synthesis between constraint and creativity.

It seems that they had everything in order to understand life and settle up another kind of technique. Why does the goal move away when we approach ?

15. The Fractured Culture

Of course it is impossible to accuretely date when, where and why it began to seem "normal" to refer exclusively reality to matter. Anyhow, it went so far as to denying any semantic fact, or any psychological reality. These soft-focus things have been left to poets and dreamers... We were educated within a fractured culture, splited up according to the two mentioned monisms. At the beginning of the Twentieth Century, Bergson (1907) stated this dismaying fact : " *Human cleverness feels at home as long as we keep it among inert objects, especially among solids... Experiment at the end can show us how life sets about obtaining some result ; then, more often, we think that its way of doing is precisely the one we wouldn't ever have thought."*

What role does the fracture play in the progress of science ? It would be a topic in itself, and I will just suggest some points. Is materialism required by experimental method ? Not plainly. The method requires a relevant scale of repetability, and possible measurement ; but it doesn't foresee the nature of phenomenon. The requirement of a proof cannot be confused with requiring a material object. The rigour of method cannot be mixed up with a thinning of reality. No method can require to think only in terms of matter, excluding functions, properties of organization, creative contributions of structures and interactions.

More than a methodological use (which would anyway be limited to the strict moment of method), mechanistic materialism runs over and appears as an ideology ; it finds its reasons elsewhere.

Was it a necessary stage of science, in order to clean mind of any old anthropomorphic or religious arguments ? If it is, may be the remedy has been worse than the disease. Clarifying methods was useful. But on purifying from anthropomorphism, we lost humanity and life. On separating from religion, we let spiritualism take away on its side any kind of semantic or symbolic fact that we are concerned about. As Koyré remarked, gods withdrew from a world where they had nothing anymore to do. Space and time, instead of being the place of presence of everything, became the general frame of absence.

"Eternal matter...inherited all the ontological attributes of deity. But only these ones : as regarding the others, God, when leaving the world, took them away with him."(Koyré, 1957)

The real foundations of materialism are to be found in an ideology, and in the fantasies which underlies it. It forms a "complex", in a freudian meaning. I will call it *"the complex of Gilgamesh"*(Bottéro, 1992), because of this mesopotamian ancester, who the first went in quest of immortality.

Consuming ernergy might give us the homeostasis of isolated and never troubled gods, and medicine might make us become immortals. *"Then you will be like a god"*(Lucretius) : a piece of matter, never divided off in yourself by difference, never removed from your own consistency, never disturbed, an immortal dead. It is an old story in Western history, and it regularly re-emerges in our civilization. It is a very remote mesopotamian root ; it has been the spring of Pharaonic Egypt ; it reappears with the Roman emperors, whose declared ambition was to be immortal and to imput immortality to whom they wished ; it is the idolatrous side of christianism (without prejudging the others), which reaches a highest point for example in the Counter-Reformation. Everything takes place as if the political or religious purpose had been transfered to science and technique when the old monarchical system felt down.

To be almighty and immortal : never to be confronted with something different, that is to use the logic of the Same ; never to confront representation with its limits, with a refuting reality. To assimilate thought into substance, and science into being. If everything is a piece of matter, then you can believe that there is nothing different from what you master ; you can imagine you have a total knowledge and a complete power on things and beings ; there is the tragic bond between mechanism and totalitarism.

Only gods have final solutions. Technology might make gods of us. A part of science serves it.

On the contrary, to meet with the reality of information and semantic facts, implies to deal with difference and plurality ; it implies to introduce time in our reasoning, that is evolution, limitation, losings and meetings. It pulverizes any omnipotence. It involves the fact to be mortal, and prevents from a narcissic representation of the world. If we are really able to understand that ...

16. The Dissident Tradition

Over this fracture of culture, a dissident tradition carried on with the frame of mind of the Enlightenment.

Although working in quite different fields, psychoanalysis and homeopathy share some features of the same worldly wisdom. Hahnemann and Freud first were both released from conformism by their criticizing the established medical practices : they thought to ask the question, whether the violence of medical intervention was vindicated by some scientific reasons. And they were brave enough to answer no, and search for something else.

Their inspiration also followed similar ways : the very invention was in getting meaning at what had been scorned as meaningless.

In the two cases, a sensitivity to significance worked : for Hahnemann, the meaning that a clinical picture could be able to get in front of a toxic one ; for Freud, the meaning that a mad talking could be able to get in front of assuming an unconscious thought. Between these levels they both found a relevant system of mediations. They both renounced almightiness : Hahnemann modestly went back to the fact that the most part of curing belonged to the patient himself ; so did Freud, when he abandonned hypnosis ; by the way, they didn't imply to trust "good nature", but right information.

They recognized the abilities of human beings as being informed, and as dealing with somatic or psychological information. And they gathered beautiful thesaurus of signifiers, even if not in the same register.
So they were "men of Enlightenment", as Thomas Mann said about Freud. They bear witness to *"this human and ethical tendancy to heal human being by drawing him out from confusion and morbid perversion."*(Thomas Mann, 1929). They maintained the tradition and demand of reason, to enlight nature in its whole. In front of the prevailing ideology, they naturally were both misunderstood and blacklisted. But at the end, though dissident and minority, the tradition is not broken off.

17. Conclusion

A living structure is material ; from one point of view, it has to be handled as a mechanical device. A human being is cognizant ; from this point of view, it has to be talked and listened to as a mindful being.

But to the question :"What is to be living ? ", Tellenbach answered :*"It is to be searching for a rhythm to agree with ."*(1961) That is the third dimension of mediations. Now, I think we have the means to overtake the split of our representations of the world : the only condition is to overtake the analytic point of view, which reduces phenomena to mechanical composition. Science has shown us a lot of phenomena where the *organization* of elements provides new and not reducible possibilities and creative synthesis.

At least, the old positivism does not arise really from science : it is more an ideology, an aristotelian and pre-scientific idea of matter. This conception is moreover conveyed by social power of techniques, by scientific pomps and appearances, than by real knowledge. The lesser experiment of physics can display the charm and diversity of matter's behaviour.

We should now be able to match our soul with our real learning. Then, at the end, reason could play in tune. May be it is time to revive this tradition and to evolve new objects of science, beyond this illusion of security provided by mechanism.
"This one who would deny the thing because it is not palpable or visible would come to deny his own substance and his own existence."(Giordano Bruno, 1584)

18. References

Bastide, M., Lagache, A. and Lemaire-Misonne, C.(1995) Le paradigme des signifiants: schème d'information applicable à l'Immunologie et à l'Homéopathie, *Revue Intern. Systémique* **9**, 237-249.
Bastide, M. and Lagache, A. (1992) *The paradigm of signifiers*, Atelier Alpha Bleue Publisher, Paris.
Bergson, H.(1957) *L'évolution créatrice* , PUF Publisher, Paris.
Bottéro, J. (1992)*L'épopée de Gilgamesh* ,Gallimard Publisher, Paris.
Bruno, G.(1584) Del Infinito Universo, Venise, in *Opera Italiana*, Leipzig, 1830.
Cassirer, E. *La Philisophie des Lumières*, Arthème Fayard Publisher, Paris.
Descartes, R.(1953) *Discours de la Méthode*, Gallimard Publisher, Paris.
Descartes, R. (1953) *Les Passions de l'Âme* , Gallimard Publisher, Paris.
Descartes, R. (1953) *Meditations Métaphysiques*, Gallimard Publisher, Paris.
Diderot, D. (1976) *D'Alembert's Dream*, Penguin Classics Publisher, London.
Jakobson, R.(1963) *Essais de linguistique générale*, Minuit Publisher, Paris.
Kant, E.(1990) *Critique de la Raison Pure*, Presses Universitaires de France Publisher, Paris.
Koyré, A. (1968) *From the closed World to the Infinite Universe*, John Hopkins Publisher, Baltimore.
Lagache, A.(1988) *Echos du Sensible*, Atelier Alpha Bleue Publisher, Paris.
Lessing, G. (1984) *Laocoon : an Essay on the limits of Paintings and Poetry*, John Hopkins Publisher, Baltimore.
Lévi-Strauss, C. (1991) *Histoire du Lynx*, Plon Publisher, Paris.

Lucretius (1985) *De Natura Rerum*, Les Belles Lettres Publisher, Paris.
Mann, T. (1970) *Freud et la Pensée Moderne*, Aubier-Flammarion Publisher, Paris,
Pascal, B. (1977) *Pensées*, Gallimard Publisher, Paris, .
Spinoza, B. (1989) *Ethics*, Everymans Classic Publisher, .
Tellenbach, H. (1979) *La Mélancolie*, Presses Universitaires de France Publisher, Paris.
Wölfflin, H. (1988) *Renaissance et Baroque*, Gérard Montfort Publisher, Paris.

WHAT IS INFORMATION ? *

From matter towards meaning

A. LAGACHE
Lycée Carnot, 145 boulevard Malesherbes
75017 Paris - France

"The highest point a man can reach is surprise".
Goethe

1. Circumstances of Understanding

We understand what we choose to understand, building up the tools to carry out: the "big book of Nature" by itself doesn't display what is to be known. Yet the whole reality is always present round about us, eternally sparkling and rolling on. But amongst it, mind is limited to choosing what he is able to evolve.

Experiment doesn't speak by itself : we need ideas in order to interpret it. About 1615, William Harvey invented the circulation. Experience had not been lacking before : for some centuries, men had bled on battlefields. But it was not interpretable, before the conceptual approach of hydraulics was usable. Thus, even if we are used of considering circulation as a very realistic idea, it had to be invented.

Every fruitful scientific activity requires a paradigm : that is a logical device of choices which makes reason operating in front of phenomena, and at the same time limits it. What is the interesting element of reality that we select, what appears to us as an object of science? What kind of interactions will we set between these objects ? What kind of coherence will we use in order to follow developing phenomena ? We have to answer these questions by choosing relevant tools.

Although these choices were unconsciously done in the history of science, they are always implicitly present, and necessary. But we don't care about when sciences work easily : the emerging part of sciences is usually a "natural" gentleman's agreement, just as well intellectual as social : a possible experimentation fits with the "normal" set of concepts available at this time.

Beyond this, abnormality and paradox arise when new objects appear, near the limits of the explicative power of a paradigm. And, by a movement of backlash, this situation on-the-limits sheds new light on the previous paradigm, until then remained tacitly accepted and now to be criticized. To a certain extend, we only know afterwards what we were doing : it is the normal historical process of knowledge.

* Communication presented at the 8th GIRI Meeting, December 1994, Jerusalem, Israel.

M. Bastide (ed.), Signals and Images, 279–291.

2. The Mechanistic Paradigm

The classical paradigm of science is, for us, the materialistic and mechanistic one. It is so usual, and has been so fruitful, that we tend to mistake it for the mere law of rationality. Since the Seventeenth Century, we learnt from physics and chemistry the ruled use of mind, including both logical deduction and control by experiment. It is not to be forgotten neither to be changed.

But this science is a method, not a metaphysical necessity : the paradigmatic choices of classical science are one way, not all the possible ways. The mechanistic paradigm must be criticized in so far as it claims to encompass every reality, and to be the only and one method.

A short examination of its main features can show these choices and their limits.

-Matter is the object of classical science : the spatial object, with a mass. It assumes the law of *situs* : one thing in one place. Each object takes up an exclusive and determined location in space. That is fine for matter! But some facts do not fulfil this condition : semantic phenomena have no location, meaning facts imply a wide organization in which no special point of space can be, by itself, the "place"of sense.

-The interactions which can take place between material objects are external and negative : symmetrical exchange of elements, or friction forces. It rules out another kind of interactions, the positive ones, through which exchanging transforms the interacting structures themselves ; like for example in the immune system, where the very meeting with an antigen irreversibly changes the background ; in this area, exchanges are creative, and different from an exchange of objects in which, at the end, there is always a zero-sum.

-In the same way, the principle of identity of substance has its limits. By saying *"One thing is and remains what it is"*, Aristotle put a great order in thought. But he made difficult to evolve a science of changing objects, like biological ones, where transformation is a constitutive piece of facts. Objects of physics fulfil the theorem of Poincaré ; on the contrary, living systems never come back to a previous state, individuality is not periodic.

-We know now that physics and chemistry deal with closed systems, in which the most balanced equilibrium is the description in terms of elements. But we know also that the living beings are open systems, and resist to analysis : their own balance involves totality, which cannot be explained by the sum of analytic elements. The concept of creative *organisation* arises from science but overtakes its previous principle.

The physico-chemical approach had been formed on a macroscopic pattern of matter. Physics itself is now ahead. But it seems that other sciences still have the same look at material objects ; in fact, a normative way for results and experiment imposes in practice an idea of materialism that physics itself has overtaken for a time. Henceforth we can wonder whether the strength of the mechanistic way of thinking is an intellectual requirement, or a social set.

3. On the Brinks

Anyway, the need to go ahead and to overtake the limits of the previous model is now actual for everybody. And many present ideas endeavour to create new ways, without loosing the best of rational requirements. But a lot of attempts remain ambiguous : the prevailing model of mechanistic thought still keeps on shaping its escape bids to go beyond itself.

3.1. ARGUING ABOUT COMPLEXITY

One can say that we are confronted to more and more complex objects, which is probably true, but according to the meaning we give to it. If it was only a question of quantitative complexity, yet science would be led to a deadlock : a more complex object would require a more complex technology, a computer as big as the earth whose product would be a new more complex object which would require a computer as big as the universe, and so on... The nonsense which arises from this description shows the process : it implies to deal with more complex objects without changing our mind, with the same concepts and techniques that we already acquired for more simple things. Let us remind history : it is not the first time that such questions are asked! The solar system "complexity" was in a similar situation at the end of the Middle Age : by the improvement of optical instruments, astronomic observations mutiplied, and their classification became too complex. Some people hastily concluded that there was no rational order in universe.

Then, around 1543, came Copernic and his hypothesis of heliocentrism... The first reactions, by the way, were that... it was far too complex to be evolved and used! About 1573, Tycho Brahé still refined observations and accumulated experimental data. But he refused the copernician assumption.

Let us then emphasize the difference, that Einstein explains, between *"psychological complexity"* and *"conceptual complexity"* : the first one comes from historical inertia and habits of thinking ; we always consider as the most simple idea what we are accustomed to using, and then mind mistakes its own acclimatization for logics. But Einstein subtly remarks that for example relativity, when applied to the right phenomena, hugely simplifies loads of problems. Light came from theory : Kepler (1609) improved the theoretical model of Universe ; and then, Galileo Galilei (1632) was able to assert : *"e pure gira !"* ; It is still going round !
Theory was good enough to organize experimental data and to prospect ahead. The right paradigm was found for the right object, and complexity was shown to be an historical effect : a temporary inadequacy between observed facts and conceptual frame. Are we in the same situation ? At least we can ask the question, whether we must reduce complexity by means of cumulative analysis, or find new concepts fitting with the new levels of organization we now deal with.

3.2. THE FORTUNE OF CHAOS

The notion of "deterministic chaos" gives rise to the same questioning. It certainly has its right implement and gets a new look at some mechanical system's behaviour. But it is too much in fashion, and I wonder if it can keep the promises that scientific litterature nowadays asks it to give. Actually, it is used beyond its strict physical definition, each time one needs to overtake a classical determinism.

Prigogine and Stengers (1988) gave a beautiful analysis of chaos, scrupulously displaying its limits. Yet the notion of chaos can describe *"the complex and unforeseenable activity, collective and erratic, the very contrary of inert disorder"* about a system. But there we find two limits, especially concerning the application of chaos to living system. First, it is a description, not an explanation; there is no new possibility to describe any new kind of interactions between elements, as we need in biology to conceive the genuine fate of molecules in the original living connecting ways. Second, and consequently, we only can describe chaos as a collective behaviour : the idea shows there its affinity with large numbers, which is the uppest limit that a mechanistic paradigm can reach without introducing qualitative changings. Then the original element that lacks in biology is missing again : the individual factor , whose need is noticeable now in so many medical and biological problems. Then we feel that we are not on the good level of interpretation for living beings, which demonstrate precisely their amazing ability in finding *their* order where we see only *our* mechanical disorder. Besides, deterministic chaos often looks like a psychological safe investment.

Sometimes chaos is used as merely reinforcing the very belief in mechanistic almightiness : as if it were necessary to make it decreasing, in order to make room for anything else. This view would suppose that mechanistic determinism is antagonistic with any other one, like for example living order ; which is obviously absurd with regard to natural laws unity : if living systems determine them in an original way, it can only be in accordance with the mechanical order, though ahead of it. Sometimes chaos simply looks like the ultimate disguise of a supplier chance, which has always been the last resort of materialism (like a new version of spontaneous generation).

It amounts to trying to transgress a limit by means of negating the first position, which is the old dialectical process. But it does not provide anything new, because the negation does not contain anything different from what it negates.

In the same kind of idea, the great development of statistics during the Twentieth Century became a marvellous tool for analysis ; but, as regards concepts, in its common use, it is only a variation on classical determinism, and doesn't introduce any new kind of relations between elements. Although the notion of determinist chaos is worthy to note, because it introduces the idea of a positive and creative irreversibility, however we must be careful not to ask it to be a panacea.

Nevertheless, from different horizons new ideas about determinism arise. They do not imply any contradiction or negation of a natural order in phenomena. But they suggest that this order can be qualitatively different from these we already know. We still assume a general and harmonizing order in nature, more than ever. But the very limits of our ways of thinking imply several paradigms to evolve it.

4. From Matter Towards Information : an Intermediate object, the Violin.

Yet cybernetics, for example, provided new possibilities. It still deals with mechanical devices, but allows to reach a new level of organization, displaying new properties. Let us take the example of a violin, as a very organized and integrated object. From a materialistic point of view, it is nothing more than about 70 pieces of wood and metal. But it presents some properties which already overtake a plain analytic description.

- Each piece is strongly linked with the others, in order to vibrate in a total unity. Everybody knows the very symbolic "soul" of the violin : the soundpost, this thin little piece of wood which joins up the back and forward sounding boards ; it ensures the free circulating of vibrations in the whole instrument. The position of the soundpost is strictly related with the quality of the sound, and the slightest changing in it is highly thrown back by low and treble registers. A violin already looks like a living system, in which the parts are so much linked that the function of the whole supplants the analytic combination : more than a combination, it becomes a solidarity. Then another feature arises, in analogy with life : a good violin has no scattering sounds ; yet the sound can

be made of different harmonics, like a voice; but like a voice too, it sounds with a recognizable unique voice: it gets individual.
- Nevertheless, this strong binding is not obtained by means of strength and stiffness. On the contrary, each linkage seems to be negociated like a compromise between lithe and sturdy joins, like a subtle mediation.
- Glues are generally a violin-maker's secret ; they include organic components ; they must merge with wood without losing their own plasticity.
- The resistance of the forward sounding board is strengthened by the harmonic-bar ; they are sticked together, but they have not exactly the same curvature : according to a very cybernetic principle, this difference of curvature creates a dynamic set of forces, which is more able to vibrate than an addition of resistance forces.
-The soundpost must not be sticked, nor the bridge : they must freely transmit the sound by a perfect adjustment. If a violin has lost its soundpost, instrument-makers say that they must "rehabilitate" it, that is to put in place very very slowly a new one : the operation can last six months. (By giving the screw a quarter of turn three times a week...).

These details already show, by analogy, the general principles of living regulation, as a mediation : a manageable difference, a mediatized distance between elements. Furthermore, we see that we are in a new field, because time arises, the incompressible and irreversible time of living organizations. If one wants to teach or to cure this kind of systems, he cannot go faster than the own internal time of the system itself.

As described, we can now understand another beautiful property of a violin : it learns to sound. The more you play a violin, the best it sounds; instead of wearing out, it improves by playing, it learns to vibrate by means of vibrating. It is the first draft of a "permanent- memory" system, that is a structure whose present qualities result from its history.

No surprise then if I conclude with noticing that a violin foreshadows the main feature of living systems : it actively reacts to environment. If you speak in the room, the violin answers, even thin-voiced. It echoes with sonorous events from surroundings.

I just played with this example in order to argue : a lot of phenomena which surrounds us every day are more than mechanical. Cybernetics is an intermediate, showing how we plainly and quickly go beyond the mechanistic paradigm, when organization increases.

The next step is the logic of information.

5. Imagery of the Passage

"In the laboratory of chemistry, the specialist had been assigned to analyse black blotches on a piece of yellowing paper. On the fixed day, the headman of department came, accompanied by a Chinese man dressed up to the nines. The specialist explained to the head of department all the characteristics concerning the nature of the sheet of paper and the blotches; besides, using the newest techniques, he was even able to determine scientifically how old it was. Thus done, he modestly concluded that he knew everything that there was to know about these blotches.
At that moment the Chinese gentleman came forward ; he bowed respectfully , praised the depth and scope of the knowledge of the analyst ; then he apologized profusely for a strange coincidence : these black blotches fulfilled exactly the esteemed chemical formulas of the scholarly eminent Doctor ; but, over and above their black and

perishable substance, they could equally be seen as antique Chinese letters whose combination stood for an old love song."
(Translated from David Shahar, 1978).

6. What is Information ? General Principle

The general principle of information is *"a difference which makes a difference".* (Bateson, 1972).

We can consider living beings like *"internal environments"*, a specific organization of matter. It constitutively deals with an external environment, on several levels of exchanges. The difference between the two environments has been overestimated, or misunderstood, because of the fantastic overevaluation of our precious self. But the two environments are made of the same material element, and they could not be defined one without the other : it is just a local specification of elements' relationships. However, the actual difference is in the set of elements,which gives a determined and relative place to each element in each type of organization ; the difference between a molecule of calcium in nature or in body is only a difference of relation with a context.

Precisely, the world of information is the determinism of context : the original role that a given element plays according to the structure in which it appears. Exchanging information will consequently be a meeting between two environments : a difference in a first environment makes a difference in a second one ; but this second difference is not the same than the previous one, because the context is not the same.

These environments must have some affinities, in order that the event in one of them can concern the second one. They must also be different, in order that some event in one of them can take a different place in the network of relations for the second one, that is to become significant. Information is a mediation, linking but also maintaining distance. *"The mere presence of a living being transforms the physical world, makes appear here **foods,** there **hiding places,** and imparts to the **stimuli** a meaning they had not."* (Merleau-Ponty, 1945). One could say that this changing of sense is only *subjective*. But precisely it is time to take into account that living beings behave as subjects.

What we must emphasize there is that a piece of information is not an object, though having generally a medium. It can also be a simple meeting, and the plain changing of context of an element. A lost jigsaw-piece in a children's room has no meaning, except making the difference between a tidy or untidy home. Putting it in its jigsaw's frame with the other pieces makes arise a new meaning : it was this pretty star which lacked in the sky in the left corner... Sense is an emigrant, the new and impalpable dimension which arises not from objects, but from the network of woven relations around objects.

7. From Signal to Representation

The term of *information* is now commonly used in scientific litterature, generally without any precise definition. Actually, very different phenomena are mixed up in this use, and we must discriminate between them.

WHAT IS INFORMATION ?

7.1. SIGNALS

The notion of *signal* is widely used, for example in studying animal's behaviour, or in general cybernetical regulation. It is very relevant in these areas but does not exceed the limits of a cybernetical logic. An element, in a range of linked elements, is first missing (quantitative level of some component, specific chemical element, etc..). Its appearing makes the system move, and produces wide chain-reactions. This element then acquires a specific role, which gets it different from another one : a pheromone amongst butterflies has a determinist power. But it works like the soundpost in the violin, on the level of things. *The last straw that breaks the camel's back* is the same as any straw, as an isolated object ; when we focuse wider, including the camel, this straw acquires a determinist power that others have not.

These are signals : already a piece of information, to a certain extend, in the sense that an element plays a part different from its material properties as an isolated substance, because it is linked with a specific context. But it would be worthy to reserve the term *information* for other facts : we remain there on a pure level of objects, given as present and material elements. (It explains why cybernetics has been often reduced to mechanism, and used in a narrow sense).

Entering the informative area introduces a new kind of differences : the image-function. Image introduces a new dimension, the notion of a representation, and consequently of an interpretation, as simplistic as they could be at the primary levels of communication.

7.2. IMAGES: THE POSITIVE HAND

The world of information begins when dealing not with the thing, but with a representation of things. For us, human beings, representation is so naturally included into symbolic language that we forget to consider other ways, more primary but strongly significant. There we must precisely build a bridge between what belongs to the body and what belongs to mind, in order to understand another kind of things : the area of semantic sensitivity, beyond mechanistic organization and short of psychological sensibleness. It is the world of mimetic representation, in which the nature becomes *"a set of private messages."*(Wiener, 1930).

As an illustration, let us follow for example the different images of hands, painted by prehistoric cavemen. They show -at least- three different ways of representation. We can organize them into a hierarchy, not suggesting that it is a chronological order that we have no means to know, but assuming that it shows a logical order, following a more and more complex way of representing: the most simple representation of something is its positive image, identical to the model. On this level, we find the *positive hands* : the form of a hand is painted and filled with colour, according to the most simple and realistic process.

Which of us, childlike, has never felt like putting his hand in the paint and applying it to the wall? It corresponds with the pleasure we all know in imprinting the bare feet in the fresh sand, rejoicing in seing our marks on the beach... (although the hollow made by the foot, inverting volume, yet suggests something more elaborated).

This kind of signifiers constitutes the most primary level of representation. It opens a first level of informative phenomena, in which images yet are signifiers, that is a concrete sign of the thing, but stand for an equivalent of the thing ; there is no transformation, just as an *alleviation* of the presence of the thing. We see this hand, and we just know that a man dropped in there. The positive hand is meaningful for us, and just says that a human presence had been. It is worth stating at this point that the informative ability of such an image is dependent up on a previous and already known lexical register : the receiver, the reader of the image must have a referent to understand it. Each image is different, and does not inform about anything else than itself ; the receiver must already have encountered the thing, or must have a memorized referent to be able to read it.

Analogically, *very low doses* in biology are not images, but behave as if they were : they seem to be dealt with by organisms in a similar way. Organisms are able to be *advised* about toxic substances, and react positively to a very low dose, up to a limited extend.

For example, we can situate the phenomenon of hormesis at this level of information. Everything happens as if the pre-treatment gave to the organism an image of the toxic substance, and provided a referent ; then the organism learns how to react to, as according to a *miniaturized* experience of toxic effect. But having being informed about some toxin does not protect against another one : the informative effect is limited to the identity between image and thing.

The ancient mesopotamians were in the same situation with their ideograms : the sign *is worth the same as* the thing. And it entails ambiguity. Mesopotamian scholars ended up as confusing images for things, dedicating their interest to worship writing, and so initiating a long tradition of intellectual idolatry. More concretely, the same thing can happen when a living structure, instead of being informed, takes an image *at the first degree*, as if it was the substance itself : images can be ambiguous. The delicate use of nosodes by homeopathic physicians can illustrate it. The substance is the very pathological object, just alleviated from an intrusive weight. It can work as a piece of information about disease, which invites to a positive regulation. On the other hand, according to the circumstances of the receiver, it can lead into a worsening, as if it were received like an excess load of disease. In this case, the image is not interpretated but literally taken as object, in a process of passive mimesis. Then, identity between thing and sign was too strong. In the case of hormesis, the external origin of substance and its distancing from natural organism protects more against confusion.

We can conclude that images must be mediations : similar to the thing, but maintaining a constitutive difference with things.

Organizing information must include a difference, a distance between object and semantic object. Something must be indicated about the difference of levels, in order to discriminate level of objects and level of information.

It gives a new look at the notion of *signifier* . A signifier, even an analogic one, can be an image ; but it must involve something about the way it takes to signify, about its own representative nature. It must include a formal feature. The same demonstration is done about signs of language: the meaning of a sentence is dependent up on the *relation of representation*, the modality of signs, which carries indications about their ways and means of being signifiers (Wittgenstein, 1921).

The composition of more complex systems of communication will include this condition. And the other prehistoric hands still show the way.

7.3. THE *INTRODUCED* OBJECT: THE OUTLINED-HAND

An interesting intermediate is given by the outlined-hands as shown in some prehistorical painting in caves. There we have the outline of a hand, like these we all designed by simply following the form of our hand put on the paper. The line of the silhouette is black. By a curious mean, upon this outline is often superimposed a spot of red colour, roughly hard by applied without regard to the form of the hand. History teaches us that explaining in terms of *primitive clumsiness is* seldom fruitful. It would be better to see there a kind of exhibition of representation in itself. The prehistoric artist delivered us a message about the very fact of representing by means of lines and colours, in a kind of intellectual *kit-form*. This outlined hand is a kind of introduction into the world of representation, a way to show to the receiver that he can be concerned about signs.

It is a kind of frame given to the object, which makes it meaningful as a sign, like in a film's picture which gives a clue by framing an object. It picks the object out of manifold trivial reality.

And it happens also in biology. The surroundings and presentation of a molecule can induce a special dealing with it. In this way the Major Histocompatibility Complex (MHC) seems to be the introducer for the immune system : it displays antigen, and so leads the system to react. The idiotypic network looks like a semantic set which reacts provided that things are presented in a semantic way, that is introduced by the MCH, representative of the self.(Bastide and Lagache, 1992) And we unfortunately know that if this introducing is missing, the system cannot react in a relevant way.

This kind of systems work by means of framing networks, which give to the elements some properties they would not have out of this frame. We really enter the complete level of active information at the stage of the *negative hand.*

7.4. PRINCIPLE OF TRACKS: THE NEGATIVE HAND

Negative hands are seen also on prehistorical cave walls; they are stenciled : the artist laid his left hand on the wall, he spread out over and around a fine pounce of colour, and then he removed his hand. Thus mankind's history is edged with two similar images : at the beginning, this negative hand paves the road for thought and arts ; at the end, for us, this Man of Hiroshima, whose corpse is transfered to the wall he protected a moment from burning, is the failure of thought and arts ; if new seeds don't grow...

Anyway, the original informative power of tracks comes from the function of absence, included into the signifier. That is why this kind of information is less ambiguous, and leads to infinite possibilities of communicating by means of analogic signifiers. The piece of information is not an object : it is the track of an object into a medium. This mediatization involves not only the designation of the content of information, but also indicates its absence as object. The signifier is settled up as indicating a level of information. We don't see a hand, we *see an absent hand*, that is the mere genius of semantic objects.

We can understand now why these negative hands are ready to step across a new stage : they become symbolic signs when the artist has painted them while folding some fingers, in a kind of finger-alphabet which refers to abstraction and linguistic communication. They go back to a function of signal, but at a new level of information.

Writing-signs followed the same evolution, from ideograms to alphabet. Letters are completely symbolic : they are no more images ; they refer to the register of verbal language as supporting the whole necessary referents. The signal-function of a letter is the element of symbolic level of written communication, capitalizing on the whole process of informative representation.

But before this break-up between signified and signifier, before entering language, there is a fascinating intermediate level of communication : the level of analogic signifiers. Evolution of writing can still help us : before becoming an alphabet, written signs went through an intermediary state of syllabication. The ideogramm, noun of a thing, has been taken as a sound, which refers not to the thing but to the vocalization of its noun. There is an analogical transfer from the reference to the thing, to the corporal event of vocalizing its noun. (Herrenschmidt, 1990). Voice is a corporal event, which binds body and mind. It is the exemplary form of a corporal signifier, as it works in aesthetics, intuitive understanding, unconscious communication and so on. We are now led to concrete communication by means of analogic signifiers. Analogic signifiers have a genuine ability of informing, provided their own structure and the patterns of communication which arise from.

They point out the content by being an image of it, a track, not so much as identical than moreover as similar to the structure of it. But they have undergone some formal shapening : alleviation, dilution, artificiality or theatricality as a *stylistic trick* , which prevent them from being mistaken for things, and indicate that they are pieces of information.

For the relevant receiver, this track of thing can be dealt with, and induces positive reactions of *active mimesis* (Lagache, 1988). One of the most beautiful achievements of human culture is to have invented their management in a therapeutic purpose, as well by theater as by medicine. This informative reaction of active mimesis actually can be described as well as by the catharsis theory of tragedy by Aristotle as by the effect of remedy in homeopathy.

Aristotle explains the cathartic power of tragedy : people are *purged of passions* by the representation of it. It would be a misinterpretation to consider catharsis like a quantitative flushing out, releasing some energy of passion for a while. It would not be so efficient!

On the contrary, the actual meaning of theater's catharsis is that tragedy *informs* about passions, calling for an upper level of information. It issues a knowledge about passions. The same *dose* of passion, if acted out in reality, would destroy people's relationship and society. The theatrical representation inverts this effect : it moves an integrative level of regulation. Without being especially conscious of it, the members of the audience receive an enlightenment which introduces a critical distance between symptomatic acting-out and regulation, a dynamic mediation. Then there is an active mimesis : information about psychological and social disorder induces to deal with it and to regulate it towards a new order. (There is the difference, says Aristotle, between art and *the spectacular*, which on the contrary leads to a passive imitation of the show, that is a *worsening* of passion : where we find again the ambiguity of images. So a criterion of good art would be its informative effect on art lovers. We also feel there the reason why Freud abandoned hypnosis, in order to evolve from return and repetition of *passion* towards an active resumption and recovery in psychoanalysis). This is a general way of art : to inform, beyond things and short of intellectual thought, by means of concrete signifiers which are elaborated tracks of objects. It invites to a sensitive experience of meaning.

In biology, homeopathic high dilutions seem to work on the same way. Even being not psychological but material, they are described as tracks of substance into a medium. Whatever they would physically consist in, they work according to the rules of analogical communication, and the paradigm of signifiers makes possible to understand their way to a certain extend and to have a look at possible experiments. (Bastide and Lagache, 1992).

Homeopathy can be devised as a ruled use of signifiers. It devolves on the law of similarity to choose the designation of the content ; the remedy is the image of disease, an anologous picture of the whole symptoms : their representation, on a corporal level of information. There is no identity between disease's poisoning and curative substance, and it is a very good requirement in order to discriminate between things and images ; the analogic relation is settled between two networks which are already corporal expressions : the whole pathogenetic symptoms according to the whole natural ones.

Nevertheless, the designation must be the most acute, in order to give rise to an actual meaningful resonance between remedy and organism.
Then it devolves on the dilution to alleviate the substance and to mediatize it so as to set it up as a signifier.

Therefore, active mimesis, this process of dynamic analogy, is worthy in many ways of therapeutics. In the Western culture, we find it in aesthetics, psychoanalysis and homeopathy. However, many traditional cultures use it in very diverse *therapeutic theaters.* Beyond its therapeutic interest, the benefit of it is to speak to our sensitivity as well as to our sense, addressing to this place whence body and mind are fused and together answer to meaning and beauty.

8. "To Whom it May Concern"

However, nothing is information by itself ; information is not an object, and doesn't *subsist in itself and for itself,* according to the classical definition of substance. That is the reason why it is often so difficult to speak about information. Our words and our conceptual habits are all patterned on the model of object. But we cannot speak about a piece of information as an isolated species, it would be absurd. What we actually observe is the informative effect on a receiver. To describe tracks tacitely implies that they have yet been recognized by someone, for whom they were significant.

Information is in the receiver. Sartre said that paintings in an art gallery don't exist during the night and closure : at these moments, they are only canvas, oil and such stuff. They exist as paintings when they are watched. In the same way, ideograms or letters don't exist as signs without a reader. The *matter of fact,* in this case, is the very meeting between a track and a relevant receiver.

The receiver is necessarily active : information arises from his reading. He creates it. Because of that, achieving to deliver information is always extremely dependent up on the circumstances of the receiver. Receivers of information are living structures, so they change in time, memorize the previous events, and carry information according to their present aims.

For example, homeopathic physicians know that their patients are not the same ones all along their life, and not the same if they received a remedy before. Organisms are permanent informed-informing structures ; they take the past into account, but always with regard to the original synthesis of the present moment. Everything is important for a living being : his past, what he is at the moment, but also the present mood and the colour of the sky today.

That is why repetability cannot be ensured in the same easy way as for mechanistic systems, concerning experiments. The informable systems are very sensitive, opened to any variation, and more parameters have to be controlled. The conditions of experiment are more demanding. It doesn't mean that there is no law. On the contrary, nothing happens by chance with information and it doesn't admit the lesser margin of error ; it is a strong determinism, and you cannot stop an informative effect when it decided to work.

Yet it means that laws are subtle, and widely unknown. They require new paradigms of reasoning, and some delicacy of reason. But there is no reason for them to remain obscure : it is just a fantastic challenge for the future.

9. The Frame of Reference: Principle of Levels

The second principle which always covers information is that we must reason upon two different levels. The materialistic thought has not this constraint. Material elements are all on the same level. They exist all in the same way, amongst the same frame of reference where elements are ruled by the same observed parameters : mass, size, speed, energy and such elements that are measurable.

Yet there is the observer, with his thought and methods. But the famous *separation of the phenomenon and the observer* (though questionable) ensures their exteriority, in such way that elements of phenomena are *pure* things. The method of measurement is supposed to ensure the best conformity between mental representation and things, so as we could *substract* the method from the result and consider reality as the material facts themselves.

But information cannot fit in this pattern. The semantic object is a double nature : it belongs to the level of things as drawing, voice, written letters, dilution, but it belongs to the level of interpretation in front of the living structures it informs. Therefore, although it does exist, information is not a substance : it is this very event which happens between levels, which goes from things to reading, the very passage itself. Sense is an emigrant : process and result are the same.

Anybody who speaks about information is leaded to think in terms of levels, even without especially intending it. It is obviously the case in the present reflections about biology. The choice of a dilution in homeopathy is often described as adressing to a certain *level* of organism, by the way suggesting the idea of an organic organization into physiological levels of information. These integrated levels would be different from a simple hierarchy of objects.

In philosophy, the analysis of thought and language by Russell and Wittgenstein is a sound example of this topic. Considering the conceptual possibility of language, Russell was led to the theory of logical classes as an informative hierarchy. And the inspiration of Wittgenstein's *Tractatus logico-philosophicus* (1921) works around the irreducible distance between thing and state-of-things, and also between what is *shown* in language, and what is *said* in fact. Then the philosophical journey consists in going over from *the forest of things* to *the point of view of eternity.*

If we want to understand how living structures are informed-informing, we must admit that they also seem to be double nature (at least !) : a sum of material elements, and an informative synthesis of them.

A level of information is a class of facts which appear in the same way, and inform us through the same channels of operating logics. It is not so much a class of objects than a class of *manner of being,* a typical way of interacting with us. It is noticeable that we cannot *simultaneously* take the chinese characters as blotches and as signs (see above the D.Shahar's story). We can consider the two aspects one after the other, but we are obliged to deal with them separately. However we cannot say that a level of information does exist *in reality.* As Einstein said, science says *how* facts work, never what they *are.*

May be this constituent difference between level of matter and level of information, or better the difference between the ways they work, is not actually a fact of the world but an epistemological constraint. Everything happens as if the brain itself refused to move on into two different logical ways together. Perhaps our traditionnal split between body and mind, matter and sense, imposes it on the thought.

My own proposal about this question would be as following : everything comes together with everything in the reality, and things rub shoulders with significance. Nature does not care about our distinctions. Reality is just an infinite wealth of possibilities. And every possibility is always close to us.

Then, understanding information according to a difference of levels between things and tracks is just to put some order in phenomena. We see that something happened, and we can organize its understanding according to a difference of levels, as rational markers. In this way, information as an event creates the difference of levels : process equals result.

Ultimately, we can conceive that several ways of thinking lead to several types of organization ; in fact, yet amongst the same nature, because there are always mediations between them, and the mediation of a mediation...

10. The Continuity of the Paradigms

This fantastic semantic capacity of living beings cannot remain for ever relegated out of science. There is no reason to think that it is an unruly world, given over to the whims of fate. On the contrary, there is a logic of it to be explored.

And there is no reason not to succeed in enlightening on it, because

"nosque partem totius naturae esse, cujus ordinem sequimur."

"so are we a part of the fullness of nature, whose measure we share."

"nous aussi, sommes une partie de l'ensemble de la nature, dont nous suivons l'ordre."

"parte somos de naturaleza entera, y su ordenanza compartimos."

"siamo anché noi una parte della natura complèssa, il mètro dalla quale seguitamo."

Spinoza, Ethics IV, Appendix, capit.32.

11. References

Bastide, M. and Lagache, A.(1992) *The paradigm of signifiers*, Atelier Alpha Bleue, Paris.
Bateson, G. (1972) *Steps to an ecology of mind*, Chandler Publishing Company, New-York.
Herrenschmidt, C.(1990) *Le Tout, l'énigme et l'illusion*, Le Débat n°62, Gallimard Publisher, Paris.
Lagache, A. (1988) *Echos du sensible*, Atelier Alpha Bleue Publisher, Paris.
Merleau-Ponty, M. (1945) *Phénoménologie de la perception*, PUF Publisher, Paris.
Prigogine, I. and Stengers, I.(1988) *Entre le temps et l'éternité*, Arthème Fayard Publisher, Paris.
Shahar, D. (1978).*Un été rue des prophètes*, Gallimard Publisher, Paris.
Wiener, N. (1930) *Cybernetics*, MIT, Cambridge, Massachussetts.
Wittgenstein, L.(1961) *Tractacus logico-philosophicus*, Gallimard Publisher, Paris.

INDEX

Printed in the United Kingdom
by Lightning Source UK Ltd.
116471UKS00007B/239